P. F. Gordon · P. Gregory
Organic Chemistry in Colour

P. F. Gordon · P. Gregory

Organic Chemistry in Colour

With 52 Figures and 59 Tables

Springer-Verlag
Berlin Heidelberg New York 1983

Dr. Paul Francis Gordon
Mr. Peter Gregory

Imperial Chemical Industries
PLC Organics Division
Hexagon House
Blackley, Manchester M9 3DA, U.K.

ISBN 3-540-11748-2 Springer-Verlag Berlin Heidelberg New York
ISBN 0-387-11748-2 Springer-Verlag New York Heidelberg Berlin

Library of Congress Cataloging in Publication Data.

Gordon, P. F. (Paul Francis), 1954 — Organic chemistry in colour. Includes bibliographies and index. 1. Dyes and dyeing — Chemistry. 2. Chemistry, Organic. I. Gregory, P. (Peter), 1946 — II. Title.
TP910.G57 1982 667.2 82-10569

Printed in GDR

2152/3020-543210

Preface

The foundations of the chemical dyestuffs industry were laid in 1856 when W. H. Perkin discovered the dye Mauveine. At approximately the same time modern chemistry was establishing itself as a major science. Thus, the chemistry of dyes became that branch of organic chemistry in which the early scientific theories were first used.

This early eminence has now been largely lost. In fact, many of our academic and teaching institutions pay little attention to this vitally important branch of organic chemistry. We believe that this book will help to rectify this unfortunate situation.

The majority of books that have been published on the subject of dyes have been technologically biased and, in our opinion, do not appeal to the mainstream organic chemist. We have, therefore, aimed at producing a book which emphasises the role of organic chemistry in dyestuffs and we have included appropriate modern theories, especially the modern molecular orbital approaches. We have assumed that the reader possesses a knowledge of the basic principles of organic chemistry;* the only other requirement is a general interest in organic chemistry.** The book should interest the newcomer to chemistry, the established academic, and the dyestuffs chemist himself.

We owe a large debt of gratitude to the many people who have helped and encouraged us in the preparation of this book. In particular, we would like to thank Professor C. W. Rees of Imperial College, London, Dr. J. Griffiths of Leeds University and Dr. R. Price and Mr. B. Parton of ICI Organics Division for reading the entire script and passing on their valuable criticisms. We would also like to thank our colleagues within Research Department for their invaluable help. A special thanks must also go to those who aided us in the preparation of the finished manuscript; to Pat Marr and especially Vera Gregory for translating our handwriting into a readable typescript, and to Derek Thorp, Andrew Gordon and Ed Marr for proof-reading. We are grateful to Miss A. Boardman, Miss D. Hutchinson and Mrs. J. Prince for their help

---

* Basic colour physics and colour perception are not included in the book. However, a bibliography for these subjects is provided in Appendix I.

** We had also intended to include a chapter on photographic dyes; however, this was omitted in order to keep the book to a reasonable length.

in photocopying the finished manuscript. Finally, we thank Dr. G. Booth who, on behalf of ICI Organics Division, gave us access to vital library and secretarial facilities.

Blackley, Manchester, U.K.
November 1982

Paul Francis Gordon
Peter Gregory

Table of Contents

Chapter 1

The Development of Dyes

1.1 Introduction

Take a look around; colour is everywhere. The clothes we wear, our surroundings, both man-made and natural, abound with colour. Indeed, from prehistoric times, man has been fascinated by colour. From the early cavemen, who adorned their walls with coloured representations of animals, through the Egyptian, Greek and Roman eras, right up to the present time, colour has been a constant companion of mankind.

Until the end of the nineteenth century, these colours were all obtained from natural sources. The majority were of vegetable origin; plants, trees and lichen, though some were obtained from insects and molluscs. Over the thousands of years that natural dyes have been used, it is significant that only a dozen or so proved to be of any practical use, reflecting the instability of nature's dyes. Today, the number of *synthetic* organic colourants exceeds 7,000 and, in 1974, the world sales of synthetic dyes amounted to a staggering £ 1,500 million! (1974 prices).

1.2 Pre-Perkin Era — The Natural Dyes

1.2.1 Introduction

In terms of numbers, the yellow dyes comprised the largest group of natural dyes but they were technically inferior to the reds, blues and blacks, having lower tinctorial strength, (*i.e.* only weakly coloured) and poor fastness properties, especially light fastness, (*i.e.* they soon faded). In contrast, the red and blue dyes had good properties, even by modern standards. Natural yellow dyes are based on chromogens[1] (mainly flavones (1), chalcones (2) and polyenes (3)) that are relatively unstable and which have been completely superseded by superior synthetic yellow chromogens. However, the anthraquinone (4) and indigoid (5) chromogens found in the natural red and blue dyes respectively still form the basis of many of the modern synthetic dyes, especially the anthraquinone derivatives (see Chap. 4).

[1] Chromogen is the term used to describe that complete arrangement of atoms which gives rise to the observed colour. The term chromophore describes the various chemical units (building blocks) from which the chromogen is built.

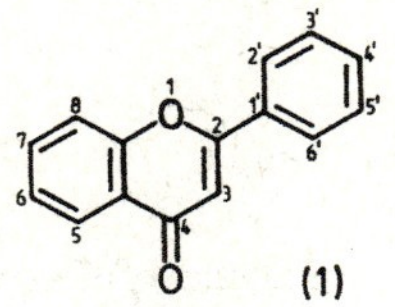

(1) 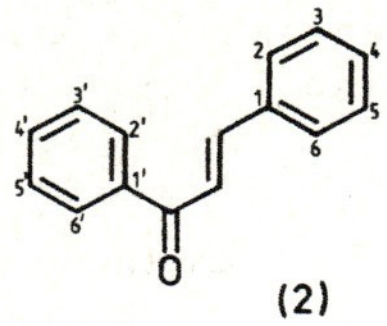 (2)

$RO_2C + CH = CR +_n CO_2R$

n = 7/9
R = H/Me

(3)

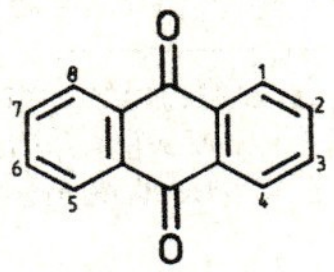

(4) 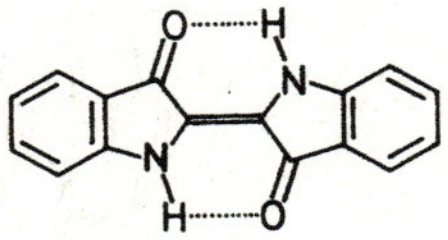 (5)

1.2.2 Yellow Dyes

All the yellow dyes were obtained from vegetable sources (see Table 1.1). The most important yellow dye in the Middle Ages was Weld.[2] It was used in conjunction with the blue dye Woad (see later) to produce the celebrated Lincoln Green, a colour made famous by Robin Hood and his merry men. Unlike most yellow dyes, Weld is based on flavone (1), not flavonol (3-hydroxy flavone) and since flavones are more resistant to atmospheric oxidation than flavonols, fabrics dyed with Weld probably displayed a higher order of light fastness than those dyed with flavonol based dyes (see Table 1.1).

Table 1.1. Some Important Natural Dyes

Colour	Class	Typical Dyes	Structure (Name)	Source
Yellow	Flavone	Weld	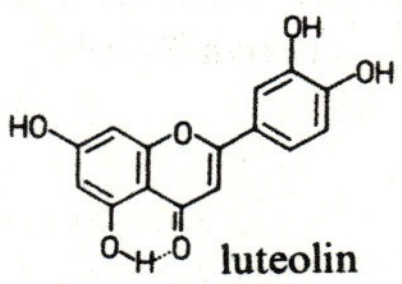 luteolin	Seeds, stems and leaves of the *Reseda Luteola L* plant (Dyer's Rocket)
	Flavonol	Quercitron	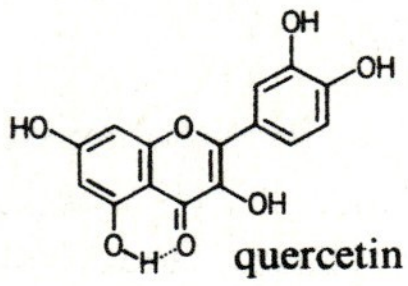 quercetin	Bark of North American oak, *Quercus tinctoria nigra*
	Chalcone	Safflower	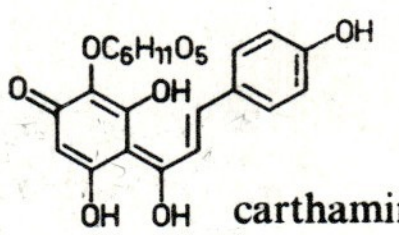 carthamin	Dried petals of *Carthamus tinctorius* (Dyer's Thistle)

[2] The convention adopted throughout this book is that all dyestuffs are denoted by a capital letter.

Table 1.1. (Continued)

Colour	Class	Typical Dyes	Structure (Name)	Source
Yellow	Polyene	Saffron	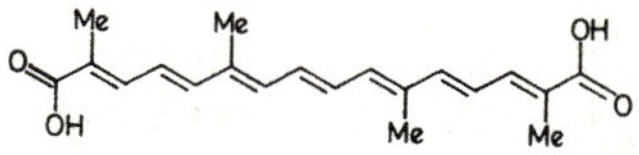 crocetin	Stigmas of *Crocus sativus* (4,000 required to give 25 g of dye)
Red	Anthraquinone	Kermes	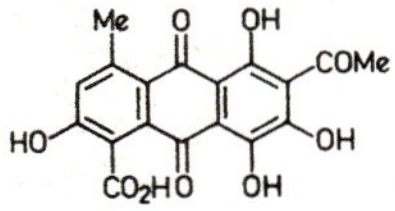 kermesic acid	Female scale insects, *Coccus ilicis*, which infect the Kermes oak
	Anthraquinone	Cochineal	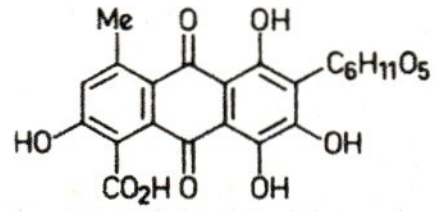 carminic acid	Female insect, *Coccus cacti*, which lives on cactus plants of the Prickly Pear family found in Mexico (200,000 → 1 kg of dye)
	Anthraquinone	Madder or Alizarin	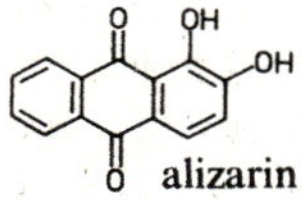 alizarin	Roots of the *Rubia tinctorum* plant. Root was known as 'alizari', hence alizarin
	Anthraquinone	Turkey Red	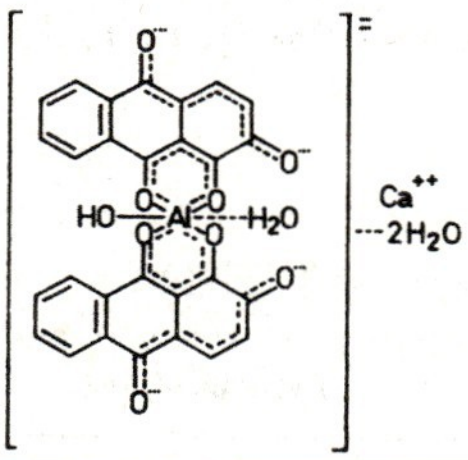	
Purple	Indigoid	Tyrian Purple	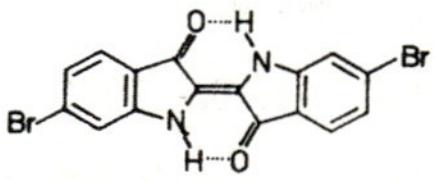 6,6′-dibromoindigo	Mollusc (*i.e.* shellfish) usually *Murex brandaris* plentiful in the Mediterranean
Blue	Indigoid	Woad; Indigo	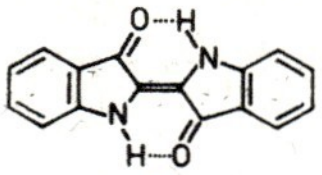 indigo	Leaves of indigo plant, *Indigofera tinctoria L*
Black	Chroman	Logwood	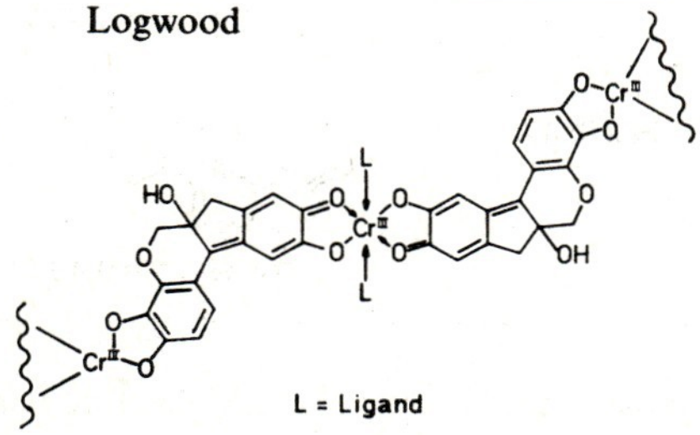 L = Ligand	Heartwood of the tree *Haematoxylon campechiancum L* found in Central America

1.2.3 Red Dyes

In contrast to the yellow dyes, three (Kermes, Cochineal and Lac) of the four natural red dyes were derived from insects rather than plants. However, the most important red dye, Madder (also known as Alizarin), was of vegetable origin. All four are hydroxy derivatives of anthraquinone (4). Synthetic anthraquinone dyes are still in widespread use today and, as a class, are noted for their outstanding light fastness and bright shades (see Chap. 4). Therefore, it is hardly surprising that the natural red dyes were far more durable to the elements than their yellow counterparts.

The most important red dye was Madder (see Table 1.1). Tunics dyed with Madder have been used by the armies of both France and England. Thus, in the 19th century, Louis Phillippe dressed his infantry in 'pantalon garance' or madder-red trousers, but perhaps more famous were the English 'red-coats'. It may be that red was selected to minimise visually the effect of blood during battle!

Alizarin was used mainly with various metallic mordants,[3] the most famous being a mixture of aluminium and calcium salts to give Turkey Red, prized for its beautiful bluish-red shade and high fastness properties. The probable structure of Turkey Red is that shown in Table 1.1 in which the aluminium ion forms a 1:2 complex with Alizarin.

1.2.4 Purple Dyes

The structure of Tyrian Purple was determined by Friedlander in 1909: using 12,000 molluscs, he obtained 1.4 g of dye which he characterised as 6,6′-dibromo-indigo (6). This fact illustrates an important feature of natural dyes, namely that vast quantities of raw material were needed to obtain just a small amount of dye.

Only recently has the complex chemistry leading to the production of Tyrian Purple been elucidated: this is outlined for the mollusc *Dicathais orbita*. Thus, the creamy coloured precursor found in the mollusc is tyrindoxyl sulphate (7); enzymic hydrolysis converts this to tyrindoxyl (8); some of this undergoes atmospheric oxidation to 6-bromo-2-methylthioindoleninone (9), and (8) and (9) form a 1:1 quinhydrone type complex, tyriverdin, which is converted by sunlight to 6,6′-dibromoindigo (6), the principal constituent of Tyrian Purple (Scheme 1.1).

Scheme 1.1

[3] A mordant is a chemical, usually a metal salt or an acid, which is applied to the fabric to be dyed prior to dyeing. During the dyeing process, it forms an insoluble complex with the dye within the fibre, which helps to retain the dye on the fibre. Mordants generally cause a bathochromic shift (*i.e.* a shift to longer wavelengths) of the colour of most dyestuffs.

Tyrian Purple derived its name from the ancient city of Tyre on the shores of the Mediterranean, where the industry originated and flourished. It was a prominent dye in the Roman era when the purple fabrics commanded a very high price. It is stated that the Romans paid as much as £ 150/kg for wool dyed with Tyrian Purple and this was so expensive that only kings and priests could afford it: 'born to the purple' is still used today to denote people of wealth or position.

1.2.5 Blue Dyes

The only natural blue dyes, Indigo and Woad, both contained the same principal colouring matter, indigo (5), although this was not realised at the time. Indeed, Indigo was thought to be of mineral origin and an English letters patent was granted in 1705 for mining it!

Indigo is one of only two natural dyes still used today although it is now produced synthetically: the other is Logwood black.

1.2.6 Black Dyes

The only important black dye is Logwood. Logwood was known in 1500, but it did not achieve any real importance until 1812 when the French chemist Chevreul discovered that it combined with metallic salts to give coloured lakes. Haematein (10), the colouring matter of Logwood, is red but in combination with chromium it gives a black shade and it is for these black shades that Logwood is renowned. Although the constitution of the metal complex has not been elucidated, it is thought that it has a macromolecular structure in which the chromium ions link the molecules together by chelation (see Table 1.1). Even today, the majority of black shades are obtained by mixing two or more dyes due to the scarcity of synthetic black dyes, and it is a measure of Logwood's success that it is still used for certain outlets, such as the dyeing of silk and leather.

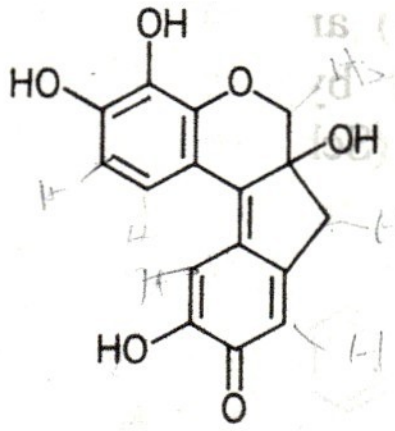

OH
HO
O
OH
HO
O

(10)

1.3 Perkin and Beyond — The Synthetic Dyes

1.3.1 Introduction

At the beginning of the nineteenth century the natural dyes, as discussed earlier, dominated the world market whilst synthetic dyes were almost unknown. In fact, the only synthetic dye of any importance, picric acid, was discovered by Woulfe in 1771, and even this only accounted for a fraction of a per cent of the world's dye production. However, the nineteenth century saw a dramatic reversal of this situation.

Within fifty years of Perkin's discovery of Mauveine in 1856, synthetic dyes accounted for over 90% of the dyes used.

These spectacular changes were initiated by a relatively new science, organic chemistry. Since organic chemistry has played such an important role in the development of the synthetic dyestuffs industry it is worthwhile summarising its progress up to 1856.

In 1800 so little was known about organic chemistry that it could scarcely be called a science. Berzelius, in 1807, had described organic compounds as materials derived from living matter, although this term now embraces virtually all carbon compounds whether natural or synthetic. Twenty years earlier, Antoine Lavoisier had devised a method for burning organic compounds and analysing the vapours that were evolved. This technique allowed him to confirm the presence of carbon in these compounds and also enabled him to detect the presence of hydrogen and nitrogen, two elements frequently found in organic compounds. Superior methods of analysis were developed from 1800 to 1831 which enabled the analyst to calculate the proportions of carbon, hydrogen, nitrogen and oxygen present in an organic compound as well as confirming their presence. Despite these advances very little was known about the structure of organic molecules. In fact the structure of benzene, a vitally important chemical to the dyestuffs industry, wasn't recognised until the second half of the nineteenth century!

Synthetic organic chemistry was in an even worse position at this time. Its progress was hindered by the presence of the 'Vital Force' theory. This theory claimed that organic compounds, as well as consisting of chemical elements, contained a 'Vital Force' which would only allow their synthesis by living organisms. However, the theory was dealt a severe blow when Wöhler, in 1828, showed that urea could be synthesised from ammonium cyanate without the intervention of any living organisms (Eq. 1.1). This observation was supported by evidence which

$$NH_4^+ \cdot CNO^- \xrightarrow{\text{heat}} NH_2-\overset{\overset{\displaystyle O}{\|}}{C}-NH_2 \qquad \text{Eq. 1.1}$$

demonstrated that organic compounds obeyed the same chemical laws as inorganic compounds; that the two branches of chemistry, organic and inorganic, have remained separate from thereon has been for the sake of convenience rather than for any 'supernatural' reason. Thus, synthetic as well as structural organic chemistry now became a focal point of interest amongst chemists.

1.3.2 Perkin's Discovery of Mauveine

The discovery of benzene by Faraday in 1824, and its discovery as a constituent of coal tar by Leigh in 1842, had passed largely unnoticed. It was really the work of A. W. Hofmann, from 1845 onwards, that focused the attention of organic chemists on benzene and other such aromatic compounds obtained from coal tar. He had been able to isolate appreciable quantities of benzene and other interesting aromatic compounds from coal tar by fractional distillation and, with the help of a team of able and enthusiastic young chemists that had collected around him, he was able to explore the chemistry of these compounds.

It was W. H. Perkin, whilst working with Hofmann, who discovered Mauveine, in 1856, and thereby started the 'dyestuff revolution'. However, he was not intentionally working towards preparing synthetic dyestuffs but towards quinine, the antimalarial drug. By consideration of the molecular formulas of allyl toluidine (11) and quinine (12) he had arrived at the relationship shown in Eq. 1.2 and thus attempted

$$\underset{(11)}{2C_{10}H_{13}N} + 3[O] \longrightarrow \underset{(12)}{C_{20}H_{24}N_2O_2} + H_2O \qquad \text{Eq. 1.2}$$

the preparation of quinine by the oxidation of allyl toluidine with potassium dichromate in sulphuric acid. If Perkin had known the structure of quinine (12a) and

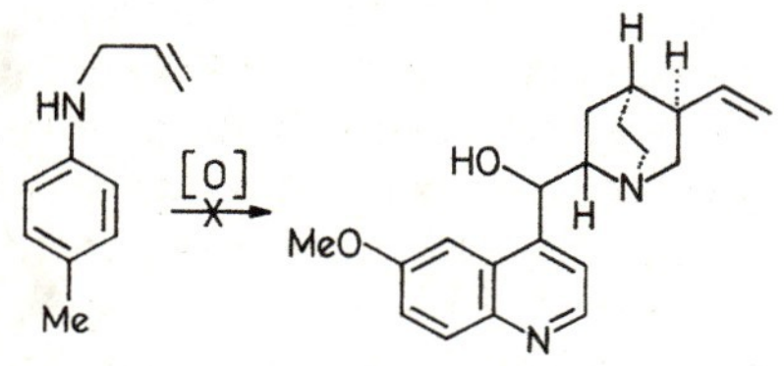

(11a) (12a)

allyl toluidine (11a) he would almost certainly have abandoned this route but, in 1856, the structure of benzene was still unknown, never mind that of quinine! Perkin tested his theory but found he obtained a very impure brown powder which didn't contain any quinine. He then turned his attention to the simplest aromatic amine, aniline, to determine if the oxidation reaction was a general one. Once again he obtained a very unpromising mixture, this time a black sludge, but on boiling his reaction mixture with ethanol he obtained a striking purple solution which deposited purple crystals on cooling. Perkin recognised that this new compound might serve as a dye, later to be called Mauveine (13), and despatched a sample to Pullar's dyehouse in Scotland. The dyers produced a very enthusiastic report about this new dye for it had superior fastness properties on silk to the available natural dyes.

Perkin had been very fortunate. Not only did his discovery arise from testing an erroneous theory but it also required the presence of substantial toluidine impurities in the starting aniline (Eq. 1.3). However, there was no luck associated with the way in

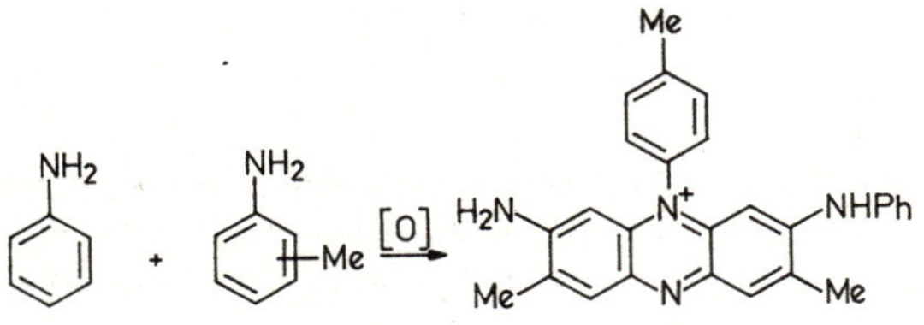

Eq. 1.3

(13)

which Perkin followed up his astute observation. Once he had established the usefulness of Mauveine as a dye, he set about preparing it on a large scale. The problems he faced were enormous. As well as having to manufacture the Mauveine, he also had to synthesise large quantities of aniline by Hofmann's route (Eq. 1.4) and in the process pioneer the techniques of large scale synthesis.

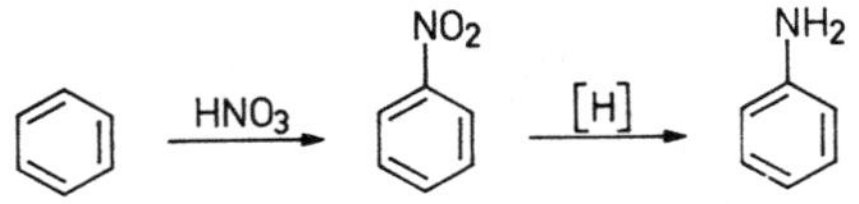

$$C_6H_6 \xrightarrow{HNO_3} C_6H_5NO_2 \xrightarrow{[H]} C_6H_5NH_2 \qquad \text{Eq. 1.4}$$

Perkin solved all these problems in a surprisingly short time and, with the help of his family, was soon manufacturing and selling Mauveine. As a dye for silk, Mauveine was an instant success. However, Perkin's dye increased greatly in importance when he discovered, with others, a method for dyeing wool with Mauveine using tannic acid. Perkin's remarkable achievements have since earned him the title of 'founder of the dyestuffs industry', and justifiably so!

Perkin's success attracted many competent chemists to the dyestuffs industry and several of the more significant discoveries made by them in the 'Post-Mauveine era' are discussed below.

1.3.3 The 'Post-Mauveine Era'

A starting point was the investigation of the chemistry of aromatic amines, particularly aniline. Thus, Verguin, in 1859, discovered Magenta (or Fuchsine) whilst studying the reaction between crude aniline and tin IV chloride, an oxidising agent. Magenta is actually a mixture of two components, homorosaniline (14) and pararosaniline (15); the composition of the mixture depends on the quantities of *ortho*- and *para*-toluidine present in the starting aniline. The effect that changing the oxidising agent had on the course of the oxidation reaction must have been more than Verguin could have expected for he had discovered a new and important class of dyes, the triphenylmethanes.

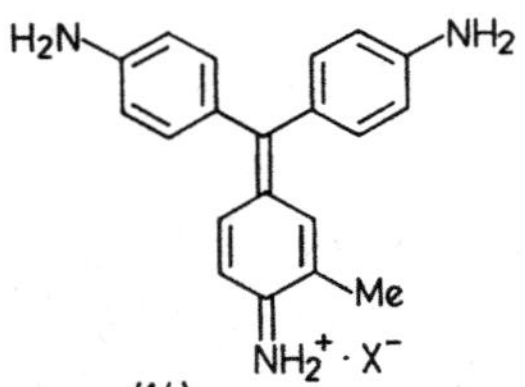

H_2N NH_2 Me $NH_2^+ \cdot X^-$ (14) H_2N NH_2 $NH_2^+ \cdot X^-$ (15)

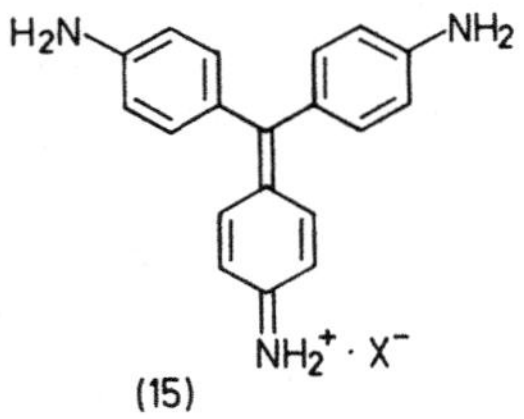

The interest which this new dye attracted quickly led to the introduction of more triphenylmethane dyes. The first of these dyes was prepared by Hofmann when he methylated the amino groups in Magenta to give a mixture of violet dyes, the so-called Hofmann Violets. The unsubstituted amino groups may be replaced by arylamino groups as well as being alkylated. Girand and de Laire discovered this reaction when they heated Magenta with pure aniline to give Rosaniline Blue (16). Unfortunately, the introduction of the hydrophobic phenyl groups resulted in poor water solubility for this otherwise useful dye. However, Nicholson, who had independently discovered the dye, was working on methods to increase its water solubility. Eventually he discovered that by treating the dyestuff (16) with concentrated sulphuric acid he was able to increase its water solubility quite markedly. He had, in fact, sulphonated Rosaniline Blue and it was the presence of the sulphonic acid groups in the dye molecule which conferred the enhanced solubility. This technique was later found to work with a wide variety of dyestuffs and is still used today.

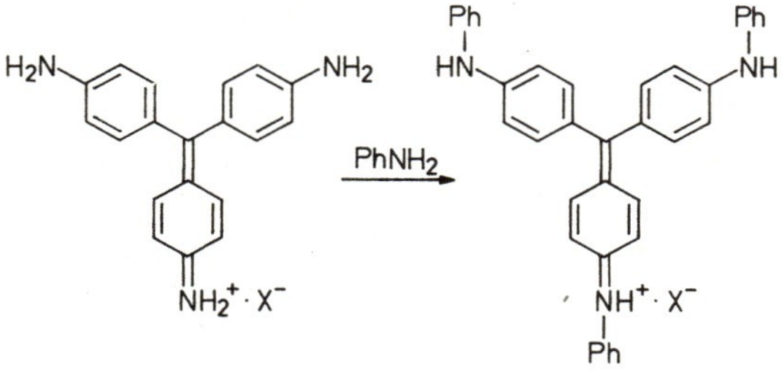

(16)

Another important class of dyestuffs, discovered in 1863, was the aniline blacks. These dyestuffs are large complex structures produced directly in the fibre by the oxidation of aniline. Here, an aniline salt is impregnated into the fibre and subsequently oxidised. The dyestuff once formed precipitates within the fibre and is characterised by its high wet fastness. Aniline blacks are still widely used today as dyes for cotton.

It was at this time that Peter Griess, working at the Royal College, discovered the diazotisation reaction. This is the singularly most important reaction carried out in the synthetic dyestuffs industry, for over 50% of current dyestuffs are azo dyes. Peter Griess found that when he treated an aromatic amine with a nitrosating

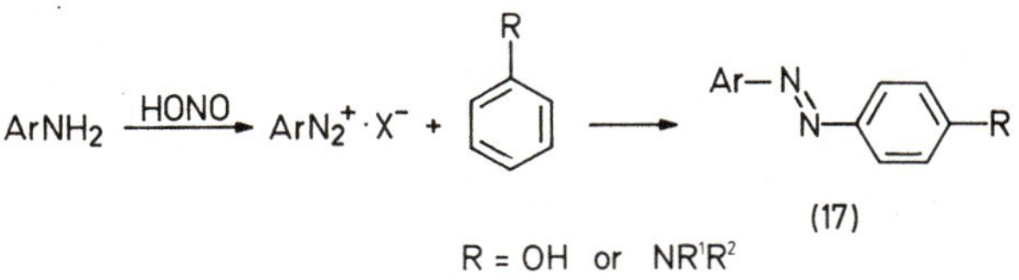

$$ArNH_2 \xrightarrow{HONO} ArN_2^+ \cdot X^- + C_6H_5R \longrightarrow Ar{-}N{=}N{-}C_6H_4{-}R \quad (17) \qquad R = OH \text{ or } NR^1R^2 \qquad \text{Eq. 1.5}$$

agent, such as nitrous acid, he obtained an unstable salt. He found this salt reacted with certain chemicals to give strongly coloured products. This unstable salt was a diazonium salt which reacts, or couples, with a phenol or an aromatic amine to give an azo compound, *i.e.* (17) (Eq. 1.5). The first commercially successful azo dye, Bismarck Brown, was discovered by Martius in 1863. Bismarck Brown is actually a mixture of dyestuffs produced from the diazotisation of *m*-phenylenediamine. A constituent of this mixture, Bismarck Brown B (18), may be prepared by tetrazotisation (double diazotisation) of *m*-phenylenediamine to produce a bis-diazonium salt which is then coupled with two moles of the parent diamine to give the disazo dyestuff.

All the excellent work carried out during the nine years since Perkin's discovery of Mauveine was done without any knowledge of the structures of the aromatic amines involved. This made rationalisation of the chemical reactions almost impossible and research tended to be a rather hit and miss affair. It was largely due to Kekulé that this unfortunate situation was resolved.

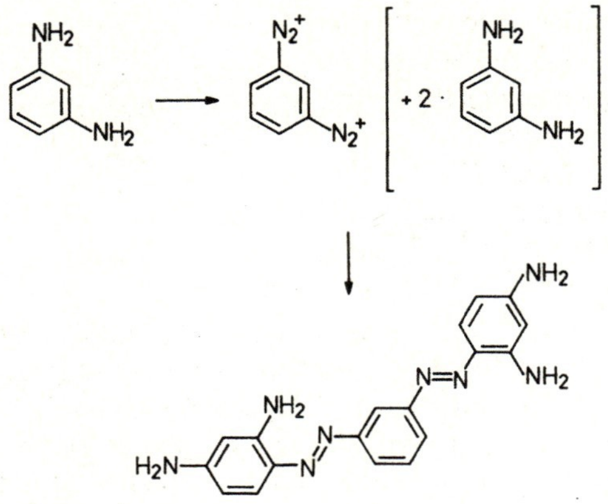

(18)

1.3.4 Kekulé's Contribution

By 1850, the empirical formula of an organic compound could be fairly easily established but the structure of these organic molecules remained a perplexing problem for the organic chemist. It was during the years 1857/58 that Kekulé published a series of papers on the theory of valency in carbon compounds. In these papers, he described the quadrivalency of carbon and postulated the equivalence of the four hydrogen atoms in methane. Kekulé also realised that the seemingly endless variety of organic compounds that existed could be attributed to the ability of carbon to form single and multiple bonds with itself and a wide variety of other elements.

Kekulé's most memorable single contribution to organic chemistry was his paper, published in 1865, on the structure of benzene. A short extract, taken from a Chemical Society publication,[4] is shown below which describes how Kekulé first visualised the benzene molecule as a ring:

> '*I was sitting writing at my textbook, but the work did not progress; my thoughts were elsewhere. I turned my chair to the fire and dozed. Again the atoms were gambolling before my eyes. This time the smaller groups kept modestly in the background. My mental eye, rendered more acute by repeated visions of this kind, could now distinguish larger structures, of manifold conformation: long rows, sometimes more closely fitted together; all turning and twisting in snakelike motion. But look! What was that? One of the snakes had seized hold of its own tail, and the form whirled mockingly before our eyes. As if*

[4] J. Chem. Soc., Trans. *1898*, 100.

by a flash of lightning I awoke; and this time also I spent the rest of the night in working out the consequences of the hypothesis. Let us learn to dream, gentlemen, then perhaps we shall find the truth . . . but let us beware of publishing our dreams before they have been put to the proof by the waking understanding.'

The solution of the structure of benzene made an enormous impact on both organic chemistry and the dyestuffs industry. Many of the apparent enigmas were seen in a new light and much of the chemistry that had gone before could now be rationalised. In this new era systematic and planned research was at last possible. Indeed, a substantial amount of this research was done in the research establishments of the newly formed dyestuffs industry. In particular, aromatic organic chemistry was thoroughly studied by dyestuffs chemists, who made a valuable contribution to our knowledge of aromatic systems.

All this newly acquired information was then put to use in the synthesis of two important natural dyestuffs, Alizarin and Indigo. These two syntheses not only illustrate the ingenuity of the early organic chemists but also bear witness to the rapid advances in organic chemistry during the last century.

1.3.5 Alizarin

It took just three years from Kekulé's disclosure of the structure of benzene to the elucidation of the structure of Alizarin by Graebe and Liebermann. During this time Graebe and Liebermann had deduced that Alizarin was 1,2-dihydroxyanthraquinone (19), a deduction arising from years of painstaking research on anthraquinone dyes. An important experiment which had determined the skeleton of Alizarin involved the distillation of natural Alizarin, obtained from the Madder plant, from zinc dust to give anthracene (20), a known compound. This confirmed the presence of a tricyclic aromatic system in Alizarin and, by careful application of the knowledge they had already gained from their work on anthraquinones, Graebe and Liebermann were able to solve the structure of this natural dye.

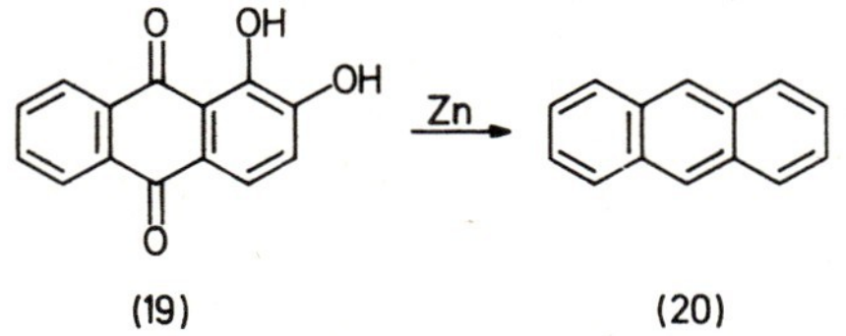

O OH OH Zn

(19) (20)

These two German chemists now undertook what they considered was an unambiguous synthesis of Alizarin from anthraquinone. This procedure of determining a chemical structure and then independently synthesising it is still used extensively in modern organic chemistry. Graebe and Liebermann envisaged a two step synthesis of Alizarin (19) from anthraquinone (4) via 1,2-dibromoanthraquinone (21; Scheme 1.2).

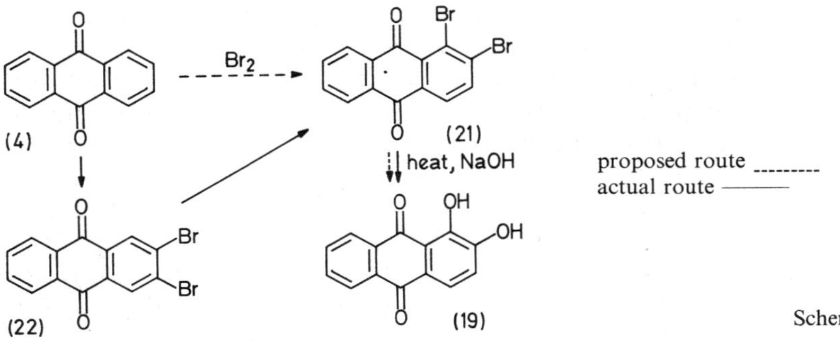

proposed route ---------
actual route ———

Scheme 1.2

The compound Graebe and Liebermann obtained on brominating anthraquinone was not the 1,2-dibromoanthraquinone (21) they expected but the 2,3-isomer (22). However, they did not perceive their mistake and continued with their synthesis by subjecting the 2,3-isomer (22), thinking it was (21), to a caustic fusion. Fortunately, the 2,3-isomer (22), during the caustic fusion reaction, rearranged to the 1,2-dibromoanthraquinone (21); the 1,2-dibromo derivative (21) was then rapidly converted to the 1,2-dihydroxyanthraquinone (19), as predicted. Graebe and Liebermann had therefore 'proven' the structure of Alizarin, although they were completely unaware of the true route.

The route that Graebe and Liebermann had discovered was not commercially viable and it was left to Caro and Perkin, working independently, to discover an efficient and cheaper route. Once again a fortuitous accident occurred which led Caro to the better route. At that time BASF, Caro's employers, had accumulated a large quantity of anthraquinone for which they had no use. Caro, in an attempt to find a use for the anthraquinone, mixed it with oxalic and sulphuric acids and began to heat the mixture; this he hoped would produce a useful new dye. However, the oxalic acid decomposed before any reaction took place but before Caro could conclude his experiment he was called away and, fortunately, neglected to switch off the gas burner under his reaction. When he returned he noticed a pink crust around the charred remnants of his reaction mixture; this pink crust, he astutely realised, was Alizarin! He quickly determined that the anthraquinone could only be sulphonated at high temperature with very strong sulphuric acid and that the sulphonated anthraquinone, once formed, could be hydrolysed to Alizarin. Caro

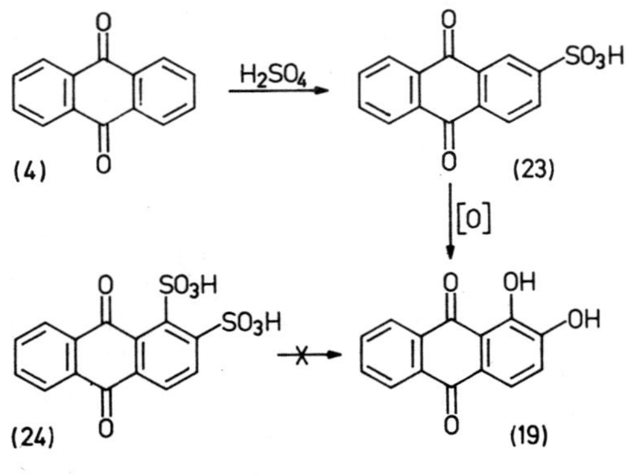

Eq. 1.6

had therefore stumbled upon a much cheaper route which would allow synthetic Alizarin to compete, commercially, with the natural product. The route Caro used is shown above (Eq. 1.6). It is notable that the Alizarin (19) is actually produced from the monosulphonic acid (23) rather than the disulphonic acid derivative (24) and requires an oxidation step; under Caro's conditions, the oxidising agent was atmospheric oxygen.

At approximately the same time Perkin, independently, discovered the same route but was beaten by one day to the patent rights by Caro. However, Perkin very quickly discovered a third and even better route. This involved chlorinating anthracene (20), which at that time was a waste product from coal tar, to obtain 9,10-dichloroanthracene (25), which could be sulphonated, oxidised and fused with sodium hydroxide to give Alizarin (19).

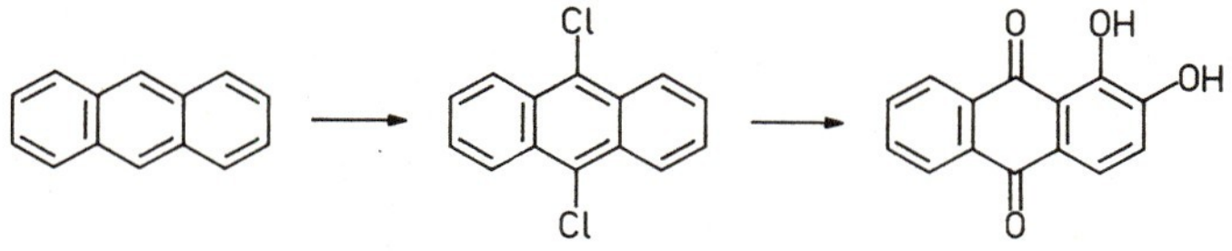

Cl Cl O OH OH O

(20) (25) (19)

The dyestuffs chemist had now produced a route by which Alizarin could not only be made cheaper but also purer than the Alizarin extracted from natural sources. Within a few years synthetic Alizarin had displaced natural Alizarin from the market place and, in Europe, 0.5 million acres of arable land used for Madder growing was used for other crops.

The dramatic success achieved in the synthesis of Alizarin prompted research into the other very important natural dyestuff, Indigo. This time, however, progress was considerably slower than in the case of Alizarin.

1.3.6 Indigo

The principal figure in the researches on Indigo was von Baeyer, although a number of chemists worked towards the determination of the structure and eventual synthesis of Indigo.

It was already known that when Indigo (5) was distilled it yielded aniline; in fact, the name aniline is derived from the Portuguese word for Indigo, *anil*. Further, von Baeyer, in 1869, obtained indole (26) from the reduction of Indigo (5). This result and other available information led him to propose the correct structure (5)

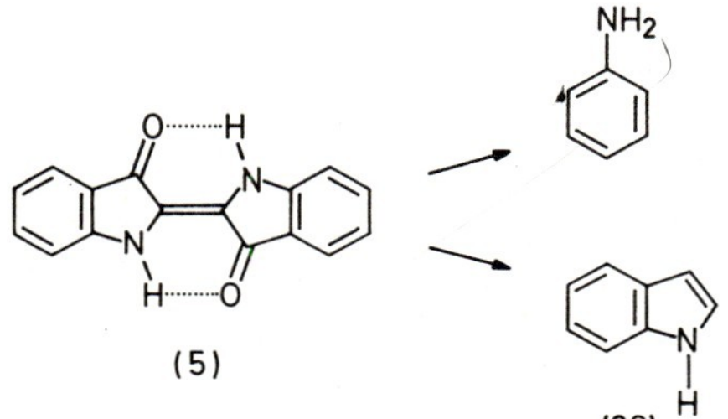

NH_2 O H N N H O N H

(5) (26)

for Indigo.[5] Eleven years later he published a synthesis of Indigo (5) from *o*-nitrocinnamic acid (27), (Scheme 1.3). However, it was not until 1897, and the outlay of £ 1 million by BASF on research, that a commercially viable route to Indigo was discovered by K. Heumann.

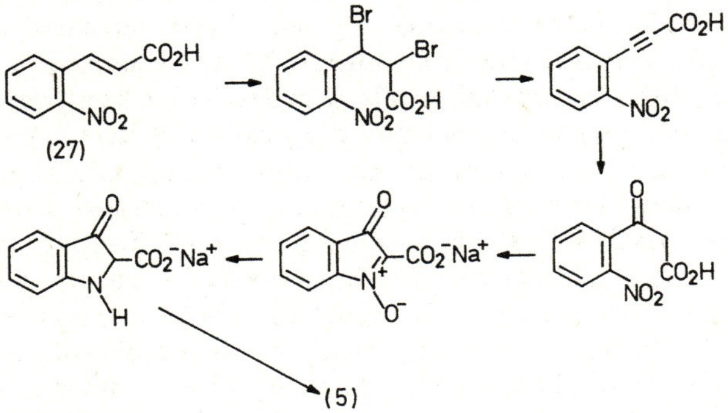

Scheme 1.3

The route that was finally chosen was a seven stage synthesis (Scheme 1.4)[6]. The first step, the conversion of naphthalene to phthalic anhydride (28), initially proved very troublesome; the reaction was very sluggish and did not always occur in high yield. However, during one manufacturing campaign a mercury thermometer, used to monitor the reaction temperature, broke and discharged its contents into the hot reaction mixture; almost immediately the naphthalene was rapidly and cleanly converted to phthalic anhydride in high yield. This occurred because the mercury reacted with the hot sulphuric acid to form mercury II sulphate, a catalyst for the reaction. This very fortunate accident was exploited by BASF and mercury II sulphate was used in all the subsequent manufactures of phthalic anhydride.

Phthalimide (29), obtained by passing ammonia through molten phthalic anhydride in a high yield process, was reacted with sodium hypochlorite to give

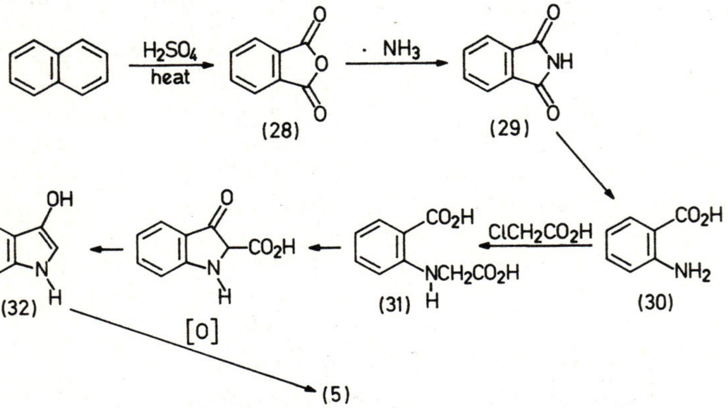

Scheme 1.4

[5] Actually, he proposed the *cis* isomer which was believed until 1928 when X-ray crystallography showed it was the *trans* isomer (5).

[6] Heumann's route was quickly superseded by a more efficient (and economical) route — see Chap. 2.

anthranilic acid (30); this step was one of the first applications of the Hofmann degradation reaction. The anthranilic acid could then be converted to indoxyl (32) in two steps; by reaction with chloroacetic acid and then a Dieckmann type cyclisation of the intermediate diacid (31). Once formed, the indoxyl (32) was readily oxidised by atmospheric oxygen to give the required product, Indigo (5).

This impressive reaction scheme is notable not only for its advanced organic chemistry but also for the technical innovations made. Many of the reagents used, such as chloroacetic acid, were not generally available and so had to be manufactured specifically for the process. The process also had to operate at maximum efficiency to make the synthetic Indigo cheaper than the natural product. Thus, by-products were utilised where possible; for example, sulphur dioxide generated in the first step was converted into sulphuric acid which was then recycled. Despite this long reaction sequence, synthetic Indigo was cheaper and of a superior quality to natural Indigo. Surprisingly, it was this superior quality of the synthetic material which caused it to be adopted so slowly by the dyers. The natural Indigo had impurities which resulted in its colour on the fabric differing from that of the synthetic material. However, the advantages of using a relatively pure dye with a predictable and reproducible shade gradually converted the rather conservative dyers; within ten years the majority of Indigo was synthetic in origin.

1.3.7 The Introduction of Novel Chromogens

Whilst synthetic processes were being developed for these two important dyes other synthetic dyes, which had no natural counterparts, were being prepared. Several new classes of dyes were discovered such as the xanthenes, phenothiazines, sulphur dyes, *etc.*, and important advances were made in the existing classes of dyes.

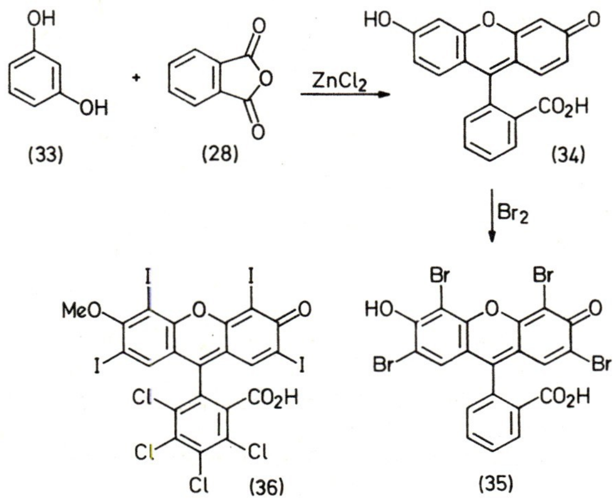

The first xanthene dye was discovered by von Baeyer when he condensed resorcinol (33) with phthalic anhydride (28) to give Fluorescein (34), which Caro later brominated to give another useful dye Eosine (35). Other members of this class of dyestuffs, which had very bright shades, were Rhodamine B, a dimethylamino analogue of (34), and Rose Bengal (36); the latter is still used today as a photo-sensitiser. Many other useful dyes structurally related to xanthene dyes were obtained in this period. Thus, azines such as safranines (37), (*e.g.* Mauveine (13)), phenothiazines (38) (*e.g.* Methylene Blue, $R^1=R^2=R^3=R^4=CH_3$), and oxazines (39), (*e.g.* CI Basic Blue 3, $R^1=R^2=R^3=R^4=C_2H_5$).

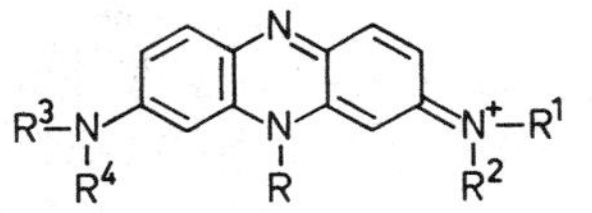

(37)

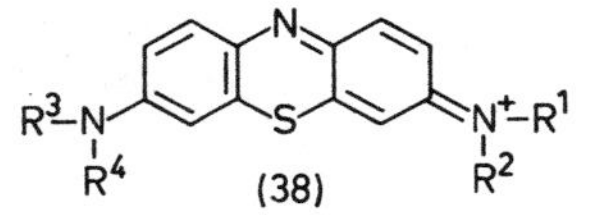

(38)

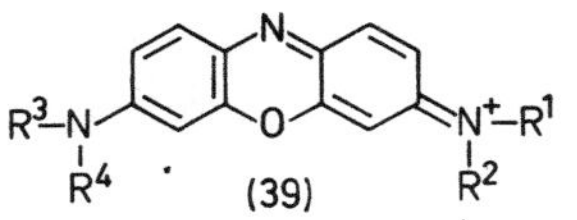

(39)

Croissant and Bretonniere, in 1873, discovered the first sulphur dye. Sulphur dyes differ from other dyes by having complex macromolecular structures that are generated within the fibre. They are usually prepared by heating sulphur with aromatic amines, phenols or aminophenols. The dyes thus obtained are water insoluble mixtures but are reduced in the dyebath to give water soluble *leuco* derivatives which impregnate the fibre. Atmospheric oxidation regenerates the sulphur dye within the fibre to give dyeings fast to washing. Sulphur Black, used on cellulose, is by far the most important sulphur dye used today and is prepared by boiling 2,4-dinitrophenol with sodium polysulphide; it is presumably a phenothiazonethioanthrone type (40).

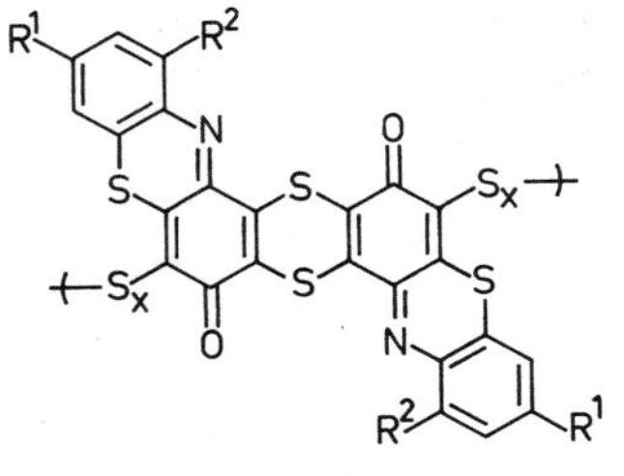

(40)

Meanwhile, significant progress had been made in azo dyes. Three examples which illustrate this progress are described below. Caro, in 1875, discovered Chrysoidine (41), a dye still used today, by diazotising aniline and coupling the diazonium compound with *m*-phenylenediamine (Eq. 1.7):

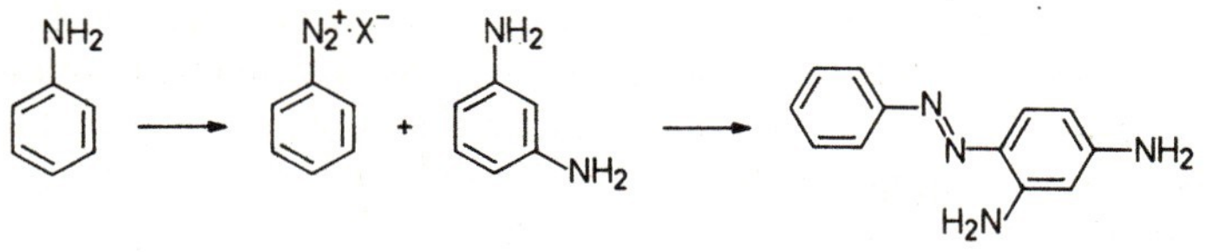

Eq. 1.7

(41)

this was the first azo dye for wool. Congo Red (42), today used as a pH indicator, was first made in 1884 by tetrazotising benzidine and coupling the bis-diazonium salt with two molecules of 4-sulpho-1-naphthylamine. Congo Red was the first dye which could be dyed directly on cellulose (dyes with this degree of affinity for cellulose are termed substantive).

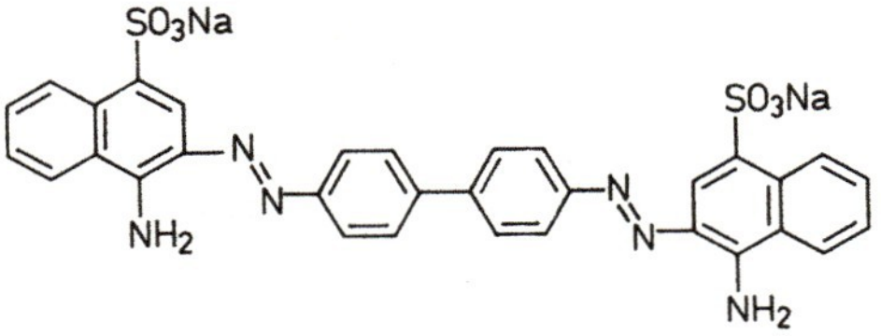

(42)

In this same period the Ingrain process was developed by Read Holliday and Sons. Here, the fabric was impregnated with a coupling component and then an ice cold solution of a diazonium salt was added. These reacted together to precipitate the dye within the fibre. The strategy behind this process is rather similar to the method of dyeing aniline blacks and vat dyes.

Just as the twentieth century dawned Böhn (1901) discovered Indanthrone (43), which had better fastness properties than any dye discovered in the preceding century. This compound set the scene for the subsequent introduction of the very important range of modern vat dyes (see Chap. 5).

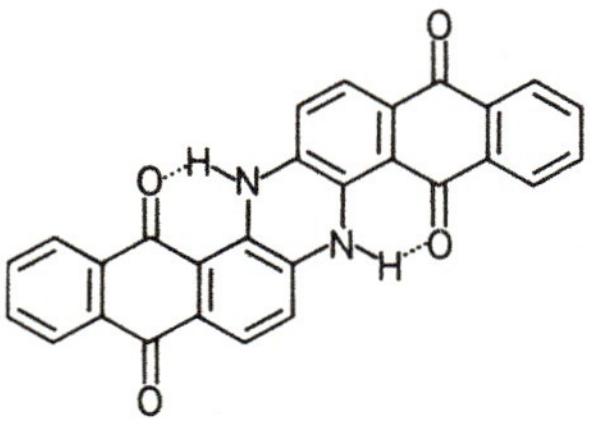

(43)

The second half of the nineteenth century had seen the introduction of many new classes of dyestuffs; since then, very few new chromogens have been added to the range of available dyestuffs. However, before these new developments are discussed it is worthwhile considering the roles played by the, then, two major synthetic dyestuffs producing countries, Britain and Germany, in the rise of the dyestuffs industry.

After Perkin's discovery of Mauveine in 1856, the British dominated the research and development effort in synthetic dyestuffs and produced a very high proportion of the early synthetic dyes. As a result, German chemists flocked to Britain to obtain experience in this new field. However, Britain's early lead was short lived and by 1875 the Germans themselves were established as the major dyestuffs manufacturers. Several of the leading German chemists, such as Hofmann and Caro, had moved back to Germany taking with them a lot of valuable experience. These chemists combined with the well trained organic chemists emerging from the German scientific institutions to give the German dyestuffs industry a solid foundation on which to build. In contrast, the British centres of learning provided little for the developing organic chemist so that the British dyestuffs industry was starved of a vital raw material — good organic chemists. Compounded with this was the total indifference of the British government to the dyestuffs industry; instead, they preferred to concentrate on the textile industry.

In the forty years after 1875, the Germans progressively built up their dyestuffs industry whilst the British remained almost stagnant. By 1914, the Germans produced 90% of the world's dyes and had a controlling interest in the British market, supplying 80% of the dyes used. At the outbreak of the First World War in 1914, all trade ceased between Britain and Germany and the British dyers were cut off from their main suppliers. The situation reached crisis point when it became apparent that the British hadn't enough dyestuffs with which to dye their soldiers' uniforms, and had to purchase the necessary dyes from the Germans! This crisis forced the British government to help rejuvenate the dyestuffs industry, which responded by increasing its output of dyes by 400% from 1913—1919. Amongst the measures the government adopted was the amalgamation of several small companies into the British Dyestuffs Corporation, later to become part of ICI, which was modelled on the large German combines such as Bayer and BASF.

The period since the First World War has seen a big expansion in the dyestuffs industry during which time the U.S.A. and Japan have become major manufacturers, as well as Britain, Germany and Switzerland. A high level of research has been maintained but the emphasis has been on optimisation and improvement of the available chromogens rather than the discovery of new types. However, this is not to say that there hasn't been any activity in the search for new chromogens. Such an example is the discovery of copper phthalocyanine (44) by several employees working at Scottish Dyes Ltd. (later ICI).

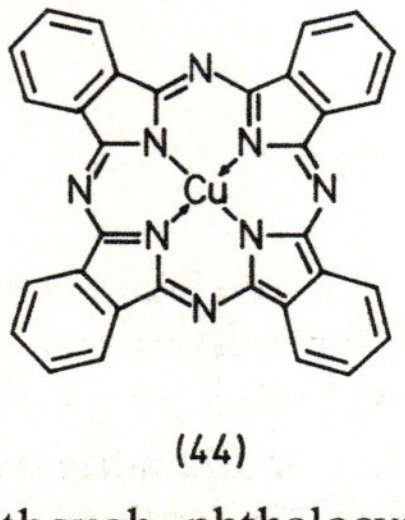

(44)

Although phthalocyanine was first discovered by Von Braun in 1907, and copper phthalocyanine twenty years later by de Diesbach, it was left to Dandridge, Drescher, Dunworth and Thomas of Scottish Dyes Ltd. to realise the potential of this new pigment. They first noticed the metal phthalocyanine as a coloured impurity in phthalimide which had been prepared by the reaction of ammonia with molten phthalic anhydride in an iron vessel. The impurity proved to be an iron phthalocyanine, the iron coming from the reaction vessel. Further research revealed that copper phthalocyanine had even better properties; its structure was later determined as (44). It is interesting to note that the phthalocyanines are structurally related to porphyrins, nature's own pigments. These phthalocyanines are valuable colouring materials with excellent properties and are used extensively today (see Chap. 5).

A second important discovery was made, in 1954, by Rattee and Stephen of ICI. They discovered a means whereby dyes containing the dichlorotriazinyl group could be made to react with cellulose under basic conditions (Scheme 1.5). The dye was thus bound to the cellulose by a covalent chemical bond and thus provided dyes of high wet fastness for cotton (further description of this technique appears in Chap. 6).

DYE∼∼NH_2 + (2,4,6-trichloro-1,3,5-triazine) ⟶ DYE∼∼N(H)–(4,6-dichloro-1,3,5-triazin-2-yl)

↓ CELLULOSE-O^-

DYE∼∼N(H)–(4-chloro-6-(O-CELLULOSE)-1,3,5-triazin-2-yl)

Scheme 1.5

Apart from these two notable discoveries, the major challenge to the dyestuffs manufacturer came from the introduction of the three main man-made fibres — nylon, polyester, and polyacrylonitrile. The problems encountered in the dyeing of nylon were quickly resolved by the use of acid dyes which, as their name suggests, contain anionic groups, especially sulphonic acid groups.

The more hydrophobic fibres, polyester and polyacrylonitrile, posed more of a problem. Eventually neutral, low molecular weight dyes (disperse dyes) were found suitable for polyester and cationic (basic) dyes for polyacrylonitrile (see Chap. 6). Though specific dye types had to be used for these fibres, they were merely modifications of dyes that were already available.

1.4 Future Trends

Two events have directed recent trends in the dyes industry — the spiralling costs of oil and energy, and much tougher toxicological and environmental constraints. As a result, the number of new dyes that are being added to manufacturers' selling ranges has decreased drastically in recent years. Instead, dye manufacturers are devoting more effort into developing better (cheaper) routes to existing dyes.

Dyes are also finding new outlets in some of the modern high technology industries such as the electronics industry. For example, dyes are being evaluated in liquid crystal displays, *e.g.* (45) and (46), lasers, *e.g.* (47) and (48), and solar cells, *e.g.* (49) and (44).

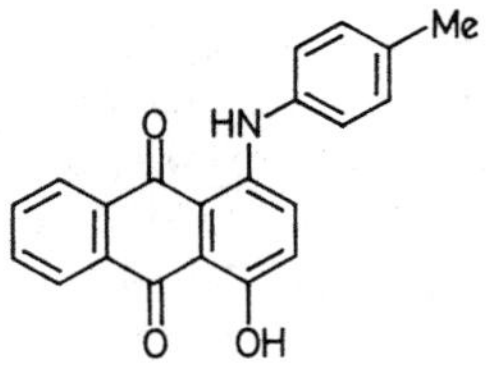

(45) 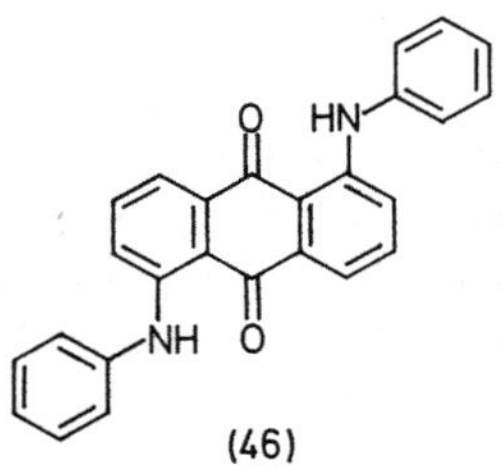 (46)

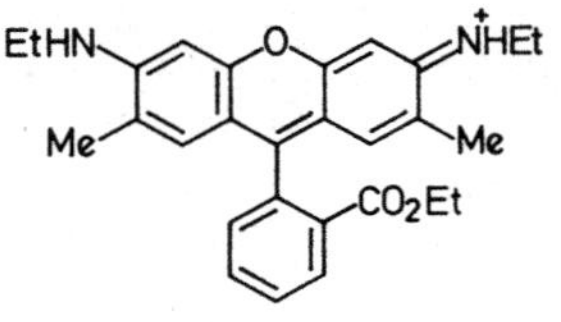

(47) 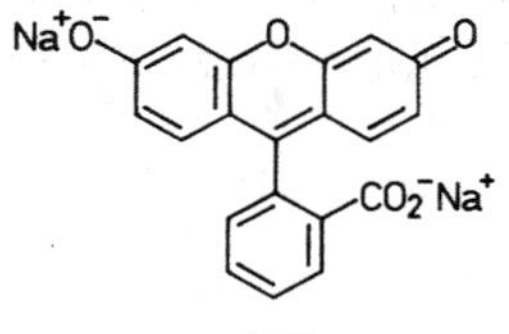 (48) 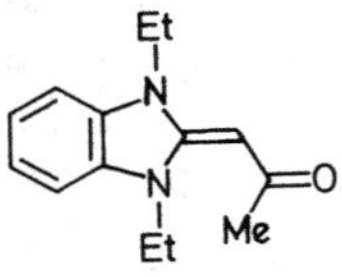 (49)

The increasingly stringent health and safety laws have prompted the development of polymeric dyes for food since here the dyes *must* be completely harmless. Polymeric anthraquinone dyes of general structure (50) look the most promising.

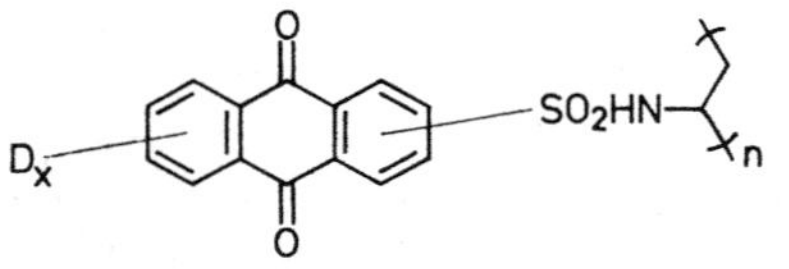

D = OH, NHR
x = 1 - 4

(50)

Finally, some members of one of the oldest class of dyes, the anthraquinone dyes, are giving exciting results as anti-cancer drugs. Important examples of these anthracycline drugs are adriamycin (51) and daunomycin (52). Because of the tremendous benefits to mankind if a successful anti-cancer drug is found, work in this area is likely to expand.

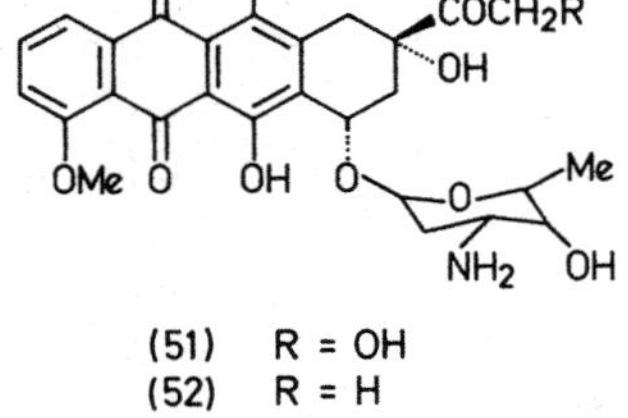

(51) R = OH
(52) R = H

1.5 Summary

Obtaining dyes from natural sources was a slow, inefficient, wasteful and very labour intensive process. The resulting natural dyes were rarely pure compounds and since the proportions in the mixture were variable, reproducibility was a serious problem. It is hardly surprising that they have been replaced by superior synthetic dyes which are produced efficiently and are pure compounds of known definitive structure, thus ensuring reproducibility.

The rise of the dyestuffs industry paralleled the early advances made in organic chemistry. It is doubtful that either would be as advanced today without the close relationship they shared, especially during the early years. The dyestuffs industry also solved many of the problems of large scale synthetic manipulations, so important to the modern organic chemical manufacturing industry. In fact, a high percentage of the present organic chemical industry is an offshoot of the dyestuffs industry; it is no coincidence that the major dyestuffs companies have successful pharmaceutical, agrochemical and other such organic chemistry intensive divisions.

Hence, just as Perkin is rightly acclaimed the Father of the dyestuffs industry, so the dyestuffs industry itself could be called the Mother of the modern organic chemical industry.

1.6 Bibliography

A. *Natural Dyes*

(a) *General Accounts with Structures*

1. Cofrancesco, A. J.: Dyes, natural. In: Encyclopedia of chemical technology, Kirk-Othmer (eds.), 2nd ed., Vol. 7, pp. 614–629. New York, London, Sydney: Interscience 1965
2. Faris, R. E.: Dyes, natural. In: Encyclopedia of chemical technology, Kirk-Othmer (eds.), 3rd ed., Vol. 8, pp. 351–373. New York, Chichester, Brisbane, Toronto: John Wiley and Sons 1979
3. Mayer, F., Cook, A. H.: The chemistry of the natural colouring matters. New York: Reinhold 1943
4. Venkataraman, K. (ed.): The chemistry of synthetic dyes. Vol. II, pp. 740–743, 820–821, 833, 1275–77. New York: Academic Press 1952
5. Baker, J. T.: Endeavour *Vol. XXXIII*, No. 118, 11 (1974)
6. Colour Index. Rowe, F. M. (ed.), 1st ed., pp. 292–300. Bradford: Society of Dyers and Colourists 1924
7. Kiel, E. G., Heertjes, P. M.: J. Soc. Dyers and Colourists *79*, 21 (1963)

(b) *General Accounts*

8. Holmyard, E. J.: Dyestuffs in the nineteenth century. In: A history of technology, Vol. V, Singer, C., Holmyard, E. J., Hall, A. R., Williams, T. I. (eds.), pp. 257–267. Oxford: Oxford University Press 1958
9. Brown, B.: Textile Colorist *65*, 487-9, 517-8 (1943)
10. Taylor, F. S.: A history of industrial chemistry, pp. 114–127. London: Heinemann 1957

B. *August Kekulé*

1. Hafner, K.: Angew. Chem. Int. Ed. Engl. *18*, 641 (1979)

C. *History of the Modern Dyestuffs and Organic Chemical Industries*

1. Ref. A8.: pp. 267–283
2. Clow, A., Clow, N. L.: The chemical revolution. London: The Patchwork Press Ltd. 1952
3. Gray, G. W.: Chimia *34*, 47 (1980)
   Pergamon Press 1966
4. Ref. A10.: pp. 214–274
5. Miall, S.: History of the British chemical industry. London: Ernest Benn Ltd. 1931
6. Johnson, A., Turner, H. A.: The Dyer and Textile Printer *1956*, 765

D. *Future Trends*

(a) *Recent Developments*

1. Clarke, A. C.: Seventies Scientists *1979*, 18
2. Feeman, J. F.: American Dyestuff Reporter *1980*, 19

(b) *Electronics Industry*

(i) *Liquid Crystal Dyes*

3. Gray, G. W.: Chimia *34*, 47 (1980)
4. Pellat, M. G., Roe, I. H. C., Constant, J.: Mol. Cryst. Liq. Cryst. *59*, 299 (1980)
5. Jones, F., Reeve, T. J.: J. Soc. Dyers and Colourists *95*, 352 (1979)

(ii) *Organic Dye Lasers*

6. Ready, J. F.: Lasers. In: Encyclopedia of chemical technology, Kirk-Othmer (eds.), Vol. 14, 3rd ed., pp. 42–81. New York, Chichester, Brisbane, Toronto: Wiley — Interscience 1981
7. Leone, S. R., Moore, C. B.: Dye lasers. In: Chemical and biochemical application of lasers, Moore (ed.), Vol. 1, pp. 12–15. New York, San Francisco, London: Academic Press 1974

(iii) *Solar Cells*

8. New Scientist *89*, 279 (1981)
9. Chamberlain, G. A., Cooney, P. J.: Nature *289*, 45 (1981)
10. Kampa, F. J., Yamashita, K., Fajer, J.: Nature *284*, 40 (1981)

(c) *Polymeric Food Dyes*

11. Leonard, W. J., Jr.: Midland Macromolecular Monographs, Vol. 5, pp. 269–292, Polymeric delivery systems. New York: Gordon and Breach 1978

(d) *Drugs*

12. Elek, S. D.: Principles of antimicrobial activity. In: Disinfection, sterilisation and preservation, Lawrence, C. A., Block, S. S. (eds.), pp. 20–25. Philadelphia: Lea and Febiger 1971
13. Remers, W. A.: The chemistry of antitumour antibiotics, Vol. 1, pp. 63–132. New York, Chichester, Brisbane, Toronto: Wiley 1979
14. Blum, R. H., Carter, S. K.: Ann. Int. Med. *80*, 249 (1974)
15. Davis, H. L., Davis, T. E.: Cancer Treat. Reps. *63*, 809 (1979)
16. Zunino, F., et al.: Biochim. Biophys. Acta *277*, 489 (1972)
17. Plumbridge, T. W., Brown, J. R.: Biochem. Biophys. Acta *479*, 441 (1977)
18. Neidle, S.: Cancer Treat. Reps. *61*, 928 (1977)
19. Murdock, K. C., et al.: J. Med. Chem. *22*, 1024 (1979)
20. Wallace, R. È., et al.: Cancer Res. *39*, 1570 (1979)

Chapter 2

Classification and Synthesis of Dyes

2.1 Introduction

The vast majority of the natural dyes used prior to the nineteenth century have been replaced by synthetic dyes discovered since then. The early advances made in organic chemistry were largely responsible for this remarkable revolution and for many years organic chemistry and dyestuffs chemistry were inextricably linked. However, as more areas of organic chemistry were investigated, *e.g.* the chemistry of natural products, so the development of organic chemistry depended to a decreasing extent upon dyestuffs research. This trend has more recently been reversed. Interest is once again focused upon dyes for, with their essentially planar π systems, they represent an interesting challenge to the application of the Molecular Orbital theories developed during the present century (see Appendix I).

Before the results of some of these studies are examined (Chaps. 3–5) the major dye classes of chemical and commercial significance will be exemplified (Sect. 2.2) and some synthetic strategy discussed (Sect. 2.3).

2.2 Classification of Dyes

Dyes may be classified either by their end use or by their chemical structure. In order that the chemistry of the dyes is emphasised, it is the latter system which has been adopted as far as possible. The classification, although by no means comprehensive, does cover the major dye classes of greatest chemical and commercial significance.

2.2.1 Azo Dyes

As already stated in Chapter 1, the azo dyes constitute, commercially, the most important class of dyestuffs. They are also the most widely studied dyes and a vast volume of literature has therefore accumulated over the past one hundred years or so.

Structure (1) is a simple representation of an azo dye: *A* and *D* are usually aromatic species (see Chap. 3). In the early dyes *A* and *D* were substituted aromatic carbocycles; however, in recent years aromatic heterocycles have been increasingly utilised.

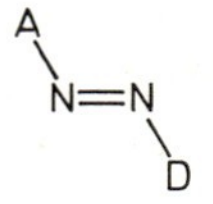

A–N=N–D (1) 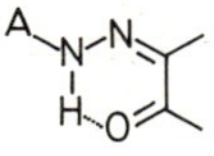 (2) $[\ldots]^{n+/-}$ (3)

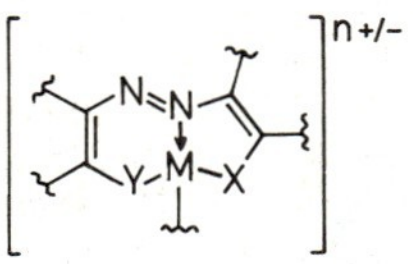

Most dyes consist of an acceptor group *A*, which is an aromatic nucleus frequently containing a chromophoric group, *e.g.* NO_2 (see Sect. 3.5.2), and a donor group, *e.g.* an aromatic nucleus *D* containing an auxochromic group, *e.g.* NR^1R^2, OH (see Sect. 3.5.2). Included in this general class are dyes which may exist as hydrazones (2), and dyes which are complexed with a metal, *e.g.* (3).

Azo dyes may contain more than one azo group, *e.g.* disazos and trisazos, which contain two and three azo groups respectively, but rarely do they contain more than four azo groups.

2.2.2 Anthraquinone Dyes

Anthraquinone dyes are second only to azo dyes in commercial importance. The commercially important anthraquinone dyes are all derivatives of 9,10-anthraquinone (4), which itself is only pale yellow. Dyes are obtained by incorporating electron-donating substituents into one or more of the eight free positions, especially the 1, 4, 5 and 8 positions. Typical of such substituents are amino, alkyl- and arylamino, hydroxy and alkoxy groups.

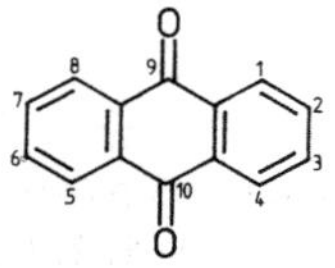

(4)

2.2.3 Vat Dyes

The majority of vat dyes (see Chap. 6 for methods of application and definition) are polycondensed aromatic carbonyl compounds, *e.g.* indanthrone (5) and benzanthrone (6). As such, they bear a close resemblance to anthraquinone dyes and many of the characteristics of anthraquinone dyes apply to these dyes.

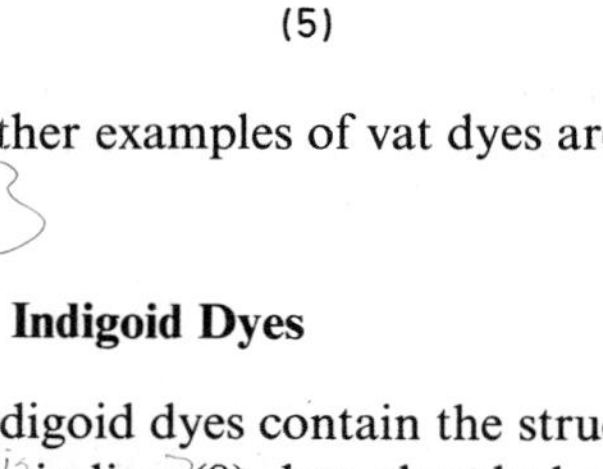

(5)

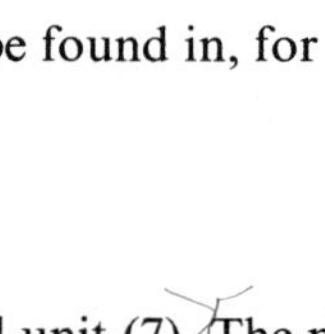

(6)

Other examples of vat dyes are to be found in, for example, indigoid and sulphur dyes.

2.2.4 Indigoid Dyes

All indigoid dyes contain the structural unit (7). The most important member of this group, indigo (8), has already been discussed (Sect. 1.3.6). However, there are also many analogues and substituted derivatives of indigo (see Sect. 5.3).

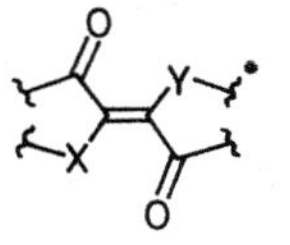

X,Y = O,S,Se,NH

(7)

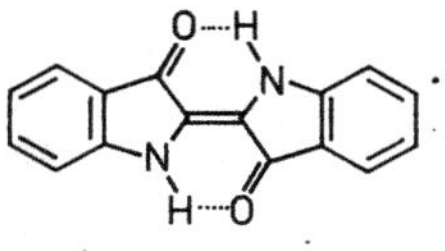

(8)

2.2.5 Polymethine Dyes

The polymethine dyes can all be represented by the general structure (9). The dyes can be cationic (z = +n), neutral (z = 0) or anionic (z = —n) according to the nature of *A* and *B*; *X* and *Y* can be either carbon or nitrogen.

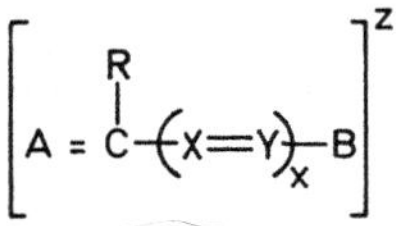

$$\left[A = \overset{R}{\overset{|}{C}} \!-\!(X{=}Y)_x\!-\! B \right]^z \qquad x = 0,1,2,\ldots \quad z = +n, 0, -n$$

(9)

Examples of cationic polymethines where *X* and *Y* are both carbon atoms are given by cyanines, *e.g.* (10), and hemicyanines, *e.g.* (11), whilst polymethines with hetero-atoms in the conjugating chain (*i.e.* *X*, *Y* = N, *etc.*) are exemplified by diazahemicyanines, *e.g.* (12), azacarbocyanines, *e.g.* (13), and diazacarbocyanines, *e.g.* (14).

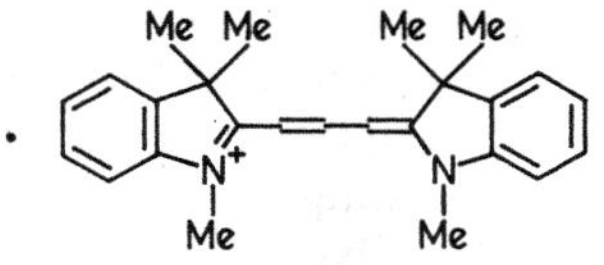

CI Basic Red 12 (10)

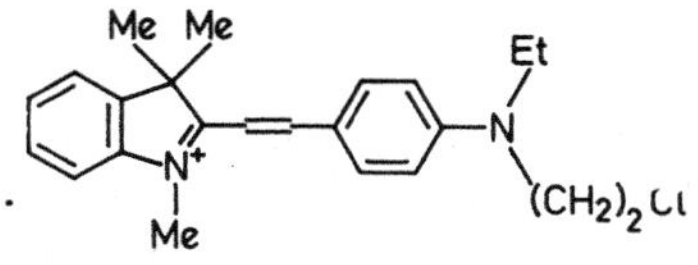

CI Basic Violet 7 (11)

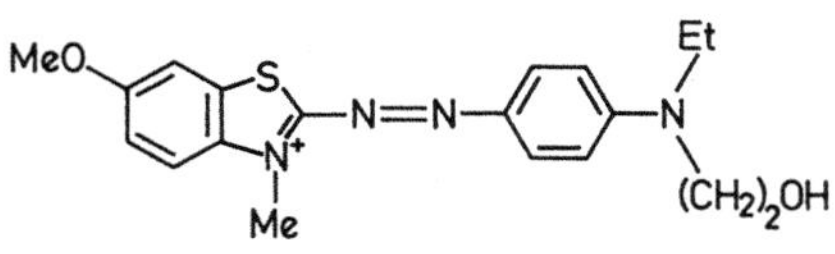

CI Basic Blue 41 (12)

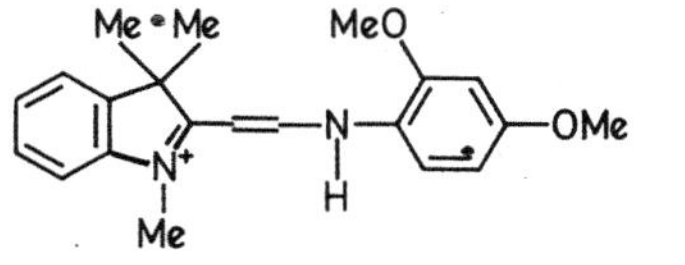

CI Basic Yellow 11 (13)

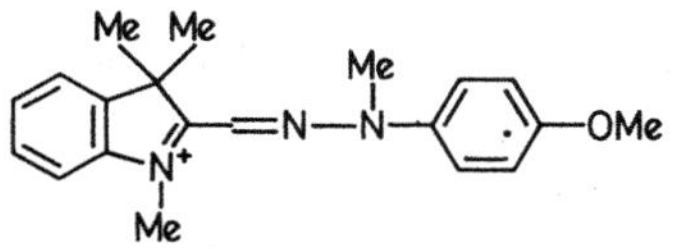

CI Basic Yellow 28 (14)

Commercially the neutral types, *e.g.* the merocyanine (15), and the anionic types, *e.g.* the oxonol (16), are much less important.

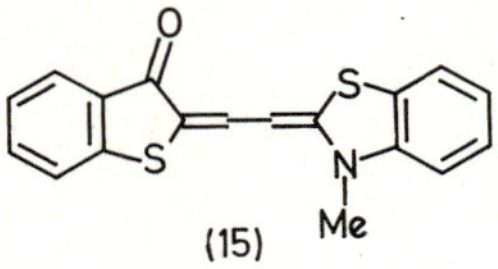

(15)

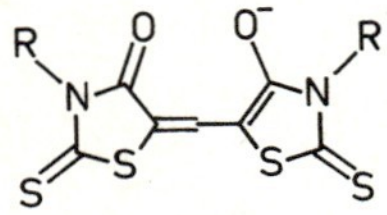

(16)

2.2.6 Aryl-Carbonium Dyes

This general class of dyes covers a great variety of structures as seen from the general formula (17).

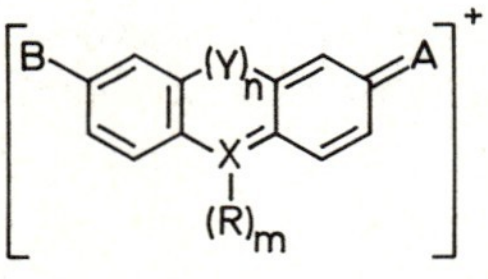

m,n = 0,1
X = C,N
Y = O,S,NR
A,B = O,S,NR
R = alkyl,aryl

(17)

Though these dyes are far less important than the azo and anthraquinone dyes, they do have some commercial significance. In fact, the first synthetic dye, Mauveine (18), belongs to this class (see Sect. 1.3.2). Dyes of this general type cover the entire shade gamut from yellow, *e.g.* (19), through red, *e.g.* (20), to blue, *e.g.* (21), and green, *e.g.* (22). To catalogue all the different types of structure that fall within this class would be unnecessarily tedious so dyes (18–22) are included as representatives of the entire class.

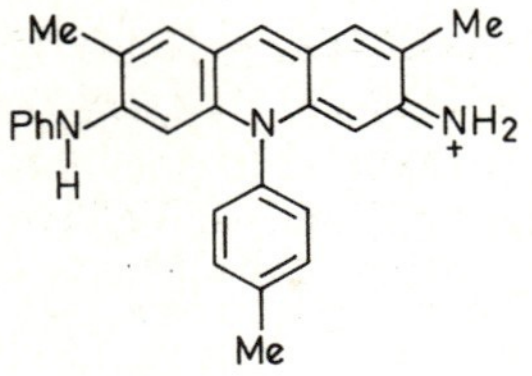

(18)

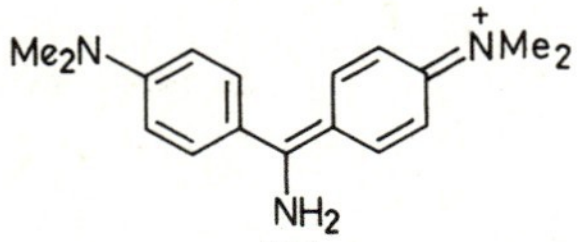

(19)

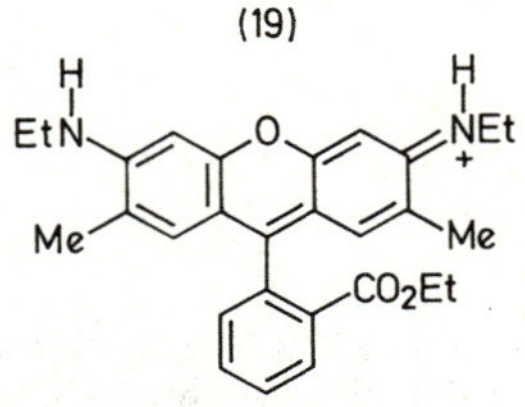

(20)

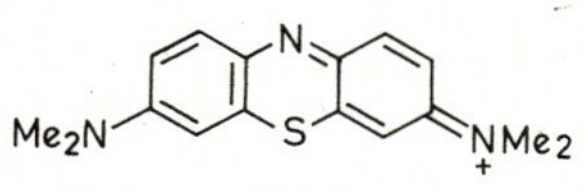

(21)

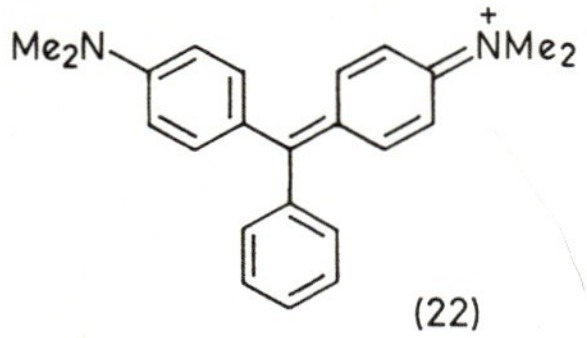

(22)

2.2.7 Phthalocyanine Dyes

Phthalocyanine dyes are a relatively new class of dyes discovered earlier this century. Since then they have become important colourants, particularly as pigments. They bear a close structural relationship to the natural pigments such as the porphyrins and in common with porphyrins they form metal complexes. Transition metals, especially copper, form the most stable complexes and for this reason the most important phthalocyanine is copper phthalocyanine (23).

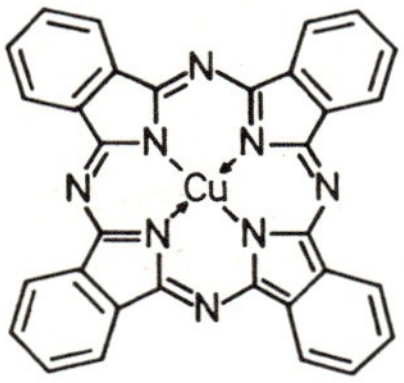

(23)

2.2.8 Nitro Dyes

Though not a very important class of dyes commercially, the nitro dyes do represent some of the simplest dye structures. Typically, they consist of two or more aromatic rings, usually benzene or naphthalene, containing at least one nitro group and one donor group, *e.g.* NH_2, OH. Martius Yellow (24) and CI Disperse Yellow 1 (25) are typical nitro dyes.

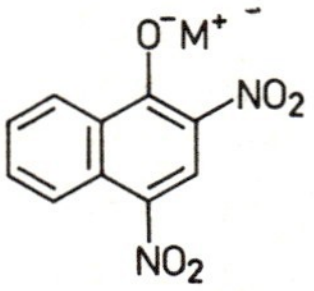

$M = NH_4^+, Na^+, Ca^{2+}$

(24)

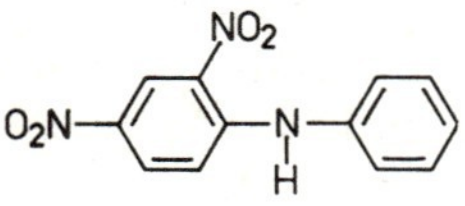

(25)

2.2.9 Miscellaneous Dye Classes

There are many other dyes which do not fall into the above eight categories. These dyes have been deliberately neglected because they are either commercially less important than those previously mentioned, *e.g.* naphtholactams (26) and coumarins (27), little studied, *e.g.* triphendioxazines (28) and formazans (29), or remain rather obscure, like the sulphur dyes (Chap. 5) whose structures are still in some doubt.

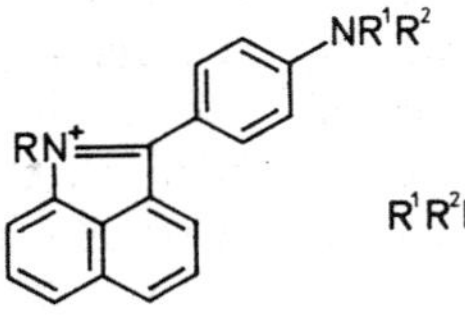

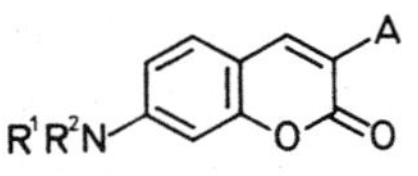

A = Acceptor

(26)

(27)

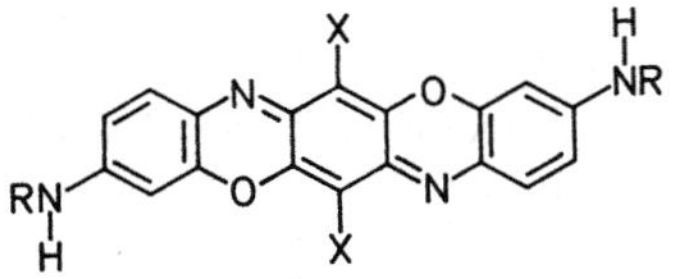

(28)

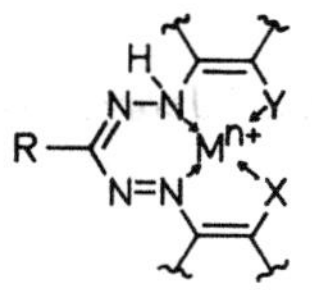

(29)

2.3 Synthesis of Dye Intermediates

The majority of dyes are prepared from intermediates which have been previously synthesised. The final step in the synthesis of the dye is usually the one in which the chromogen is actually assembled, as in the diazotisation and coupling of aromatic amines to give azo dyes. This is so for the majority of dyes. However, a notable exception is the anthraquinone class of dyes. Here the anthraquinone skeleton is usually assembled first and the required substituents introduced afterwards. Thus, the synthesis of dye intermediates will be considered first (Sects. 2.3.1, 2.3.2) whilst the methods by which the different chromogens are assembled will be considered later (Sect. 2.4). In the case of the anthraquinone dyes the synthesis of the anthraquinone skeleton is discussed at the same time as the routes by which the various substituents are introduced. There will therefore be a slight overlap of section 2.4.2 with section 2.3.1.

2.3.1 Synthesis of Aromatic Carbocycles

Benzene and naphthalene are by far the most important aromatic carbocycles used in the dyestuffs industry: both can be obtained from either petroleum or coal tar by fractional distillation. The hundreds of benzene and naphthalene intermediates used can be prepared from these parent compounds by the sequential introduction of a variety of substituents, *e.g.* NO_2, NR^1R^2, Cl, SO_3H, CN, SO_2R, *etc.* These substituents are introduced into the aromatic ring by one of two routes; either by electrophilic substitution or by nucleophilic substitution.

In general, aromatic rings, because of their inherently high electron density, are much more susceptible to electrophilic attack than to nucleophilic attack. Nucleophilic attack only occurs under forcing conditions unless the aromatic ring already contains a powerful electron-withdrawing group, *e.g.* NO_2. In this case, nucleophilic attack is greatly facilitated because of the reduced electron density at the ring carbon atoms.

Electrophilic Substitution

The commonest mechanism for electrophilic attack at an aromatic system involves the initial attack of an electrophile $E^{\oplus}$ to give an intermediate containing a tetrahedral carbon atom; loss of $X^{\oplus}$, usually a proton, from the intermediate then gives the product (Eq. 2.1).

X → (+E⁺) → E X (+) → (−X⁺) → E

Eq. 2.1

The S_E1 mechanism (Eq. 2.2), in which the leaving group departs before the electrophile attacks, is less frequently encountered.

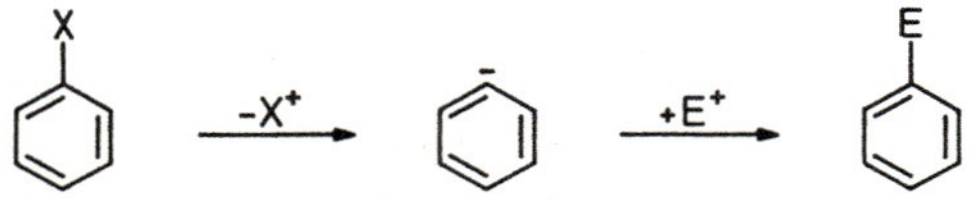

X → (−X⁺) → − → (+E⁺) → E

Eq. 2.2

The electrophile can be either cationic, *e.g.* $NO_2^{\oplus}$, or a system capable of polarisation, *e.g.* $Cl^{\delta+}—Cl^{\delta-}$. The rate determining step in these reactions is usually either the formation of the electrophile or its attack at the aromatic ring.

In an unsubstituted benzene ring the position of attack presents no problems by virtue of the symmetry of the benzene ring. However, problems arise when electrophilic attack occurs at a benzene ring already containing a group for there are now three possible sites of attack (30).

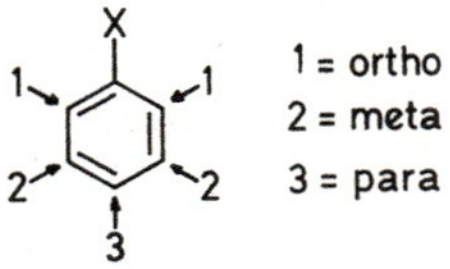

X; 1, 1, 2, 2, 3

1 = ortho
2 = meta
3 = para

(30)

Fortunately, the position of attack may be predicted with a fair degree of accuracy because the various *X* groups fall into two main categories; those which direct the incoming electrophile to the *ortho* and *para* positions, and those which direct it to the *meta* position. The directing effect of the substituent is not exclusive and so *meta* directing groups may give some *ortho* and *para* products and *vice-versa*. However, the directing effect is usually sufficient to give one 'type' of product.

Thus, nitration of nitrobenzene gives approximately 93% of *meta*, 6% of *ortho* and 1% of *para* dinitrobenzene.

The majority of *ortho-para* directing groups are also activating substituents (with respect to hydrogen) whilst *meta* directing substituents are deactivating. Groups which fall into the first category are: $—O^{\ominus}$, $—NR^1R^2$, —OR, —SR, alkyl, aryl, $CO_2^{\ominus}$, and those in the second category are: $—NR_3^{\oplus}$, NO_2, CN, SO_3H, COR, CO_2R and CCl_3. The halogens, F, Cl, Br, I, are peculiar in that they deactivate the benzene ring to electrophilic attack yet direct the incoming substituent to the *ortho* and *para* positions.

Two successful explanations for these observations have been proposed; the first is based on the stability of the intermediate formed after attack. The energy of each of the three possible intermediates (31–33) is estimated by considering the respective orbital energies.[1] The lowest energy intermediate would be expected to form in preference to the other two; attack is therefore predicted at the position which gives the most stable intermediate. This system works well and does lead to agreement with experimental evidence in most cases.

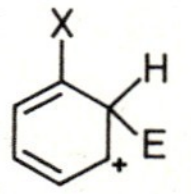

(31) 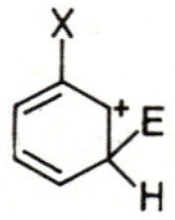 (32) 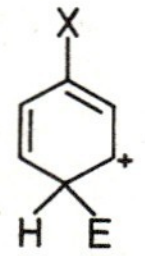 (33)

The second explanation is based on the Frontier Orbital (FMO) approach and considers the interaction between the HOMO of the substituted benzene and the LUMO of the electrophile. Thus, in contrast to the first method it considers the energy changes as the reactants *begin* to interact with each other. In particular, the orbital coefficients of the HOMO at the five vacant positions in the substituted benzene ring are estimated. The position with the highest orbital coefficient is predicted to be the one at which electrophilic attack takes place. This method works well provided that the HOMO of the benzene ring is sufficiently separated from the lower lying orbitals. Where this is not the case, it is normal to take account both of the HOMO and also of the orbitals closest in energy to the HOMO by taking a 'weighted average'.

The ability of a substituent to activate or deactivate the benzene ring to electrophilic attack may also be estimated by comparing the energy of the HOMO of the substituted benzene with the HOMO energy of benzene.[2] The FMO method works well for the substituents mentioned earlier. It is particularly elegant in its explanation for the behaviour of the halogens since it shows high orbital coefficients

[1] The reader may carry out a simple exercise by considering the number of canonical forms which may be written for each intermediate. It will be seen that in (31) and (33) the positive charge may be placed on the carbon atom bearing the substituent *X* whilst in (32) it cannot. Thus, a substituent which stabilises a positive charge should favour (31) and (33) and *vice-versa*.

[2] Those with a HOMO higher in energy than benzene would be expected to activate the ring and *vice-versa*.

at the *ortho* and *para* positions yet the HOMO for the halobenzenes is shown to be lower than the HOMO for benzene.

When the benzene ring contains more than one group it is usually harder to predict where the incoming substituent will enter. However, a few simple rules have been formulated for disubstituted benzene rings: —

(1) If the two groups favour attack at one position then the electrophile will attack there (see below). Thus, *p*-chlorobenzoic acid reacts with electrophiles *ortho* to the chloro group and *meta* to the carboxylic acid, *viz*:

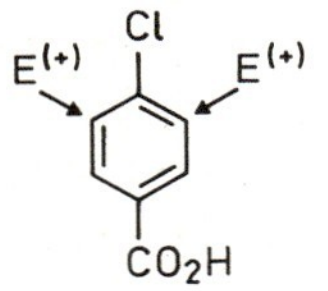

Cl
$E^{(+)}$ $E^{(+)}$
CO_2H

(2) If a strongly activating group competes with a group which deactivates or only weakly activates the benzene ring, then the position of attack is controlled by the strongly activating group.
(3) For steric reasons, an electrophile is least likely to attack the position between two groups in a *meta* aspect to each other.
(4) In the situation shown in (34) attack will take place at position *A* rather than *B*.

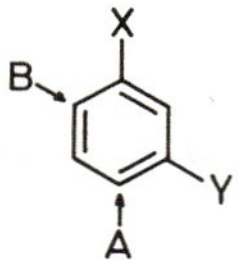

X
B
Y
A

X electron donor
Y electron acceptor

(34)

The other important carbocyclic ring system used in dyes is naphthalene (35). Here the preferred position of attack is at the 1-position. However, 2-substituted naphthalenes are thermodynamically more stable and under equilibrating conditions the 2-isomer is formed in preference to the 1-isomer. Thus, 1-naphthalenesulphonic acid is formed when naphthalene is sulphonated at lower temperatures whereas 2-naphthalenesulphonic acid is formed at higher temperatures (Eq. 2.3). This kinetic versus thermodynamic control is found only in the sulphonation reaction and does not occur for other electrophilic reactions.

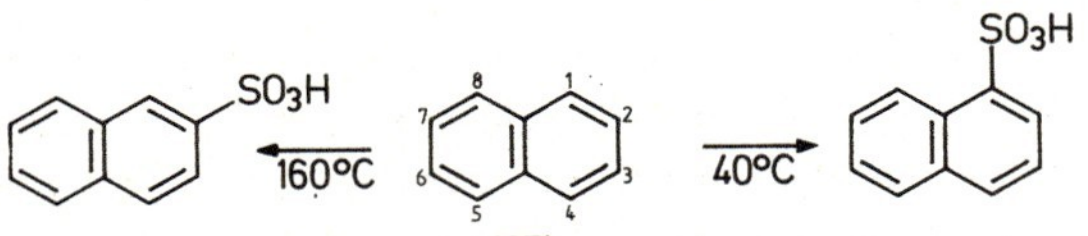

SO_3H ← 160°C (35) 40°C → SO_3H Eq. 2.3

Activating substituents in one of the rings of naphthalene promote substitution in that ring — see (36) and (37). These effects may be rationalised by similar arguments to those put forward in the case of benzene and many analogies may be drawn between the two systems.

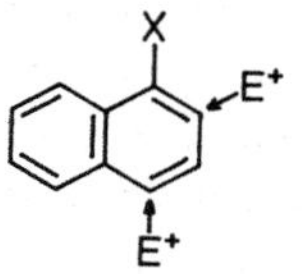

(36)

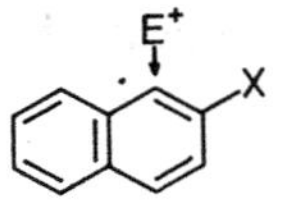

(37)

Nucleophilic Substitution

As already pointed out, the unsubstituted benzene ring is not susceptible to nucleophilic attack. However, if the benzene ring contains a good leaving group, (*e.g.* Cl) and a strong electron-withdrawing substituent, (*e.g.* NO_2) in an *ortho* or *para* position, then nucleophilic substitution is greatly facilitated.

The most important mechanism for nucleophilic aromatic substitution is the S_NAr mechanism (Eq. 2.4). The first step is usually rate determining since this is the step in which the aromaticity is lost. The S_N1 mechanism, in which the substituent leaves before the incoming nucleophile attacks, is less frequently encountered, although it is known to occur in the substitution of the diazonium group ($—N^{\oplus}{\equiv}N$).

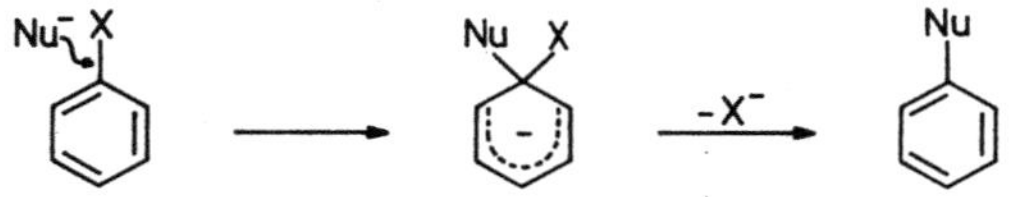

Eq. 2.4

A few other rather special mechanisms for nucleophilic substitution at benzene have been observed. One of these is the proposed intermediacy of benzynes which are then attacked by nucleophiles to give the substituted product. If the nucleophile attacks at a position other than that from which the leaving group originated, it is called *cine* substitution. A feature of this type of reaction is the need for a very strong base, (*e.g.* $NaNH_2$) — Eq. 2.5.

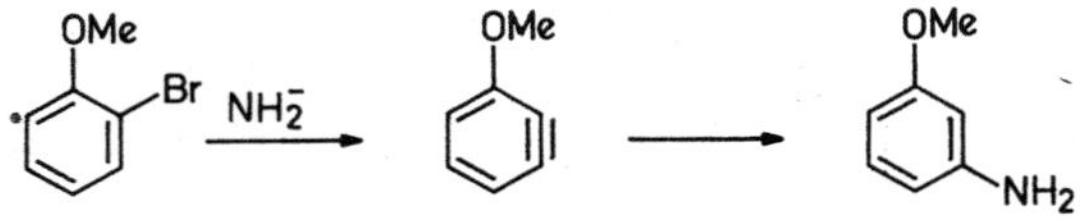

Eq. 2.5

Orientation in nucleophilic aromatic substitution is not problematic since the nucleophile usually replaces the leaving group, although there are exceptions (see Eq. 2.5). As in nucleophilic aliphatic substitution the common leaving groups are the halides, sulphonates and ammonium salts. In addition, nitro, alkoxy, sulphone and sulphonic acid groups are frequently encountered as leaving groups in aromatic substitution. In fact, the nitro group is found to be an excellent leaving group. An approximate order of leaving group ability is: F > NO_2 > OTs > > SOPh > halogens > N_3 > $NR_3^{\oplus}$ > OR, SR, NR_2. However, the order depends to some extent upon the nucleophile.

To predict a similar order for nucleophiles is somewhat more difficult because nucleophilic attack depends upon the substrate, solvent and reaction conditions. However, the following series gives an approximate idea of such an order: $NH_2^{\ominus} >$ $> PhNH^{\ominus} > ArS^{\ominus} > RO^{\ominus} > R_2NH > ArO^{\ominus} > OH^{\ominus} > ArNH_2 > I^{\ominus} > Br^{\ominus} >$ $> Cl^{\ominus} > H_2O >$ ROH. A notable omission from this list is the cyanide ion, which is a poor nucleophile for aromatic systems and only attacks the aromatic nucleus in exceptional circumstances.

Reactions

The methods of introduction of the most important groups encountered in dyestuffs is now presented.

(a) Nitro Groups

The introduction of nitro groups into aromatic hydrocarbons is one of the most important processes encountered in the dyestuffs industry. Not only is the nitro group important in its own right, as a chromophore, but it also provides a ready access to the amino group by reduction.

The nitro group is almost always introduced into the aromatic ring by direct nitration using either concentrated nitric acid or a mixture of concentrated nitric and sulphuric acids. The active nitrating species is the nitronium ion ($NO_2^{\oplus}$) which is formed according to Eq. 2.6a and b respectively. Since the equilibrium lies far to the right in Eq. 2.6b, most nitrations are carried out using a mixture of concentrated nitric and sulphuric acids.

$$\text{(a)} \quad 2HNO_3 \rightleftharpoons NO_2^+ + NO_3^- + H_2O$$

$$\text{(b)} \quad HNO_3 + 2H_2SO_4 \rightleftharpoons NO_2^+ + H_3O^+ + 2HSO_4^- \qquad \text{Eq. 2.6}$$

Mononitration is fairly easily accomplished and on the industrial scale the nitration of benzene to give nitrobenzene is a high yield process carried out at ~50 °C (Eq. 2.7). The introduction of a second nitro group to give *m*-dinitrobenzene requires a higher temperature and stronger reagents; this is consistent with the deactivating effect of the first nitro group (Eq. 2.7).

$$C_6H_6 \xrightarrow{'NO_2^+'} C_6H_5NO_2 \xrightarrow{'NO_2^+'} m\text{-}C_6H_4(NO_2)_2 \qquad \text{Eq. 2.7}$$

As mentioned earlier, the formation of *m*-dinitrobenzene is accompanied by small amounts of the *ortho* and *para* isomers. On the industrial scale these impurities are removed by treating the dinitrated benzenes with sodium sulphite (Eq. 2.8a, b). The sulphite attacks both the *ortho* and *para* dinitrobenzenes to give the corresponding water soluble nitrosulphonates which are easily separable from the water insoluble nitrobenzene. This interesting method of purification demonstrates the remarkable effect the nitro group has in facilitating nucleophilic attack at the *ortho* and *para* positions in a benzene ring, and also underlines the ease with which suitably activated nitro groups may be replaced.

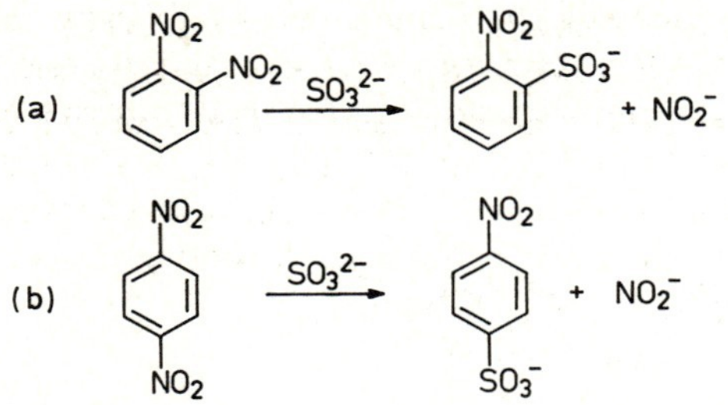

(a) o-dinitrobenzene $\xrightarrow{SO_3^{2-}}$ o-nitrobenzenesulphonate + NO_2^-

(b) p-dinitrobenzene $\xrightarrow{SO_3^{2-}}$ p-nitrobenzenesulphonate + NO_2^-

Eq. 2.8

Naphthalene nitrates under even milder conditions than benzene to give 1-nitronaphthalene, a major source of 1-naphthylamine. By increasing the severity of the reaction conditions both dinitration and trinitration are possible, although mixtures of nitrated naphthalenes are thus obtained. Even tetranitration is possible under very forcing conditions (*e.g.* Scheme 2.1).

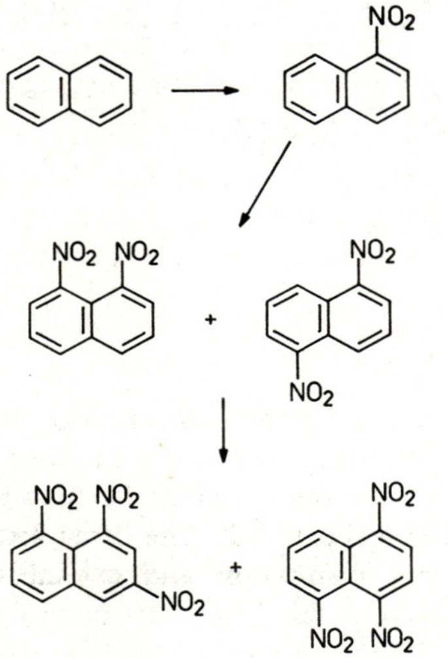

Scheme 2.1

Nitration of the naphthalenesulphonic acids provides access to some important dyestuff intermediates. If the 1-position is free then the nitro group enters there (*e.g.* 38). The importance of these nitronaphthalenes lies in their ability to be reduced to the corresponding amino compounds which are used widely as coupling components for azo dyes.

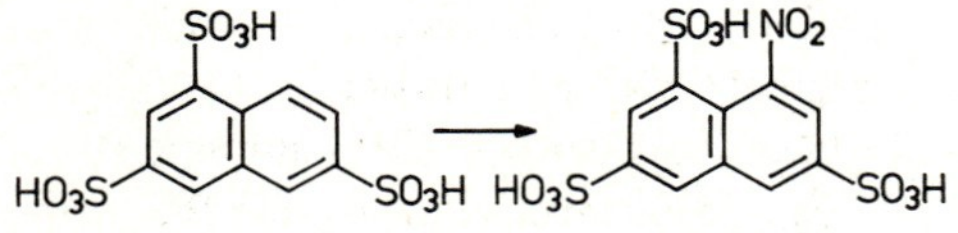

(38)

When the aromatic hydrocarbon contains an activating group, *e.g.* CH_3, NH_2, OH, then milder nitration conditions can be employed. However, care has to be exercised when nitrating these hydrocarbon derivatives in order to prevent oxidation. Thus, toluene is nitrated industrially to give *ortho* and *para*-nitrotoluene at ~20 °C to avoid it being oxidised to the corresponding carboxylic acid, whilst groups are usually protected by acetylation. An elegant method of protecting *o-aminophenols* is by preforming the benzoxazolidone (39) which is then nitrated and hydrolysed to the required nitroaminophenol.

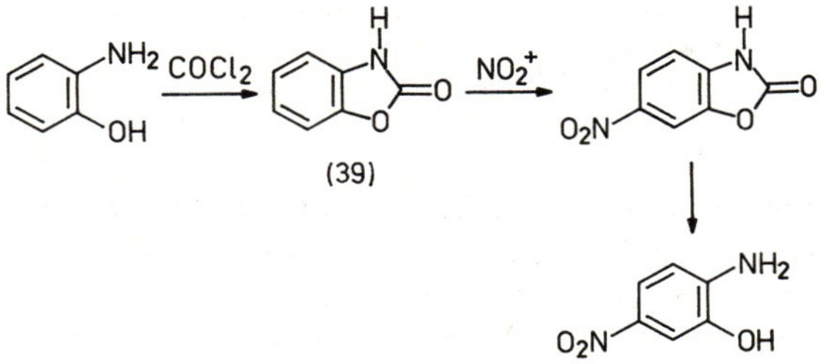

NH_2 $COCl_2$ OH (39) NO_2^+ O_2N NH_2 O_2N OH

Protonation of an amino group also protects it from oxidation but the resulting ammonium salt is now, of course, both *meta* directing and deactivating. This strategy has been used to advantage in the nitration of *o*-anisidine to give the nitroanisidine (40) in high yield. The methoxy and ammonium groups reinforce attack at the position *para* to the methoxy group and *meta* to the ammonium group.

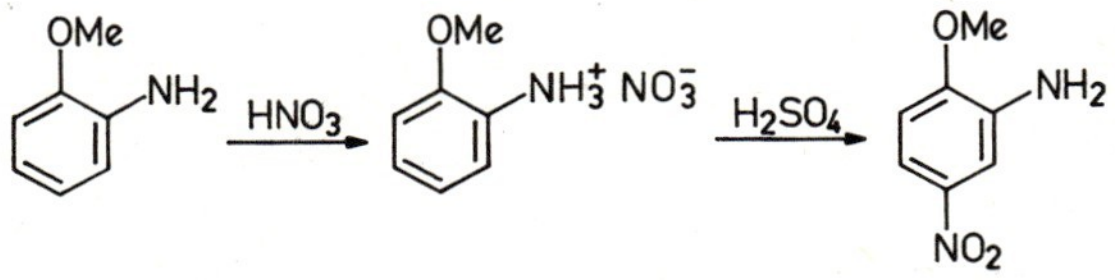

OMe NH_2 HNO_3 OMe $NH_3^+ NO_3^-$ H_2SO_4 OMe NH_2 NO_2

(40)

Chloronitrobenzenes are also important dyestuff intermediates since they may be reduced to chloroanilines or aminated to give nitroanilines. The two possible orientations of the chloro and nitro group may be obtained simply by altering the sequence of chlorination/nitration as depicted in Scheme 2.2. The *ortho* and *para* chloronitrobenzenes are separated by fractional distillation and crystallisation. Further nitration of either of these compounds gives 2,4-dinitrochlorobenzene (Scheme 2.3). (N.B. 2,6-dinitrochlorobenzene is formed in only minor amounts from 2-chloronitrobenzene).

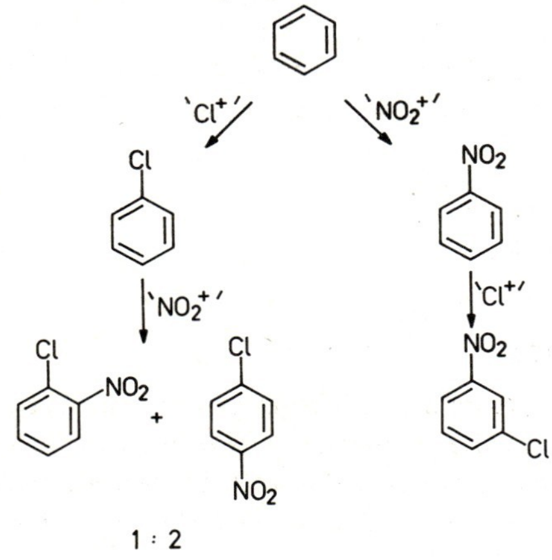

'Cl^+' 'NO_2^+' Cl NO_2 'NO_2^+' 'Cl^+' Cl NO_2 + Cl NO_2 NO_2 Cl

1 : 2

Scheme 2.2

The chlorine atom in 2,4-dinitrochlorobenzene is highly activated towards nucleophilic attack by the two nitro groups and provides a convenient source of 2,4-dinitroaniline by amination.

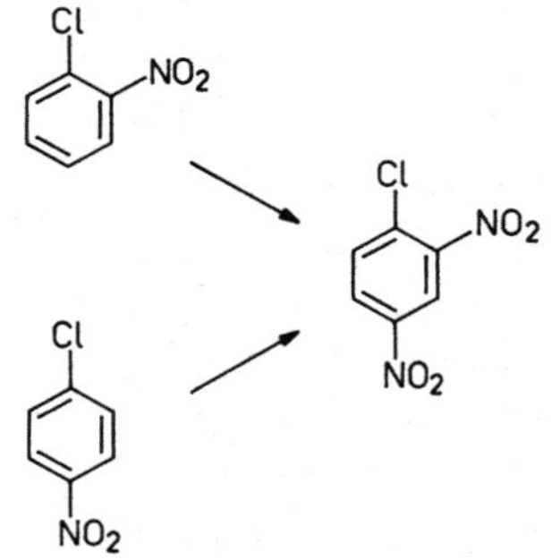

Scheme 2.3

(b) Sulphonic Acid Groups

The sulphonic acid group is used extensively in the dyestuffs industry for its water solubilising properties and for its ability to act as a good leaving group in nucleophilic substitutions. It is used almost exclusively for these purposes since it has only a minor effect on the colour of a dye.

The sulphonic acid group can be introduced into the aromatic ring by a variety of reagents, *e.g.* H_2SO_4, oleum, SO_3, and $ClSO_3H$, and under a variety of conditions. As for nitration the reaction conditions for sulphonation depend on the reactivity of the substrate, the reagent, and the number of groups that are to be introduced.

It has been difficult to establish the active sulphonating species for it is thought that this varies according to the reagent in use. Thus, in sulphuric acid the electrophile is thought to be $H_3SO_4^{\oplus}$ whilst in concentrated sulphuric acid (>85%) it appears to be $H_2S_2O_7$. In oleum ($H_2SO_4 + SO_3$) an even more active electrophile is encountered, $H_3S_2O_7^{\oplus}$, and at higher concentrations of SO_3 the electrophile is $H_2S_4O_{13}$. All these reagents are merely SO_3 carriers and not surprisingly free SO_3 is found to be the most active sulphonating agent. However, the latter has to be used in aprotic solvents. As a result sulphuric acid, oleum and chlorosulphonic acid are the preferred sulphonating agents on the manufacturing scale. A general mechanism for sulphonation is shown in Scheme 2.4.

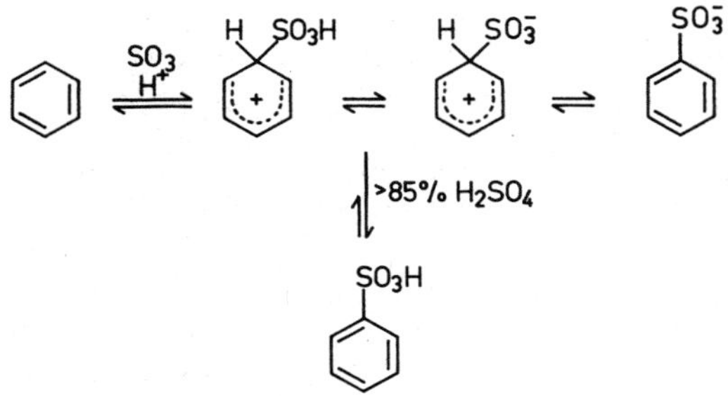

Scheme 2.4

The sulphonic acid group can be introduced into molecules containing a variety of other groups such as halogen, hydroxy, acylamino, *etc.*, without interfering with this functionality. Sulphonation is retarded by the presence of deactivating substituents, especially nitro groups. For instance, under normal conditions 2,4-dinitrobenzene cannot be sulphonated. Sulphonation of benzene gives initially the monosulphonated benzene: this is then converted to the *m*-disulphonic acid under more severe conditions or longer reaction times. Sulphonation of toluene leads to a mixture of *ortho* and *para* isomers which can be sulphonated further to toluene-2,4-disulphonic acid.

As in the nitration of naphthalene, sulphonation gives the 1-substituted naphthalene. However, because the reverse reaction (desulphonation) is appreciably fast at higher temperatures, the thermodynamically controlled product, naphthalene-2-sulphonic acid, can also be obtained. Thus, it is possible to obtain either of the two possible isomers of naphthalene sulphonic acid. Under kinetically controlled conditions naphthalene-1-sulphonic acid is obtained whilst thermodynamic control gives naphthalene-2-sulphonic acid (Scheme 2.5). Similarly, 2-naphthol can give either the 1-sulphonic acid (41) or the 6-sulphonic acid (42).

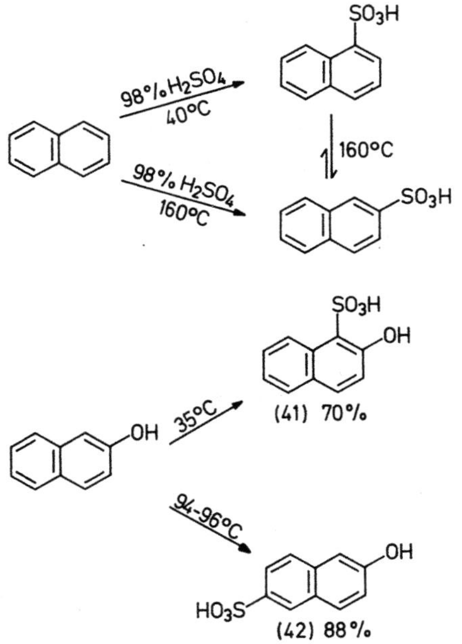

Scheme 2.5

As is the case with benzene, prolonged heating and stronger reagents lead to the introduction of further sulphonic acid groups into the naphthalene ring. In this way disulphonated and trisulphonated derivatives can be obtained. By careful manipulation of the reaction conditions a high yield of just one isomer is possible, *e.g.* in the preparation of 1,3,6-naphthalenetrisulphonic acid (43). When sulphonic acid groups are introduced into a naphthalene ring they always enter at a vacant position which is not *ortho*, *para*, or *peri* to another sulphonic acid group. The result of this interesting experimental observation is that only one tetrasulphonated naphthalene (44) is obtained and that pentasulphonated naphthalenes cannot be obtained by sulphonation.

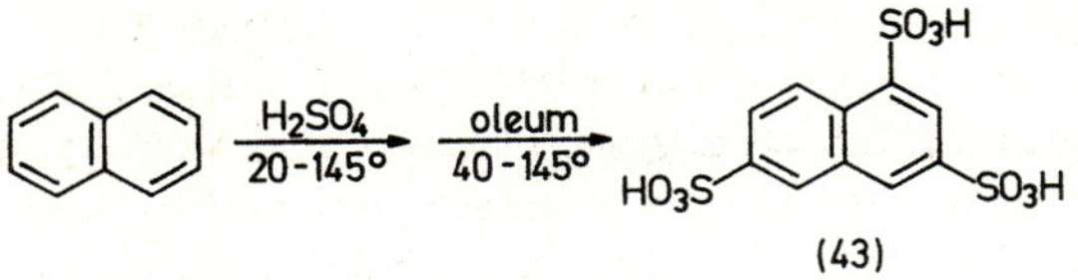

(43)

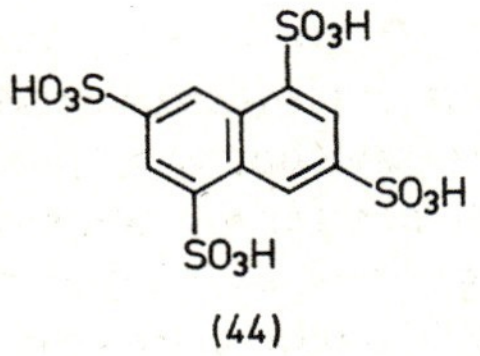

(44)

'Bake sulphonation' is an important variant on the normal sulphonation procedure. The reaction is restricted to aromatic amines, the sulphate salts of which are prepared and heated (dry) at a temperature of approximately 200 °C *in vacuo*. The sulphonic acid group migrates to the *ortho* or *para* positions of the amine (Eq. 2.9): this tendency is also apparent in polynuclear systems so that 1-naphthylamine gives 1-naphthylamine-4-sulphonic acid. The reaction is therefore useful for the introduction of a sulphonic acid group into a specific position. An example where bake sulphonation complements conventional sulphonating procedures is given by the sulphonation of 2,4-dimethylaniline. Conventional sulphonating conditions result in the sulphonic acid group entering the 5-position (directed of course *ortho* and *para* to the methyl groups and *meta* to the $NH_3^{\oplus}$ group) to give the aniline-3-sulphonic acid (45), whereas bake sulphonation gives the aniline-2-sulphonic acid (46).

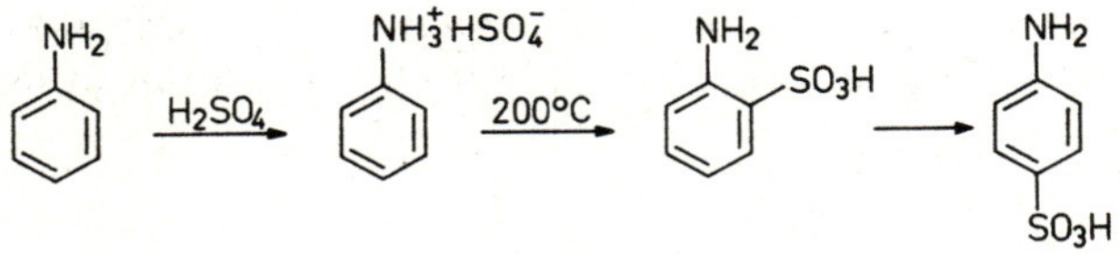

Eq. 2.9

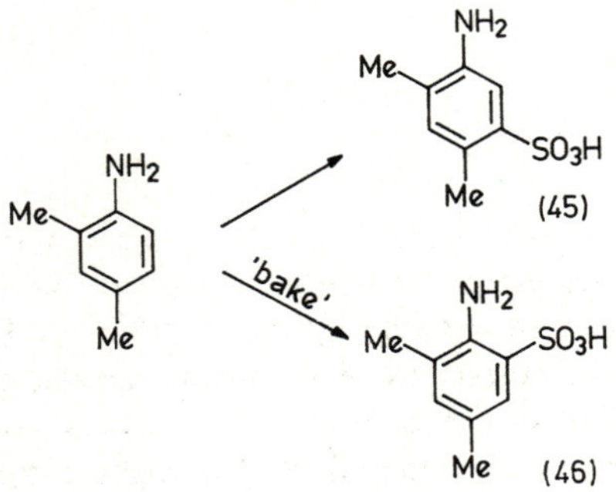

(45)

(46)

In cases where a large excess of acid is undesirable chlorosulphonic acid is employed (Eq. 2.10). An excess of chlorosulphonic acid leads to the introduction of a chlorosulphonyl group which is a useful intermediate for the preparation of sulphonamides and sulphonate esters.

$$ArH + ClSO_3H \longrightarrow ArSO_3H \xrightarrow{ClSO_3H} ArSO_2Cl \begin{cases} \xrightarrow{HNR^1R^2} ArSO_2NR^1R^2 \\ \xrightarrow{ROH} ArSO_3R \end{cases}$$ Eq. 2.10

It is possible to introduce sulphonic acid groups by alternative methods but these are little used in the dyestuffs industry. However, one worth mentioning is sulphitation because it provides an example of the introduction of a sulphonic acid group by nucleophilic substitution. The process involves treating an active halogen compound with sodium sulphite. Equation 2.11 illustrates such a reaction. As mentioned earlier, this reaction is used in the purification of *m*-dinitrobenzene.

$$Cl\text{-}C_6H_4\text{-}NO_2 \xrightarrow{Na_2SO_3} {}^-O_3S\,Na^+\text{-}C_6H_4\text{-}NO_2 + NaCl$$ Eq. 2.11

(c) Halogens

In common with nitro and sulphonic acid groups the halogens, F, Cl, Br and I, fulfil two functions in dyestuffs; first as substituents in the final dyes and second as useful synthetic intermediates. The two most important halogens are chlorine and bromine.

Chlorine and bromine are usually introduced into benzenoid aromatics by direct halogenation. In contrast, they are frequently introduced into the naphthalene nucleus using the Sandmeyer (Eq. 2.12a) or Gattermann (Eq. 2.12b) reactions. When chlorine and bromine are introduced directly via the elemental forms, *i.e.*

(a) $$C_{10}H_7N_2^+ \xrightarrow[X^-]{CuX} C_{10}H_7X + N_2$$

Eq. 2.12

(b) $$C_{10}H_7N_2^+ \xrightarrow{Cu/HX} C_{10}H_7X + N_2$$

chlorine gas or bromine liquid, a catalyst is required which is usually the corresponding iron III halide ($FeCl_3$ or $FeBr_3$). The iron catalyst is added either directly or is formed *in situ* by the addition of iron to the reaction mixture ($2\,Fe + 3\,X_2 \rightarrow 2\,FeX_3$). In the presence of catalyst halogenation proceeds by electrophilic attack of the halonium ion ($FeX_3 + X_2 \rightarrow X^{\oplus} + FeX_4^{\ominus}$) at the aromatic nucleus to give the corresponding halobenzene. Thus, benzene is chlorinated by passing a stream of chlorine gas into benzene containing a catalytic amount of $FeCl_3$ at approximately 30 °C. Similarly toluene is chlorinated to give a mixture of *ortho* and *para*-chlorotoluenes which is separated by fractional distillation. Predictably, deactivated benzenes, *e.g.* nitrobenzene, require higher temperatures (> 50 °C) for reaction to occur.

In contrast, activated systems such as phenols and anilines require no catalyst and the reaction proceeds readily at room temperature. In fact, phenols are so active that dilute sodium hypochlorite (NaOCl) solutions must be used to obtain the mono-chlorinated derivative whilst aromatic amines have to be deactivated by acetylation to avoid over chlorination (Scheme 2.6).

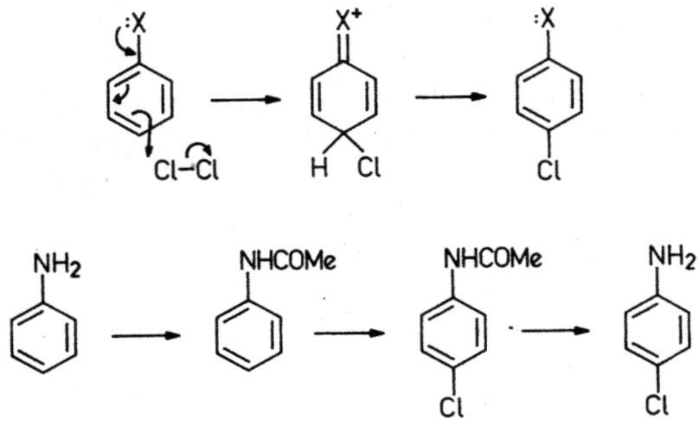

Scheme 2.6

If the aromatic hydrocarbon is substituted by alkyl groups then, under the right conditions, substitution can occur in the side chain (Scheme 2.7).

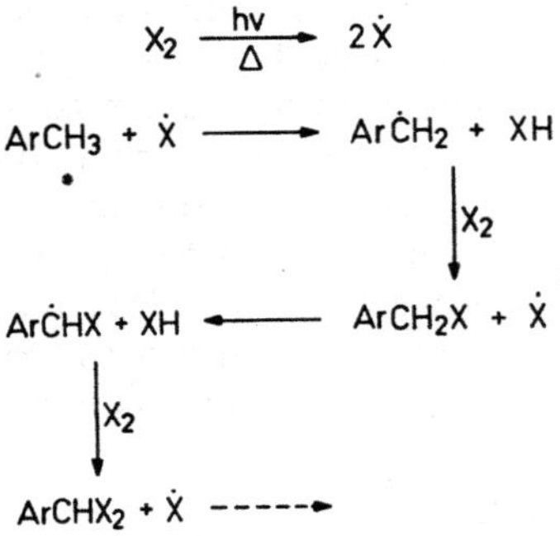

$$X_2 \xrightarrow[\Delta]{h\nu} 2\dot{X}$$

$$ArCH_3 + \dot{X} \longrightarrow Ar\dot{C}H_2 + XH$$

$$Ar\dot{C}H_2 \xrightarrow{X_2} ArCH_2X + \dot{X}$$

$$ArCH_2X + \dot{X} \longrightarrow Ar\dot{C}HX + XH$$

$$Ar\dot{C}HX \xrightarrow{X_2} ArCHX_2 + \dot{X} \dashrightarrow$$

Scheme 2.7

The mechanism involves radicals and therefore the conditions for ionic nuclear halogenation must be avoided, *e.g.* no Lewis acid catalysts (*viz.* FeX_3) must be present. As a result these reactions are carried out in iron free containers (enamel containers).

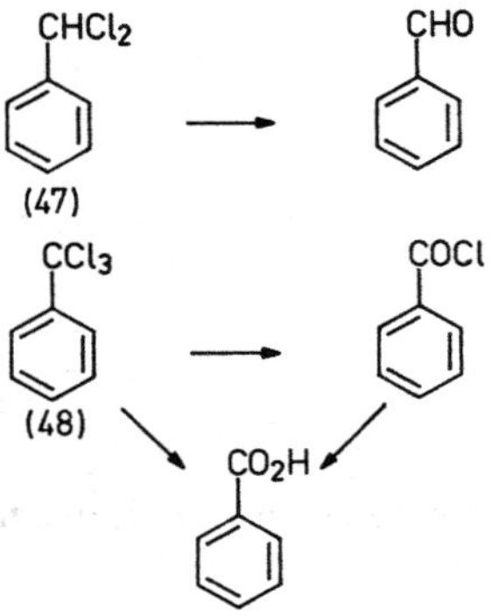

(47) $C_6H_5CHCl_2 \longrightarrow C_6H_5CHO$

(48) $C_6H_5CCl_3 \longrightarrow C_6H_5COCl$; (48) $\longrightarrow C_6H_5CO_2H \longleftarrow C_6H_5COCl$

Higher temperatures, which facilitate homolytic cleavage of the halogen, are generally required than for the ionic reaction and the presence of radical initiators, *e.g.* ultra-violet light or benzoyl peroxide, is beneficial.

Benzyl chloride, a useful alkylating agent, may thus be obtained from toluene. Further chlorination of benzyl chloride gives a mixture of the dichloro derivative (47) and the trichloro derivative (48). The latter two compounds are important sources of benzaldehyde and benzoyl chloride/benzoic acid respectively.

Fluorination and iodination reactions are used very little in dyestuffs synthesis. The only fluorinated species worthy of mention is the trifluoromethyl group which can be obtained from the trichloromethyl group by the action of hydrogen fluoride (Eq. 2.13) or antimony pentafluoride.

$$C_6H_5CCl_3 \xrightarrow{HF \text{ or } SbF_5} C_6H_5CF_3$$

Eq. 2.13

(d) Hydroxy Groups

Hydroxy groups are found in a wide variety of dyes; their importance lies primarily in the desirable properties they confer upon the dyes (see Chaps. 3–5) and rather less in their synthetic utility. The hydroxy group is rarely introduced directly into the parent aromatic nucleus but rather via nucleophilic attack at a position containing a leaving group, especially chloro, bromo, and sulphonic acid groups.

Phenol, for instance, can be prepared by heating chlorobenzene with aqueous sodium hydroxide at high temperature (> 300 °C) and pressure (> 200 atmosphere). This route is used for the industrial preparation of phenol. An alternative industrial route to phenol is by the oxidation of cumene (49), obtained from two readily available petroleum products, 1-propene and benzene (Scheme 2.8). Acetone, an important organic solvent, is also obtained. However, this latter route is restricted to phenol whilst the former route is more typical of industrial hydroxylations.

$$C_6H_6 + MeCH{=}CH_2 \xrightarrow{AlCl_3} C_6H_5CHMe_2\ (49) \xrightarrow{O_2/NaOH} C_6H_5CMe_2OOH \xrightarrow{H^+} C_6H_5CMe_2O{-}\overset{+}{O}H_2 \longrightarrow C_6H_5O\overset{+}{C}Me_2 \longrightarrow C_6H_5OCMe_2OH \longrightarrow C_6H_5OH + Me_2CO$$

Scheme 2.8

Sulphonic acid groups can be displaced from benzene and naphthalene rings by the hydroxide ion. For instance, 2-naphthol can be prepared from 2-naphthalenesulphonic acid (Eq. 2.14), and the selective replacement of just one sulphonic acid group from the trisulphonated naphthylamine (50) gives another important naphthol, H-acid (51).

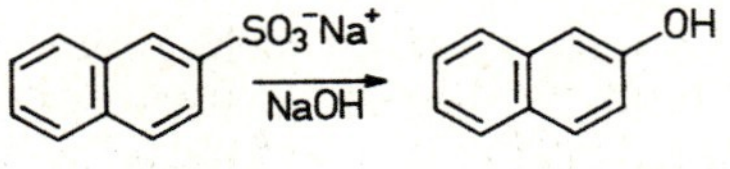

Eq. 2.14

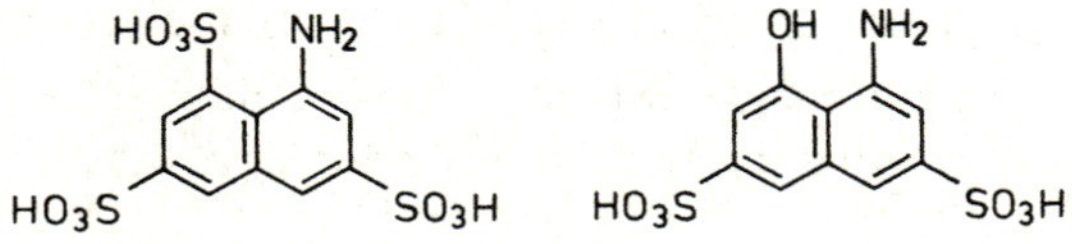

$SO_3^-Na^+$ NaOH OH

HO_3S NH_2 HO_3S SO_3H (50)

OH NH_2 HO_3S SO_3H (51)

Displacement of two sulphonic acid groups from a ring gives the corresponding dihydroxy compound, as in the preparation of resorcinol (52). The introduction of the first hydroxy group slows the rate at which the second group is introduced; this effect is particularly noticeable in the production of *ortho* and *para* dihydroxy benzenes. For example, *o*-dichlorobenzene requires a high temperature (275 °C), pressure (110 atmospheres) and a catalyst (CuCl) to give catechol.

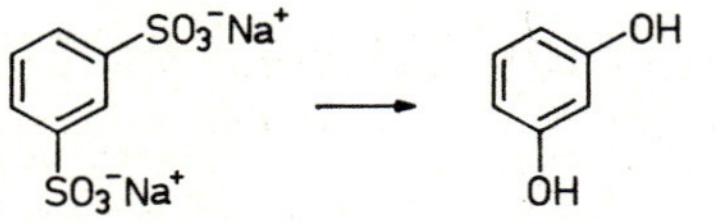

$SO_3^-Na^+$ $SO_3^-Na^+$ → OH OH (52)

In contrast, strong electron-withdrawing groups (*ortho* and *para*) greatly facilitate the displacement of chlorine from an aromatic compound. Thus, *p*-chloronitrobenzene gives *p*-nitrophenol with a 15% aqueous solution of sodium hydroxide at 160 °C and 2,4-dinitrophenol is obtained from the corresponding chlorobenzene with 5% aqueous sodium hydroxide at 100 °C.

A widely used laboratory method for the introduction of hydroxy groups into a benzene ring is to hydrolyse a diazonium group. However, this process has found little application on the industrial scale — replacement of halides and sulphonic acid groups is preferred.

(e) Amino Groups

The amino group is the single most important group in dyestuffs chemistry. In azo dyes it not only provides access to the azo group (by diazotisation and coupling) but it is also the most widely used auxochrome, as is evident by its presence in so many coupling components (see Chap. 3). It is also the most important donor group found in anthraquinones (see Chap. 4) and is present in nitro, arylmethane and polymethine dyes (see Chap. 5).

As was the case with the hydroxy group, the amino group is almost always introduced via the intermediacy of another group. Probably the most popular method is by the reduction of a nitro group; the method normally employed is known as the Béchamp method. In this method the nitro compound, iron and a small quantity of an acid are mixed in water with an initial heating period to initiate the reaction. The reduction proceeds according to Eq. 2.15 and works for many nitro compounds. Thus,

$$2\,ArNO_2 + 5\,Fe + 4\,H_2O \longrightarrow 2\,ArNH_2 + Fe_3O_4 + 2\,Fe(OH)_2 \qquad \text{Eq. 2.15}$$

aniline, chloroanilines, aminophenols, anilinesulphonic acids and various other anilines and naphthylamines are thereby obtained. However, iron is not the sole reducing agent employed and sodium sulphide and sodium bisulphite find applications. Sodium sulphide is particularly useful for it will selectively reduce nitro groups in the presence of azo groups and will even reduce just one nitro group in a polynitro compound. Thus, *m*-dinitrobenzene can be reduced to *m*-nitroaniline, and picric acid (53) can be reduced to the dinitroaminophenol (54). Good yields are obtained in both cases.

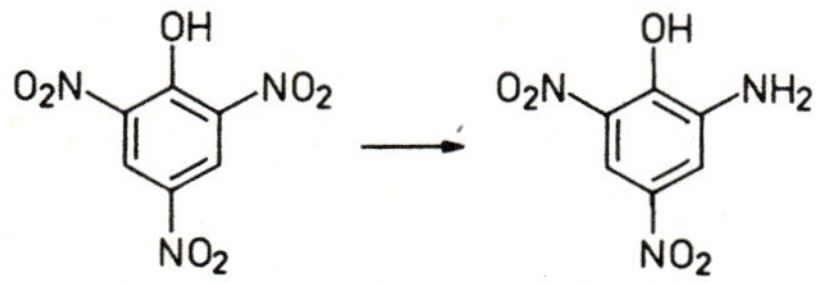

OH, O_2N, NO_2, NO_2 (53) → OH, O_2N, NH_2, NO_2 (54)

Sodium bisulphite is useful when a sulphonated amine is required for it is capable of reducing a nitro group and introducing a sulphonic acid group at the same time (Eq. 2.16).

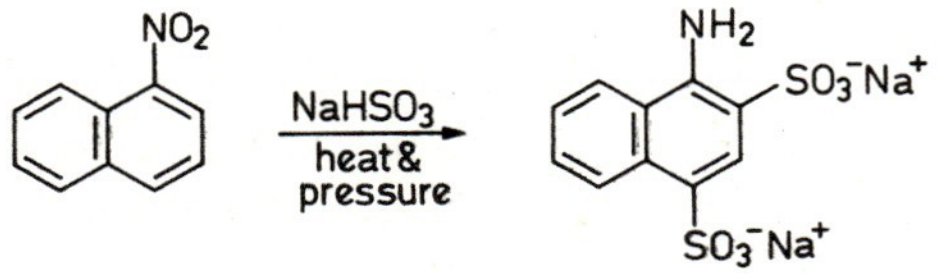

NO_2 (naphthalene) —$NaHSO_3$, heat & pressure→ NH_2, $SO_3^-Na^+$, $SO_3^-Na^+$ Eq. 2.16

As for hydroxylations, aminations can be effected by nucleophilic substitution. A halogen atom or a sulphonic acid group are the usual leaving groups, Eq. 2.17.

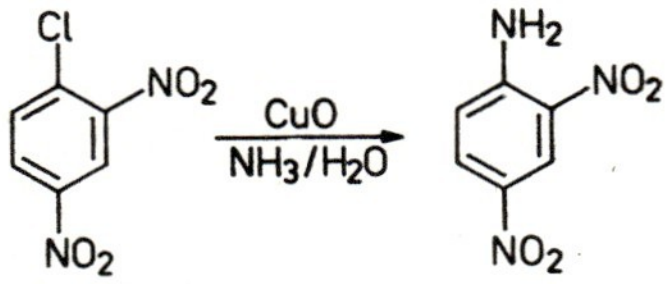

Cl, NO_2, NO_2 —CuO, NH_3/H_2O→ NH_2, NO_2, NO_2 Eq. 2.17

In the naphthalene series a very important method of introducing an amino group is via the Bucherer reaction. Scheme 2.9 illustrates the reaction for the conversion of 2-naphthol to 2-naphthylamine, once an important dyestuffs intermediate. However, 2-naphthylamine has been found to be a potent carcinogen in humans and is therefore no longer prepared directly. Fortunately, its sulphonated derivatives are non-carcinogenic and naphtholsulphonic acids are therefore converted to naphthylamine sulphonic acids via the Bucherer reaction.

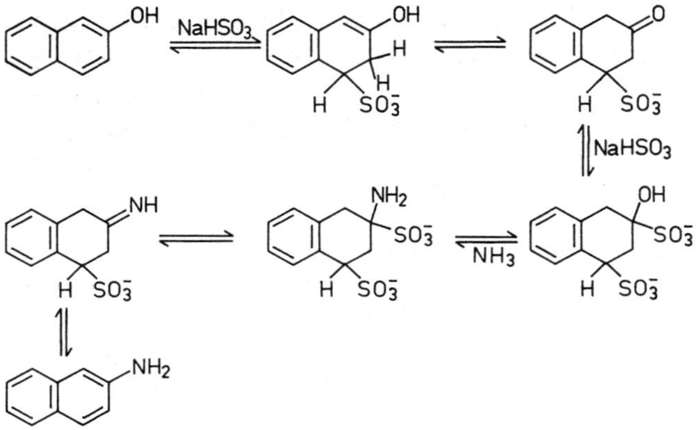

Scheme 2.9

The benzidine rearrangement provides access to some important dyestuff intermediates. As the name implies, these intermediates are all related to benzidine (55). Although the mechanism is not known for certain, that depicted in Scheme 2.10 is consistent with our present knowledge of the reaction. Particularly noteworthy is the intramolecular nature of the reaction as indicated by the lack of 'crossover products'.[3]

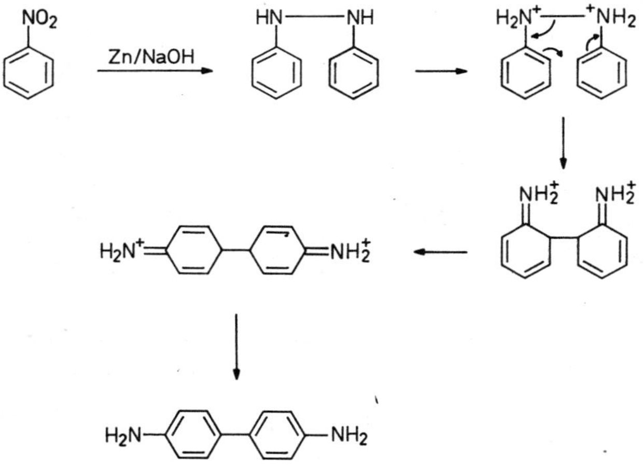

(55)

Scheme 2.10

(f) Miscellaneous Reactions

The above examples list some of the most important functional groups introduced into aromatic systems and the methods by which this is accomplished. However, there are many more functional groups and types of reaction that have not been discussed, *e.g.* alkylation, arylation, acylation, formylation, and nitrosation, which provide important intermediates, *e.g.* salicylic acid (Scheme 2.17) and phthalic

[3] Crossover products are products which are derived by the scrambling of structural subunits from two similar precursors which independently would give only one type of product. Such scrambling is indicative of an intermolecular reaction.

anhydride (56 → Scheme 2.15) or exemplify interesting chemistry, *e.g.* copper cyanide displacements, Scheme 2.16. These have been omitted, sometimes rather arbitrarily, to keep the length of this chapter within reasonable bounds. However, Schemes 2.11 to 2.17 have been included in an attempt to show how the chemistry and intermediates are interrelated and by reading the sections that follow the reader should build up a coherent picture of dyestuffs synthesis.

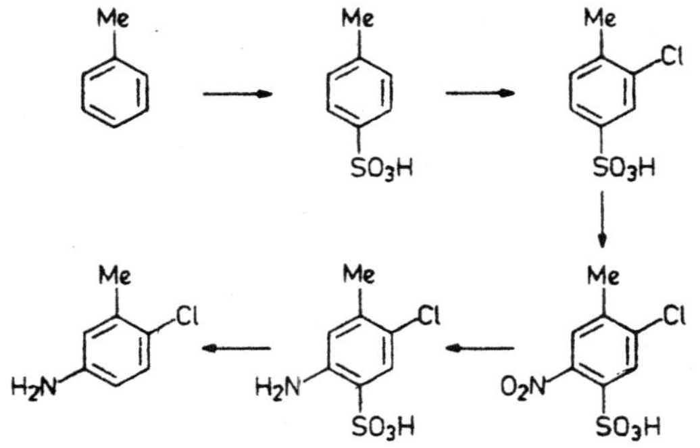

Me; Me, SO_3H; Me, Cl, SO_3H; Me, Cl, O_2N, SO_3H; Me, Cl, H_2N, SO_3H; Me, Cl, H_2N

Sulphonation-desulphonation

Scheme 2.11

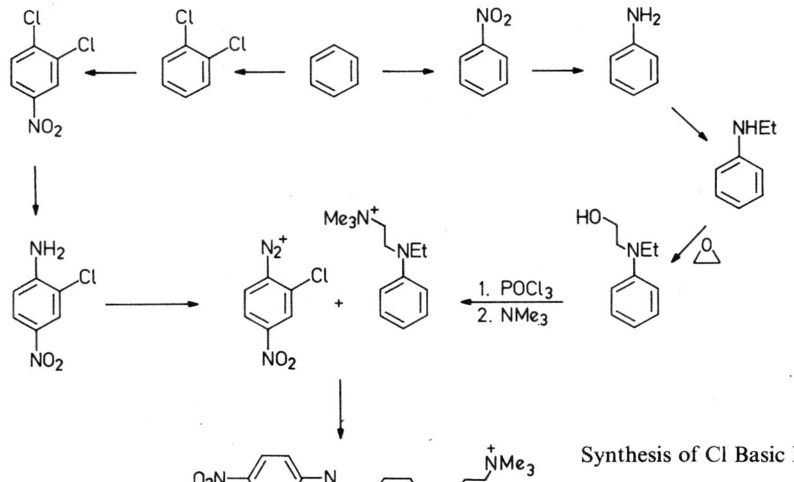

Cl, Cl, NO_2; Cl, Cl; NO_2; NH_2; NHEt; HO, NEt; NH_2, Cl, NO_2; N_2^+, Cl, NO_2; + Me_3N^+, NEt; 1. $POCl_3$ 2. NMe_3; O_2N, N, N, Cl, N, Et, $\overset{+}{N}Me_3$

Synthesis of Cl Basic Red 18

Scheme 2.12

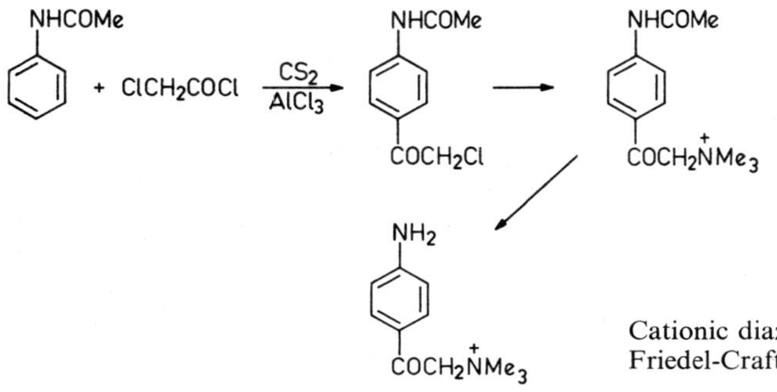

NHCOMe + $ClCH_2COCl$ $\xrightarrow[AlCl_3]{CS_2}$ NHCOMe, $COCH_2Cl$ → NHCOMe, $COCH_2\overset{+}{N}Me_3$ → NH_2, $COCH_2\overset{+}{N}Me_3$

Scheme 2.13

Cationic diazo component via the Friedel-Crafts reaction

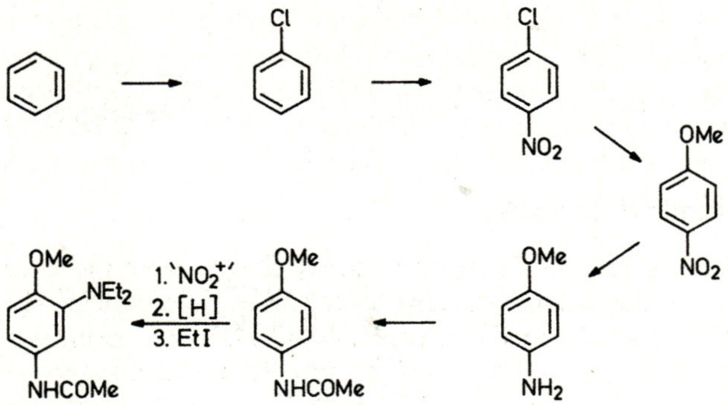

Scheme 2.14

Synthesis of a coupling component used in disperse dyes

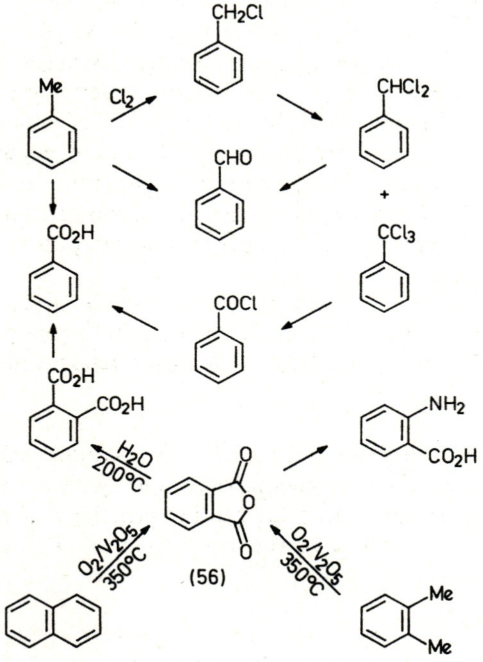

Scheme 2.15

Some useful intermediates

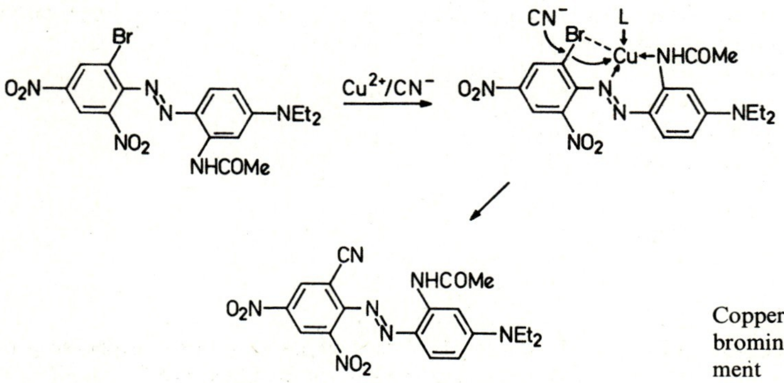

Scheme 2.16

Copper promoted bromine displacement

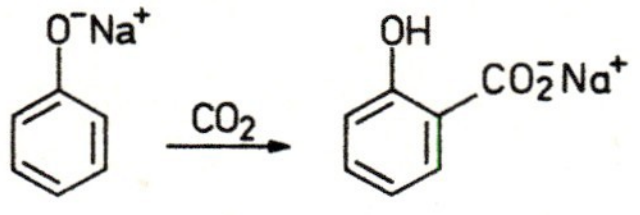

Scheme 2.17
Kolbé-Schmidt Reaction

2.3.2 Synthesis of Aromatic Heterocycles

In the early days of the dyestuffs industry the majority of dyes were prepared from benzene and naphthalene derived intermediates, including heterocyclic[5] dyestuffs such as Mauveine and Indigo. However, a recent trend has seen the introduction of an increasing number of heterocycles as dye precursors and nowhere has the impact been felt more than in azo dyes. Thus, hitherto unobtainable shades, especially blues and greens, and properties such as brightness have become attainable and this has increased even further the commercial importance of azo dyes. It is for this reason that the majority of the heterocyclic intermediates discussed in this section find their application in azo dyes. Some of the heterocyclic intermediates are still derived from benzene, *e.g.* indoles and benzothiazoles, but many, *e.g.* pyrazolones and pyridines, are synthesised from acyclic precursors. In contrast to the benzenoid intermediates it is unusual to find a heterocyclic intermediate that is synthesised via the parent heterocycle. Generally, this is because the parent heterocycle is not readily available in bulk from sources such as coal tar and petroleum, as are benzene and naphthalene.

The most important heterocycles are those with five or six membered rings and these rings may be fused to other rings, especially a benzene ring. Nitrogen, sulphur and to a lesser extent oxygen are the most frequently encountered heteroatoms.

(a) Synthesis of 5-Membered Rings

The synthesis of a number of 5-membered ring heterocycles will be discussed, both unfused and fused to other ring systems. In general, only one synthesis will be described in each case. However, it is hoped that the general principles of heterocyclic synthesis will emerge from these examples which represent, in the majority of cases, the most important commercial routes.

Pyrazolones

Pyrazolones, *e.g.* (57), are used as coupling components since they couple readily at the 4-position under alkaline conditions to give important azo dyes in the yellow-orange shade area (see Chap. 3).

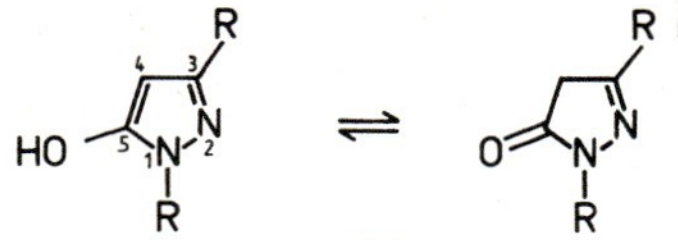

(57)

[5] An aromatic heterocycle is an aromatic system which contains one or more heteroatoms, especially nitrogen, sulphur and oxygen, as a member of the aromatic ring.

The most important synthesis of pyrazolones involves the condensation of a hydrazine with a β-ketoester. Commercially important pyrazolones carry an aryl substituent at the 1-position, mainly because the hydrazine precursors are prepared from readily available and comparatively inexpensive diazonium salts by reduction. In the first step of the synthesis the hydrazine is condensed with the β-ketoester to give a hydrazone: heating with sodium carbonate then effects cyclisation to the pyrazolone. In practice the condensation and cyclisation reactions are usually done in 'one pot' without isolating the hydrazone intermediate. The example in Scheme 2.18 illustrates the condensation of phenylhydrazine with ethyl acetoacetate.

$PhNH_2 \longrightarrow PhN_2^+ \xrightarrow{NaHSO_3} Ph{-}NH{-}NH_2$

$EtO{-}CO{-}CH_2{-}CO{-}Me$

$PhNH{-}N{=}C(Me){-}CH_2CO_2Et$

Scheme 2.18

Wide variations in the 1-substituent are possible merely by choosing a different aniline as precursor to the hydrazine. In this way pyrazolones with chlorophenyl and phenylsulphonic acid groups may be prepared and such pyrazolones are presently in use. The substituent at the 3-position is limited to alkyl, aryl, and carboxylic acid (including derivatives) groups by the ready availability of the corresponding β-ketoesters. In fact, the two commonest β-ketoesters in pyrazolone synthesis are ethyl acetoacetate ($CH_3COCH_2CO_2Et$) and diethyl oxalacetate ($EtO_2CCOCH_2CO_2Et$).

Indoles

Indoles (58) are used as coupling components to give azo dyes; the structurally related indolenines (59) are also employed as intermediates in polymethine dyes (see later). Azo dyes are obtained when indole (58) is coupled at the 3-position with a diazonium salt. In contrast, the most important polymethine dyes (for textiles) are obtained by reaction of a 3,3-disubstituted indolenine (59) at the 2-substituent.

R = H, aryl, alkyl
R, R^1 = aryl, alkyl
R^2 = alkyl

(58) (59)

The majority of commercial indoles are synthesised by way of the Fischer-Indole synthesis, which involves the rearrangement of phenylhydrazones to indoles and their derivatives. The hydrazones may be obtained either by condensation of a hydrazine with a ketone, Eq. 2.18b, or by reaction of a diazonium salt with a β-ketoester (Japp-Klingemann reaction), (Eq. 2.18a). Such hydrazones must contain at least one α-hydrogen for rearrangement to occur and acid catalysts such as $ZnCl_2$, HCl, polyphosphoric acid and H_2SO_4 are usually used to effect the rearrangement (Scheme 2.19).

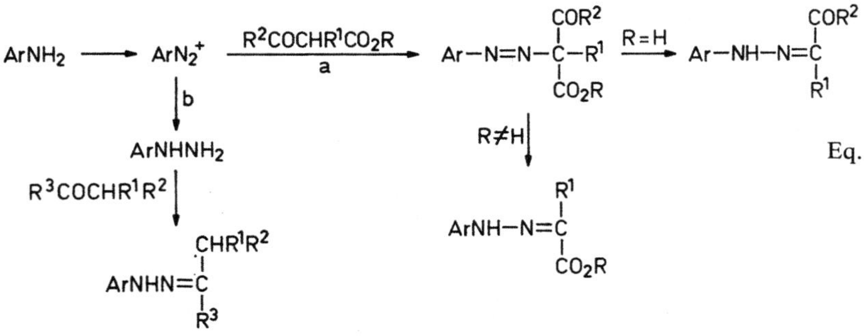

ArNH$_2$ → ArN$_2^+$ —($R^2COCHR^1CO_2R$, a)→ Ar—N=N—C(COR^2)(R^1)(CO_2R) —(R=H)→ Ar—NH—N=C(COR^2)(R^1)

R≠H → ArNH—N=C(R^1)(CO_2R)

ArN$_2^+$ —b→ ArNHNH$_2$ —($R^3COCHR^1R^2$)→ ArNHN=C(CHR^1R^2)(R^3)

Eq. 2.18

H_3O^+

(I)

(II)

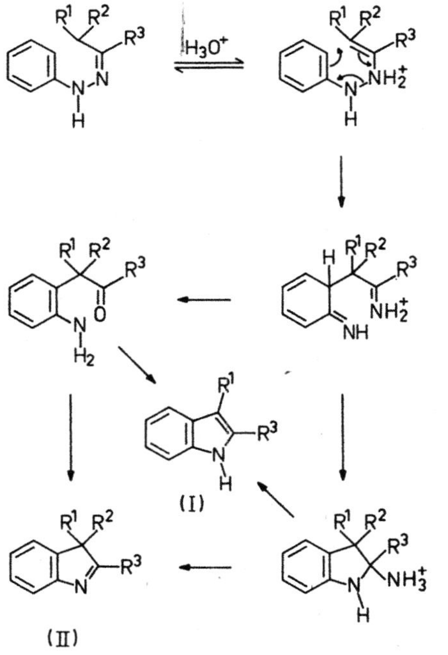

Scheme 2.19

For azo dyes only indoles of type I (R^1 = H), where R^3 is usually an aryl group (from an acetophenone derivative) or a methyl group (from acetone), are of any importance. Indolenines II are used in polymethine dyes where they are alkylated to further activate the 2-substituent. The most important of these is Fischer's Base (60).

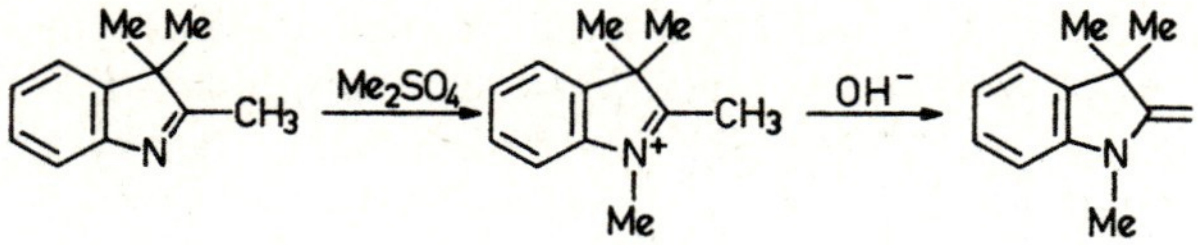

(60)

Aminothiazoles

In contrast to the pyrazolones and indoles just described, aminothiazoles are used as diazo components. As such they provide dyes which are more bathochromic than their benzene analogues. Thus, aminothiazoles are used chiefly to provide dyes in the red-blue shade areas.

The most convenient synthesis of 2-aminothiazoles is by the condensation of thiourea with an α-chlorocarbonyl compound: for instance, 2-aminothiazole (61) is prepared by condensing thiourea with α-chloracetaldehyde — both readily available intermediates.

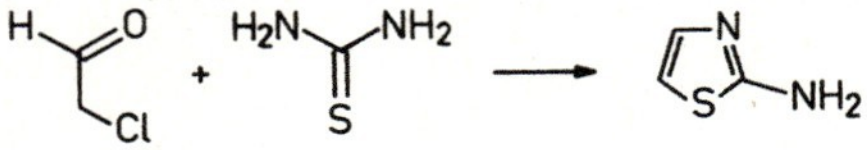

(61)

Substituents can be introduced into the thiazole ring either by using suitably substituted precursors or by direct introduction via electrophilic attack. An interesting example of the latter method is the preparation of 2-amino-5-nitrothiazole from the nitrate salt of 2-aminothiazole.

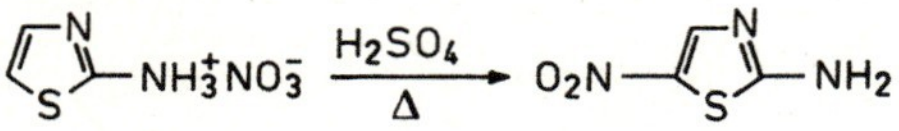

Aminobenzothiazoles

Aminobenzothiazoles are prepared somewhat differently to the thiazoles. Here the thiazole ring is annelated on to a benzene ring, usually via an aniline derivative. Thus, 2-amino-6-methoxybenzothiazole (63) is obtained from *p*-anisidine and thiocyanogen. The thiocyanogen, which is formed *in situ* from bromine and potassium thiocyanate (the Kaufmann reaction), reacts immediately with *p*-anisidine to give the thiocyanato derivative (62); this spontaneously ring closes to the aminobenzothiazole (63). If the 4-position of the aniline is free the thiocyanate group enters at this position as well as at the 2-position to give the 6-thiocyanatobenzothiazole (64). The latter compound is used as a diazo component in its own right and also as an intermediate for other useful sulphur-containing benzothiazole diazo components, *e.g.* (65).

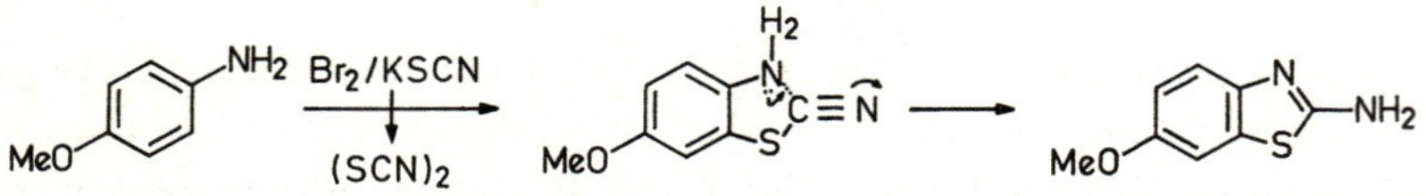

(62) (63)

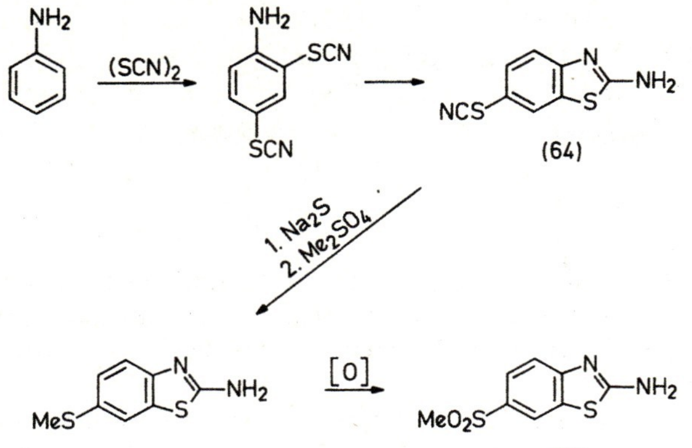

(64)

(65)

An alternative to the above synthesis is used to prepare 6-unsubstituted benzothiazoles. The aromatic amine is reacted first with potassium thiocyanate to give the 1-arylthiourea which is then cyclised with bromine to the corresponding 2-aminobenzothiazole, *e.g.* (Eq. 2.19).

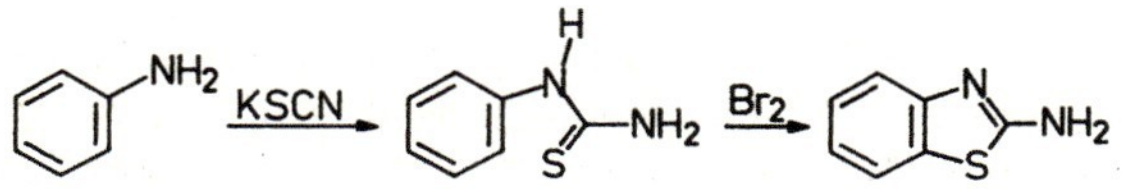

Eq. 2.19

Dyes derived from aminothiazoles and aminobenzothiazoles are used on a variety of fibres. For instance, anionic[6] (benzo)thiazole azo dyes are used on nylon fibres, cationic[6] (benzo)thiazole azo dyes are used on polyacrylonitrile, and neutral (benzo)thiazole dyes are used on polyester fibres.

Benzoisothiazoles

5-Aminoisothiazoles are relatively difficult to prepare cheaply. However, the benzo-homologues are far easier to prepare and are important diazo components. Aminobenzoisothiazoles give dyes which are even more bathochromic than the corresponding dyes from aminobenzothiazoles. Aminobenzoisothiazoles can be prepared from *o*-nitroanilines by the standard chemical transformations outlined in Scheme 2.20.

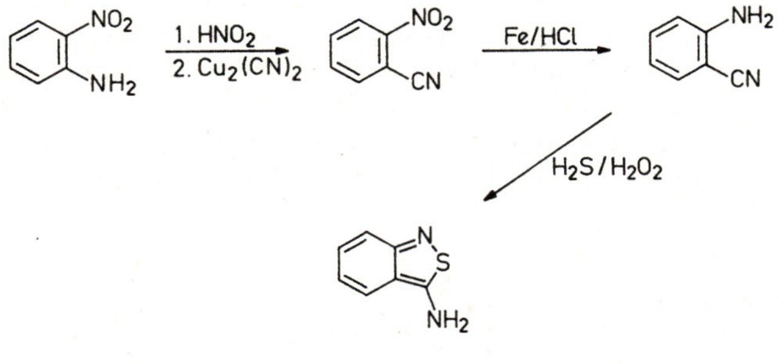

Scheme 2.20

[6] Typical anionic dyes contain $CO_2^{\ominus}$ or $SO_3^{\ominus}$ groups whilst a typical cationic dye would contain a $NR_3^{\oplus}$ group (also see diazahemicyanines).

Thus, *o*-nitroaniline is diazotised to give the diazonium salt and this is converted to the *o*-cyanonitrobenzene with copper I cyanide (Sandmeyer reaction). The *o*-cyanonitrobenzene is then reduced to the *o*-cyanoaniline. This is treated first with hydrogen sulphide to give the thioamide (66) and then with hydrogen peroxide to effect cyclisation — for postulated mechanism see Scheme 2.21.

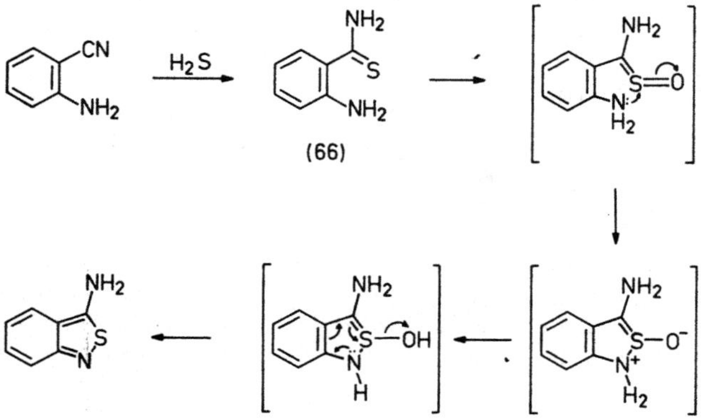

CN NH2 H2S NH2 S NH2 (66) NH2 S=O N H2

NH2 S N NH2 S—OH N H NH2 S—O⁻ N+ H2

Scheme 2.21

The most important commerical benzoisothiazole is the nitrobenzoisothiazole (67), prepared by an analogous route but starting from 2-cyano-4-nitroaniline.

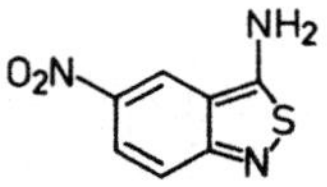

O2N NH2 S N

(67)

Triazoles

The triazoles are fairly representative of five-membered heterocycles containing three heteroatoms.[7] Heterocycles containing more heteroatoms are not generally found in dyestuffs and even triazoles are not of widespread importance, although they do provide some useful red dyes, *e.g.* (68), for polyacrylonitrile fibres (see Sect. 5.5.3).

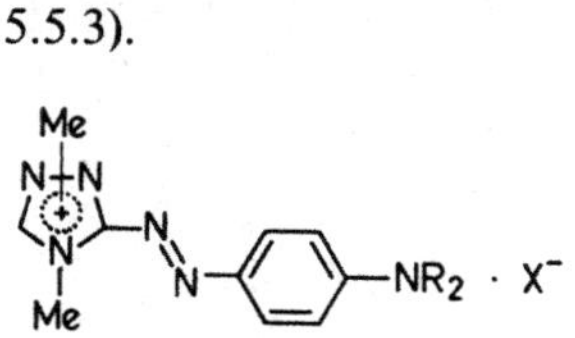

Me N N N N N Me NR2 · X⁻

(68)

[7] The only other commercially important heterocycles containing three hetero atoms are 5-amino-1,2,4-thiadiazoles, (*e.g.* 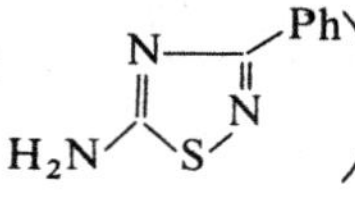 N—Ph N H2N S) and the isomeric 2-amino-1,3,4-thiadiazoles, (*e.g.* N—N MeS S NH2).

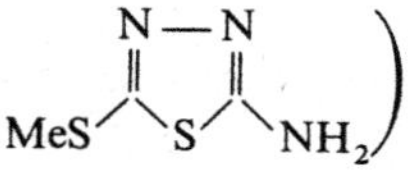

The most important triazole is 3-amino-1,2,4-triazole itself, used in the synthesis of the above diazahemicyanine dye (68), (see Sect. 2.4). The preparation of 3-amino-1,2,4-triazole (69) is simple, using readily available and quite inexpensive starting materials. Thus, cyanamide is reacted with hydrazine to give aminoguanidine (70) which is then condensed with formic acid. It can be seen that substituents in the 5-position are introduced merely by altering the carboxylic acid used.

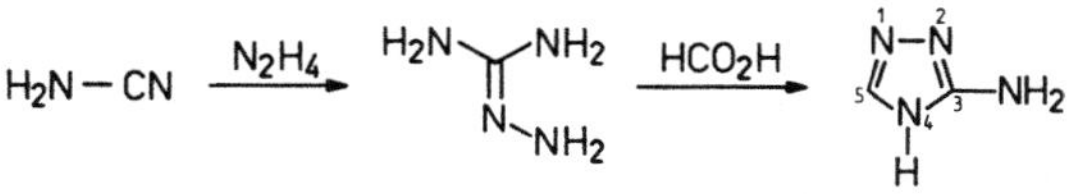

H_2N-CN —(N_2H_4)→ (70) —(HCO_2H)→ (69)

Thiophenes

Thiophenes have only recently been introduced as useful heterocycles for dyestuffs. The important thiophenes, *i.e.* 2-aminothiophenes, are used as diazo components for azo dyes and are capable of producing very bathochromic dyes, *e.g.* greens, having excellent properties. Part of the reason for their late arrival on the commercial scene has been the difficulty encountered in attaining a good synthetic route. However, this problem has been largely overcome by ICI.[8] The synthetic route is very flexible so that a variety of protected 2-aminothiophenes can be obtained merely by altering the reaction conditions or alternatively by stopping the reaction at an intermediate stage (see Scheme 2.22).

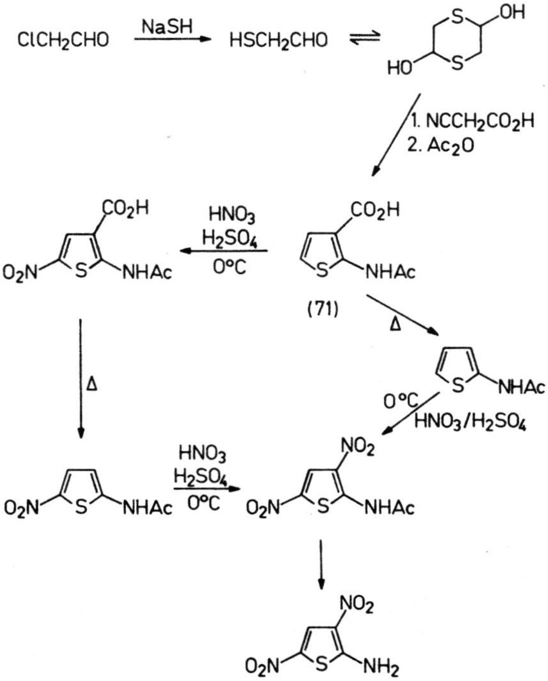

$ClCH_2CHO$ —(NaSH)→ $HSCH_2CHO$ ⇌ dithiane diol —(1. $NCCH_2CO_2H$, 2. Ac_2O)→ (71); (71) —(HNO_3, H_2SO_4, 0°C)→ nitro acid; (71) —(Δ)→ 2-acetamidothiophene —(0°C, HNO_3/H_2SO_4)→ dinitro; nitro acid —(Δ)→ O_2N-thiophene-NHAc —(HNO_3, H_2SO_4, 0°C)→ dinitro-NHAc → (72)

Scheme 2.22

[8] GB 1419074 and 1419075 (ICI).

Thus, the initially formed thiophene (71) can be converted to the useful dinitrothiophene (72) by either of two routes, as shown.

(b) 6-Membered Rings

Pyridine Derivatives

Pyridine itself has little importance as a dyestuff intermediate. However, its 2,6-dihydroxy and 2,6-diamino derivatives have achieved prominence in recent years as coupling components for azo dyes, particularly in the yellow-red shade areas.

In solution the 2,6-dihydroxypyridine tautomer (73) is usually the minor tautomer and the 6-hydroxy-2-pyridone (74) the major tautomer. Thus, within this class lie the 1-substituted-6-hydroxy-2-pyridones (75). These pyridones couple at a vacant

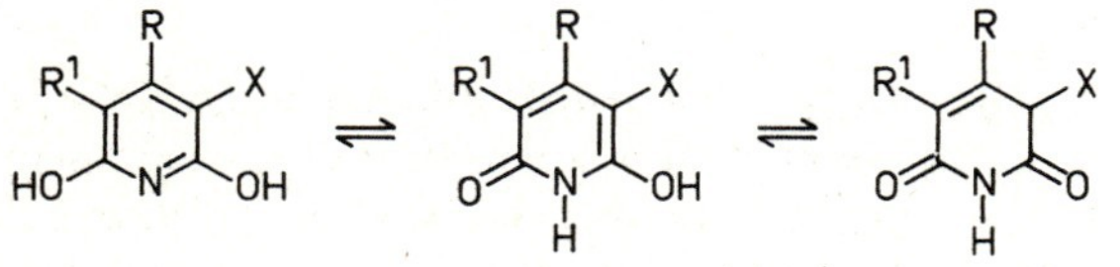

(73) (74)

3/5 position with diazonium salts to give the hydrazone form of the azo dye, *e.g.* (76) (see Sect. 3.3).

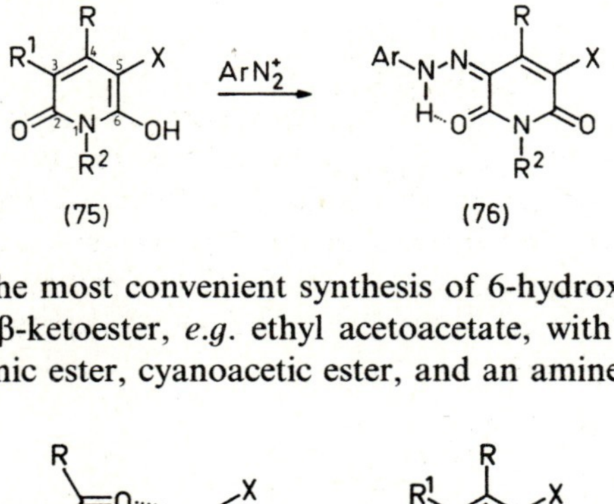

(75) ArN_2^+ (76)

The most convenient synthesis of 6-hydroxy-2-pyridones is by the condensation of a β-ketoester, *e.g.* ethyl acetoacetate, with an active methylene compound, *e.g.* malonic ester, cyanoacetic ester, and an amine (Eq. 2.20).

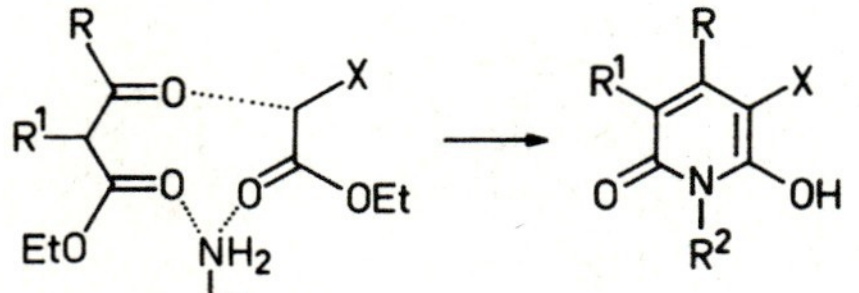

$X = CN, CO_2R, NR_3^+$ Eq. 2.20

The amine can be omitted if an acetamide is used (Eq. 2.21) and in some cases this modification results in a higher yield. R is usually alkyl or less frequently aryl, whilst R^1 is H, and the R^2 group can be alkyl, aryl and even an amino or a hydroxy group. By virtue of the synthetic method *X* is an electron-withdrawing group, *e.g.* CN, CO_2R, although simple derivatives, *e.g.* $X = CONH_2$, H, are readily obtained by hydrolysis. Dyes from pyridones in which *X* is an ammonium group (from $R_3N^{\oplus}CH_2CONH_2$) are used for polyacrylonitrile fibres.

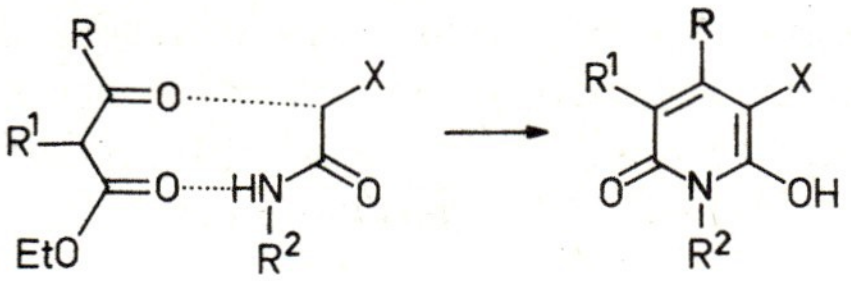

Eq. 2.21

An interesting situation is encountered in the synthesis of the pyridone (77) using malononitrile (78) as the active methylene compound. Here an amine is not required since the nitrogen is provided by the nitrile which undergoes a facile cyclisation from the nitrile ester intermediate (79).

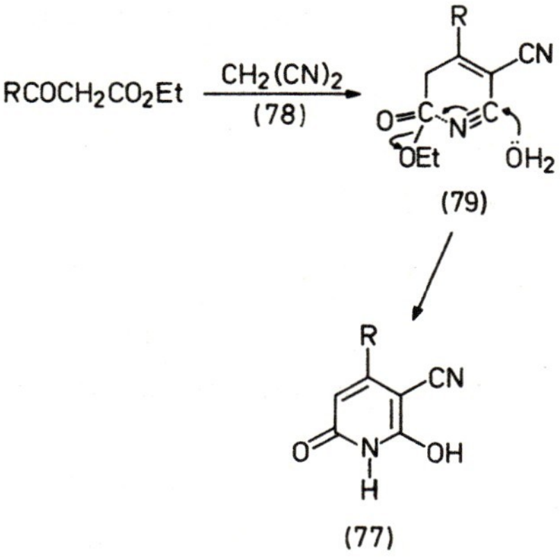

$$\mathrm{RCOCH_2CO_2Et} \xrightarrow[(78)]{\mathrm{CH_2(CN)_2}} (79) \rightarrow (77)$$

Diaminopyridines are easily obtained from the corresponding 6-hydroxy-2-pyridones by the action of phosphorus halides, (*e.g.* $POCl_3$), Eq. 2.22, and then amination. The first step converts the hydroxypyridone to the dichloropyridine; this reaction proceeds even with 1-alkylsubstituted hydroxypyridones. In these cases the alkyl group is lost under the reaction conditions, usually as an alkyl halide, from an intermediate such as (80). Under mild conditions it is possible to isolate the monochloro derivative (81).

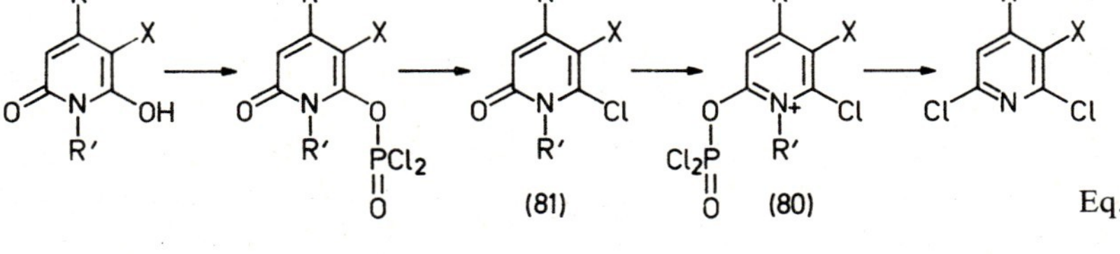

(81) (80) Eq. 2.22

The dichloropyridine is then treated with two moles of an amine to give the diaminopyridine (Eq. 2.23).

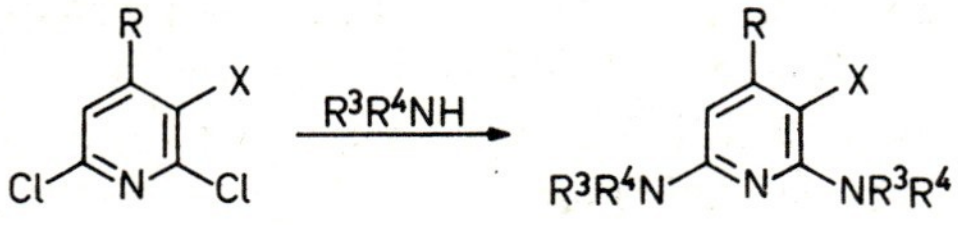

$$\xrightarrow{R^3R^4NH}$$

Eq. 2.23

If X is an electron-withdrawing group, *e.g.* CN, then the reaction is greatly facilitated. In fact, two different amines may be introduced into the pyridine nucleus; however, if the dichloropyridine is unsymmetrical (82) then mixtures of diaminopyridines, *e.g.* (83) and (84), are obtained, although steric factors may ensure that (83) is the major, if not the exclusive, isomer.

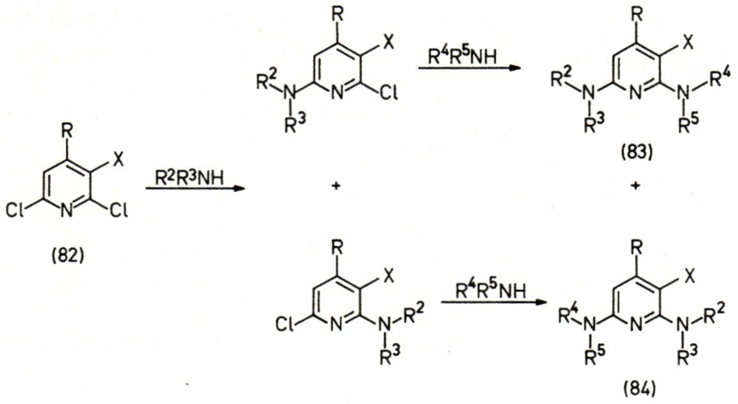

R^2R^3NH R^4R^5NH

(82) (83) (84)

Triazines

The most commercially important triazine is 2,4,6-trichloro-*s*-triazine (cyanuric chloride, 85). Cyanuric chloride has not achieved prominence because of its value as part of a chromogen but because of its use for attaching dyestuffs to cellulose, *i.e.* as a reactive group (see Chap. 6). This innovation was first introduced by ICI in 1956 and since then other active halogen compounds have been introduced.

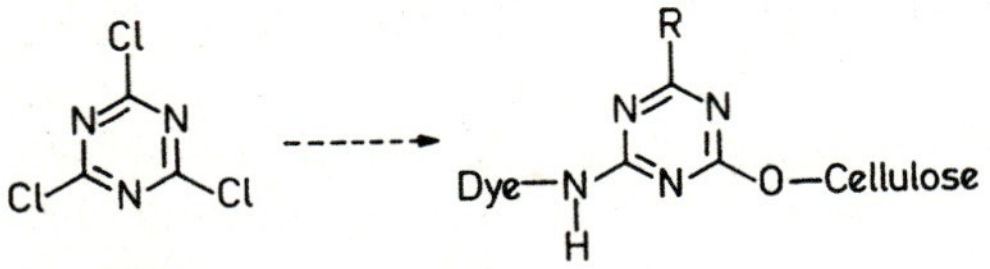

(85)

On the large scale cyanuric chloride is produced by the trimerisation of cyanogen chloride. The cyanogen chloride is produced by chlorination of hydrogen cyanide and is trimerised by passing it over charcoal or charcoal impregnated with an alkaline earth metal chloride at a high temperature (250–480 °C) (Scheme 2.23).

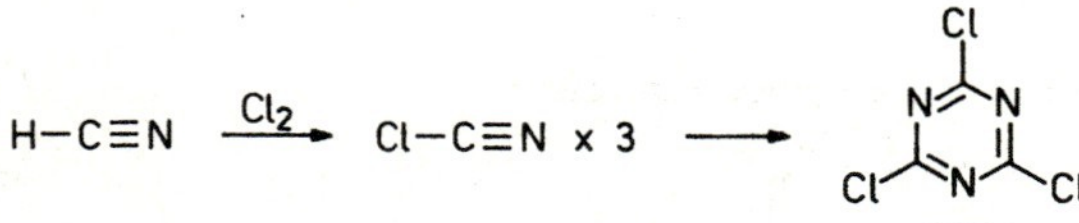

$$H{-}C{\equiv}N \xrightarrow{Cl_2} Cl{-}C{\equiv}N \times 3 \longrightarrow$$ (85)

Scheme 2.23

2.4 Synthesis of Dyes

This section contains those reactions in which the actual chromogen is assembled. Many of the intermediates that have been discussed in the last section will be seen in the following six sections.

2.4.1 Azo Dyes

The emphasis in this section is on the mechanistic aspects of the diazotisation and coupling reaction which is the single most important method used for the preparation of azo dyes. Diazotisation of an aromatic amine gives the diazonium salt (86), a relatively weak electrophile. Attack at a suitable substrate, the coupling component (EH), then gives the azo dye (Eq. 2.24).

$$ArNH_2 \longrightarrow \underset{(86)}{ArN_2^+} \xrightarrow{EH} Ar-N{=}N-E + H^+ \qquad \text{Eq. 2.24}$$

The intermediates required are synthesised as described in the previous section. They are most commonly aromatic amines, from which both diazo and coupling components can be obtained, and aromatic hydroxy compounds (both carbocyclic and heterocyclic).

Diazotisation

The diazotisation reaction essentially involves the nitrosation of a primary amine followed by its deprotonation and dehydration (Eq. 2.25).

$$RNH_2 \xrightleftharpoons{NOX} R-\overset{+}{N}H_2-N{=}O \rightleftharpoons R-NH-N{=}O \rightleftharpoons R-N{=}N-OH \xrightleftharpoons{HA} R-N^+{\equiv}N\ A^- \qquad \text{Eq. 2.25}$$

The study of the kinetics of the overall reaction is greatly simplified since the rate determining step occurs either with the generation of the nitrosating species or in the nitrosation of the amine. This conclusion has been reached following studies of the nitrosation of N-methylaniline, which shows very similar kinetics to the diazotisation reaction yet cannot undergo the later steps. In aqueous solution the nitrosating agents most frequently encountered are: — nitrous anhydride (N_2O_3), nitrosyl halide ($O=N-X$), the nitrous acidium ion ($O=N-OH_2^{\oplus}$) and the nitrosonium ion ($NO^{\oplus}$). The nitrosating species present in any particular reaction depends primarily upon the acidity of the reaction mixture. At low acidities (0.1 M $H_3O^{\oplus}$) the rate of diazotisation of primary aromatic amines is given by Eq. 2.26. However, if the acidity is reduced a point is reached where the rate of diazotisation is no longer dependent on the amine concentration, *i.e.* Eq. 2.27 is satisfied.

$$\text{Rate} = k\,[HNO_2]^2\,[ArNH_2] \qquad k = \text{rate constant} \qquad \text{Eq. 2.26}$$

$$\text{Rate} = k\,[HNO_2]^2 \qquad \text{Eq. 2.27}$$

At this point the rate determining step is the formation of the nitrosating agent, nitrous anhydride (87), which reacts as soon as it is formed with the free amine (the latter is present in large concentration at these higher pH values).

$$2\,HONO \rightleftharpoons \underset{(87)}{N_2O_3} + H_2O$$

That the rate of reaction is dependent on the rate of formation of the nitrous anhydride has been further demonstrated by oxygen-18 exchange experiments (Eq. 2.28); these show similar kinetics to the diazotisation reaction.

$$H_2O^{18} + N_2O_3 \rightleftharpoons HONO + HO^{18}{-}NO \qquad \text{Eq. 2.28}$$

With non-basic amines, *e.g.* *p*-nitroaniline, the rate relationship never adopts the form shown by Eq. 2.27, even at low acidities, because of the poor nucleophilicity of the amine nitrogen atom.

As the acidity of the diazotisation medium is increased (0.1 M → 6.5 M $H_3O^{\oplus}$), the nitrosating agent changes and a new rate relationship applies (Eq. 2.29).

$$\text{Rate} = k_1[ArNH_2][HNO_2]h_0 + k_2[ArNH_3^+][HNO_2]h_0$$
$$h_0 = \text{Hammett acidity function} \qquad \text{Eq. 2.29}$$

The nitrosating species now appears to be the nitrous acidium ion (88). The relative contribution from the two terms for any amine depends on the acidity of the reaction medium and the basicity of the amine; the first term is important at low acidities, (*i.e.* in weakly acidic media) and with non-basic amines, *e.g.* nitroanilines, whereas the second term is important at high acidities (*i.e.* strongly acid media) and with basic amines.

$$H_3O^+ + HONO \longrightarrow \underset{(88)}{H_2O^+NO} + H_2O$$

Surprisingly, Eq. 2.29 reveals that it is the protonated amine which is the nucleophile in the absence of an appreciable concentration of free amine. A mechanism (Scheme 2.24) has been proposed to account for this and this has received further support from a study of substituent effects on the rate of reaction.

NH_3^+ (Z) ⇌['NO+'] NH_3^+ ($\to NO^+$, Z) →[B^-] $H_2N^+\cdots H\cdots B^-$, $\cdots NO^+$ (Z) → $H_2N^+{-}NO$ (Z)

Scheme 2.24

At very high acid concentrations the nitrosating species is the reactive nitrosonium ion ($NO^{\oplus}$). The rate of reaction now decreases with increasing acidity (Eq. 2.30). This feature points to a further change in the rate determining step which now appears to be deprotonation of the nitrosated species (89).

$$\text{Rate} = k\,[ArNH_3^+]\,[HNO_2]\,h_0^{-2} \qquad \text{Eq. 2.30}$$

$$NO^+ + ArNH_3^+ \xrightleftharpoons{\text{fast}} \underset{(89)}{ArNH_2^+NO} + H^+ \xrightleftharpoons{B} ArNHNO + BH^+$$

Two pieces of evidence support this postulate; the first is the high primary isotope effect observed for the deuterated analogue of the primary amine ($ArNH_2D^{\oplus}$), and the second is that *different* aromatic amines react at approximately the *same* rate at high acid concentration.

When the diazotisation of an aromatic amine is carried out in the presence of halide ions ($X^{\ominus}$) a marked increase in the rate of diazotisation is noted over diazotisations carried out in halide free solutions. This catalysis is particularly important at low nitrous acid concentrations and is given by chloride, bromide and iodide ions but not by fluoride ions. The rate is as shown in Eq. 2.31. It is entirely consistent with the nitrosation of the free amine by nitrosyl halide (Scheme 2.25) and at low nitrous acid concentration takes place in preference to nitrosation by nitrous anhydride.

$$\text{Rate} = k\,[ArNH_2]\,[H^+]\,[HNO_2]\,[X^-] \qquad \text{Eq. 2.31}$$

$$HNO_2 + H^+ + X^- \xrightleftharpoons{\text{fast}} NOX + H_2O$$

$$NOX + ArNH_2 \xrightleftharpoons{\text{slow}} ArNH_2^+NO + X^- \downarrow \qquad \text{Scheme 2.25}$$

As far as possible diazotisations of aromatic amines are carried out in dilute aqueous mineral acids. The amine is usually dissolved in approximately 2.5 molar equivalents of acid and diazotised at 0–5 °C with concentrated sodium nitrite solution. Diazotisations in concentrated acid are used for non-basic amines such as polynitroanilines and for most heteroaromatic amines, *e.g.* aminodinitrothiophenes and aminobenzoisothiazoles, since hydrolysis of the diazonium salt occurs in dilute acid. Here, the acid of choice is concentrated sulphuric acid, used either alone or in admixture with glacial acetic acid or phosphoric acid. A particularly important reagent combination is nitrosylsulphuric acid ($ON.HSO_4$)[9] which is used extensively as a nitrosating agent for non-basic amines. Concentrated nitric acid is not favoured because of the danger from explosions and competing nitration and oxidation reactions. Concentrated hydrochloric acid is avoided because of its oxidation to chlorine in nitrous acid.

Diazotisations in organic solvents, *e.g.* alcohols, ketones, amides, *etc.*, with organic nitrites, *e.g.* pentyl nitrite, are usually used only for the preparation of solid diazonium salts in the laboratory. Unless suitably stabilised, most solid diazonium salts are explosive; hence, on the industrial scale, the diazonium salt is used without isolation.

The Coupling Reaction

As mentioned in the last section the diazonium salt is not isolated but reacted *in situ* with a suitable substrate called the coupling component. In aqueous solution the diazonium salt (86) exists in equilibrium with its diazohydroxide (90) and diazotate (91), the position of the equilibrium depending upon the pH of the solution. At high pH values the equilibrium is shifted to the right whilst at low pH values the reverse is true.

$$\underset{(86)}{ArN_2^+} \underset{H_3O^+}{\overset{OH^-}{\rightleftharpoons}} \underset{(90)}{Ar{-}N{=}N{-}OH} \underset{H_3O^+}{\overset{OH^-}{\rightleftharpoons}} \underset{(91)}{Ar{-}N{=}N{-}O^-}$$

The most common coupling components are aromatic amines (92) and phenols (93). These also exist in equilibrium, the amine with its ammonium salt (94) and the phenol with its phenolate (95). The position of the equilibrium is once again pH dependent and, at high pH, lies to the left for the amine and to the right for the phenol. In the coupling reaction it is the phenolate (95) and the free amine (92) which actually participate in the reaction with the diazonium salt, Eq. 2.32.

$$\underset{(92)}{Ar{-}NR_2} \rightleftharpoons \underset{(94)}{Ar{-}\overset{+}{N}HR_2}$$

$$\underset{(93)}{Ar{-}OH} \rightleftharpoons \underset{(95)}{Ar{-}O^-}$$

[9] Nitrosylsulphuric acid can be prepared from sodium nitrite and concentrated sulphuric acid. The reagent is isolable as a crystalline solid.

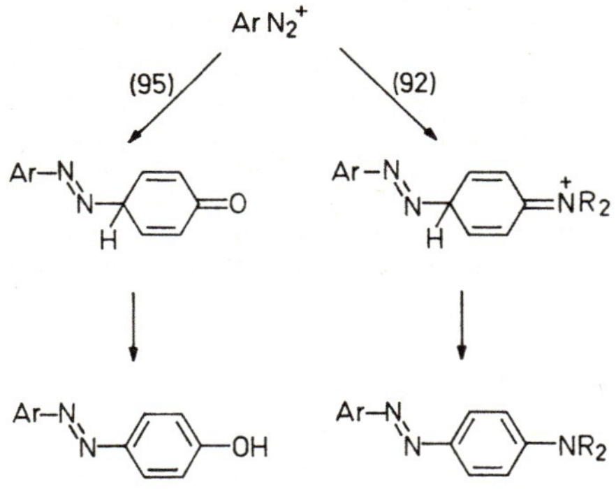

Ar N$_2^+$ (95) (92)

Obviously the pH of the reaction mixture plays an important role in the coupling reaction and a careful balance has to be struck between the equilibria governing the coupling component and the diazo component. The pH for coupling is chosen in the range pH 4–10 and depends upon the reactants involved. Phenols and naphthols are usually coupled at pH $\geqq$ 7 to ensure the presence of the phenolate or naphtholate anion (*i.e.* pH of solution $\geqq$ pK_a of phenol or naphthol). However, the pH of the solution must not be allowed to rise too high (pH > 10) otherwise diazotate formation occurs and, even more serious, the diazonium salt decomposes. On the other hand, the much slower coupling aromatic amines are coupled at a lower pH, typically pH 4–7, (*i.e.* such that pH > pK_a for $ArNH_3^{\oplus}$). This pH range strikes a balance between the needs of the diazo component and those of the coupling component. Thus, it ensures an appreciable concentration of free amine whilst maintaining the stability of the diazonium salt (at pH > 7 diazonium salt decomposition seriously competes with the coupling reaction for aromatic amines). The rate of coupling is given by Eq. 2.32 where [C] is the concentration of the coupling species and $[ArN_2^{\oplus}]$ is the concentration of the diazonium salt.

$$\text{Rate} = k\,[ArN_2^+]\,[C] \qquad \text{Eq. 2.32}$$

In some coupling reactions the rate of coupling can be markedly increased by the addition of an organic base such as pyridine. The effect is noticeable in slow coupling reactions and it has been used in the dyestuffs industry for many years. The effect has been explained by assuming that the pyridine promotes deprotonation of the intermediate (96) to give the corresponding azo dyes. Of course, the effect is only noticed when the deprotonation step has a comparable rate to the reverse of the intermediate forming step.

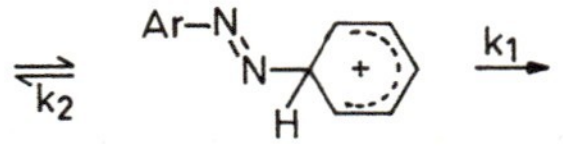

k_2 k_1

(96)

Consistent with the coupling reaction being an electrophilic attack by the diazonium salt, coupling takes place at a position of high electron density. Thus, phenols and amines couple at the *para* position, whilst 1-naphthols couple at the 2- and 4-positions, and 2-naphthols couple at the 1-position, in the absence of steric constraints. The position of coupling can be influenced by steric factors: for instance, the 1-naphthol (97) couples only at the 2-position and not at all at the 4-position. Examples of some common coupling components are shown in Scheme 2.26.

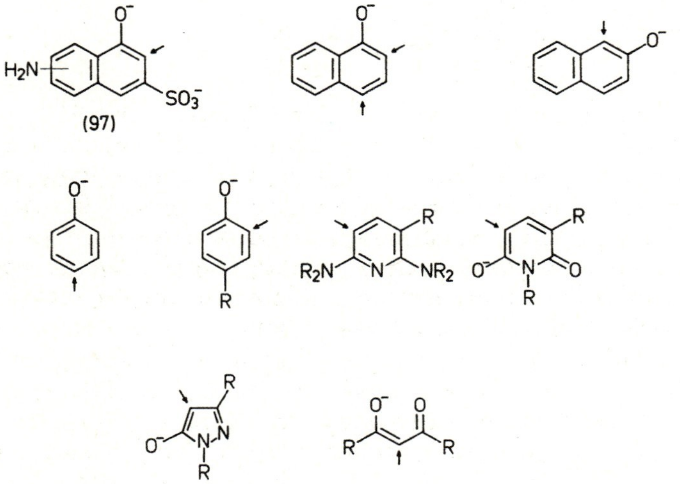

(97)

Scheme 2.26

Of course, coupling between the diazonium salt and the coupling component not only depends upon the pH of the reaction medium but also upon the reactivity of both the diazo and coupling components. The electrophilicity of an aromatic diazonium ion is increased by the introduction of electron-withdrawing groups. Thus, *p*-nitrobenzenediazonium chloride couples approximately 10^5 times faster with 1-naphtholate than does *p*-methoxybenzenediazonium chloride, whilst 2,4,6-trinitrobenzenediazonium chloride even couples with 1,3,5-trimethylbenzene (98)! In contrast, the activity of the coupling component is increased by the introduction of electron-releasing groups. In fact, a hydroxy or amino group must normally be present in an aromatic ring before coupling takes place at all, except with the most active diazo components, *e.g.* 2,4,6-trinitrobenzenediazonium chloride noted above. The introduction of another electron-releasing group increases further the rate of reaction. For instance, *ortho*-methoxyphenol couples approximately 10 times faster than phenol.

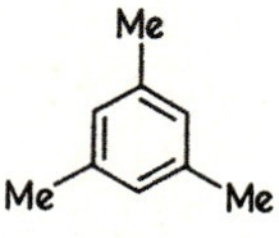

(98)

An intriguing situation is encountered in the coupling reaction of aminonaphthols (99). In these cases two different donors are present in separate rings and conditions can be chosen so that coupling may be induced to take place exclusively in either ring. For instance, 1-amino-8-naphthol (100) couples either at the 4-position (pH < 7) or at the 5-position (pH > 7).

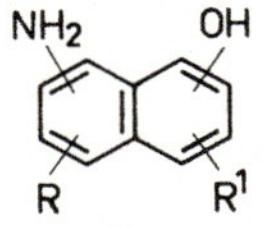
NH2
OH
R
R1
(99)

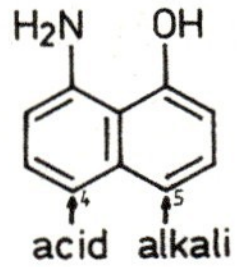
H2N
OH
4
5
acid
alkali
(100)

This pH dependence is as a consequence of two factors. Below pH 7 only the free amine is available for coupling since the naphtholate is not present in solution (*i.e.* its pK_a > pH). At high pH both the free amine and naphtholate are now present but the naphtholate couples much faster than the amine. This means that monoazo compounds can be prepared readily and, more importantly, that disazo compounds from two different components are accessible.[10] Several important dyes are manufactured from aminonaphtholsulphonic acid coupling components, *e.g.* H-acid (51) and J-Acid (101), using these principles.

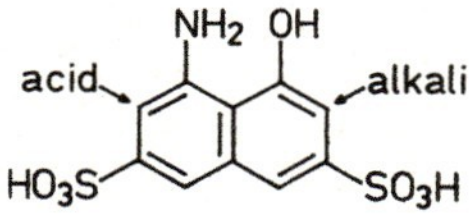
NH2
OH
acid
alkali
HO3S
SO3H
(51)

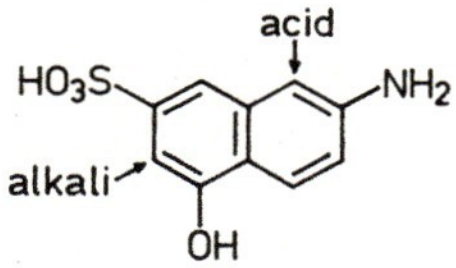
acid
HO3S
NH2
alkali
OH
(101)

As well as the aromatic systems already discussed, aliphatic molecules can also form azo dyes with diazonium salts. In the dyestuffs industry the use of such systems is limited to β-ketoacid derivatives, *e.g.* (102). The coupling reaction takes place in alkaline solution to give yellow dyes which are used for either fibre or pigment outlets (see Sect. 3.3.3).

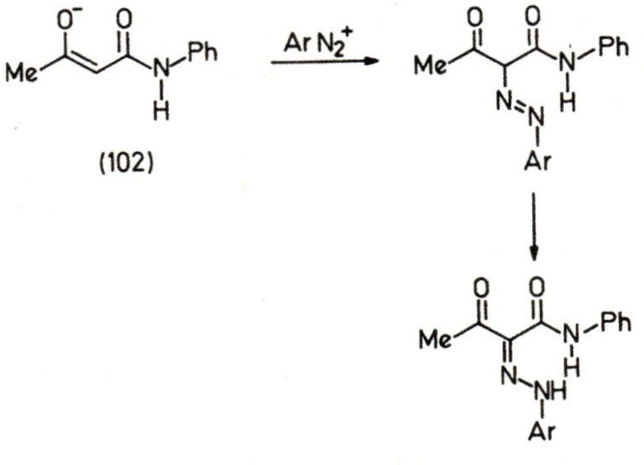
O-
O
Me
N
Ph
H
(102)
Ar N2+
O
O
Me
N
Ph
N
N
H
Ar
O
O
Me
N
Ph
H
N
NH
Ar

[10] In such cases coupling is carried out at low pH first and then at high pH.

Metal Complex Azo Dyes

Nowadays, the most important metallised azo dyes are those in which the azo group is chelated directly to the metal. Dyes are metallised for one or both of two reasons; the first is to achieve a better fibre affinity (see Sect. 1.2.2) and the second is to improve a specific property of the dye, especially light fastness (see Chaps. 3 and 6).

For complexation to be effective the azo dye requires substituents at the *ortho*-position of the aromatic rings, *A* and *B*. One substituent is sufficient but for the most stable complexes two substituents are preferred, *e.g.* (103). These substituents are usually hydroxy groups, OH, and carboxy groups, CO_2H. The metals used are the transition metal ions, Cu II, Co III and Cr III (see Sect. 3.4).

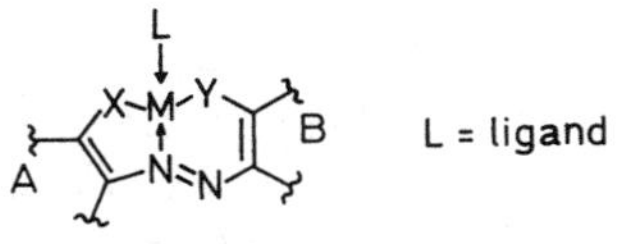

L = ligand

(103)

In many cases the azo dye is prepared first and then metallised before application to the fibre; alternatively it can be impregnated into the fibre with subsequent metallisation (see Chap. 6). However, the synthesis of metallised *o,o'*-dihydroxyazo dyes present some problems which stem from the difficulties encountered in preparing *o,o'*-dihydroxyazo dyes by conventional diazotisation and coupling of *o*-aminophenol and 2-aminonaphthol diazo components. The problem with 2-aminonaphthols is their tendency to be oxidised by nitrous acid to 2-naphthoquinones. In contrast, *o*-aminophenols diazotise easily but once formed the *o*-hydroxybenzenediazonium salts are very slow to couple because of the ready formation of the internal betaine (104), and hence the resultant low electrophilicity of the diazonium group, *i.e.* see (105). As a result the copper complexes of *o,o'*-dihydroxyazo dyes are prepared by alternative procedures.

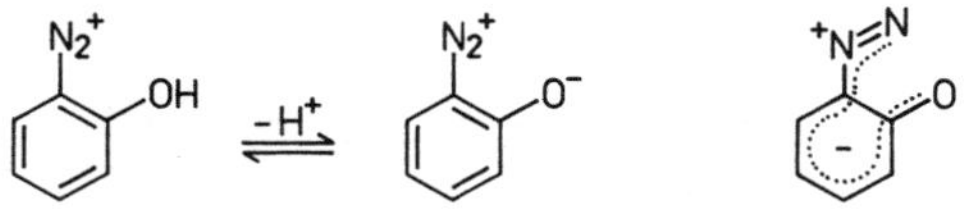

(104) (105)

The first method involves the preparation of the azo dye (106) from, for instance, an *o*-methoxyaniline and a suitably substituted phenol. The azo dye (106), when heated in the presence of cuprammonium sulphate and an alkanolamine, undergoes a copper promoted dealkylation to (107). This method, which is also applicable to naphthols, is known as dealkylative coppering.

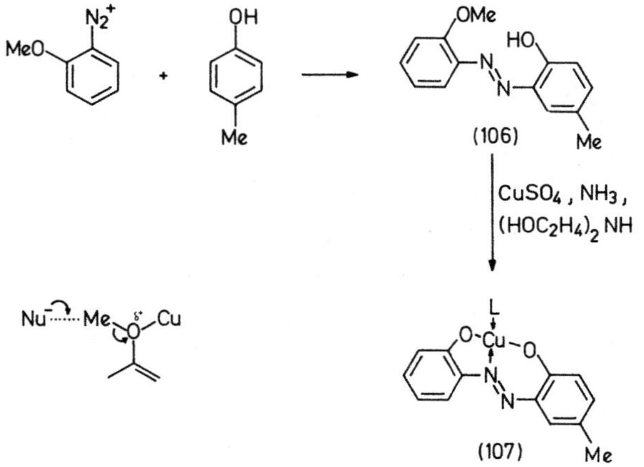

(106)

$CuSO_4$, NH_3, $(HOC_2H_4)_2NH$

(107)

The second method utilises the ready replacement of an *o*-halogen, usually chlorine. Again the replacement of the chlorine is promoted by the presence of the copper chelate (Scheme 2.27). Hydrolytic coppering is the name given to this reaction.

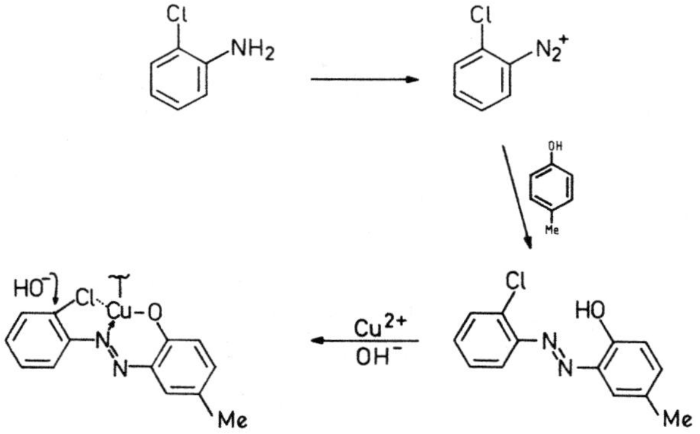

Scheme 2.27

The third procedure, known as oxidative coppering, requires the presence of an oxidising reagent, *e.g.* O_2 or H_2O_2. Thus, hydrogen at the *ortho* position is replaced by a hydroxy group; once again the copper ion plays an important role in facilitating the oxidation (Scheme 2.28).

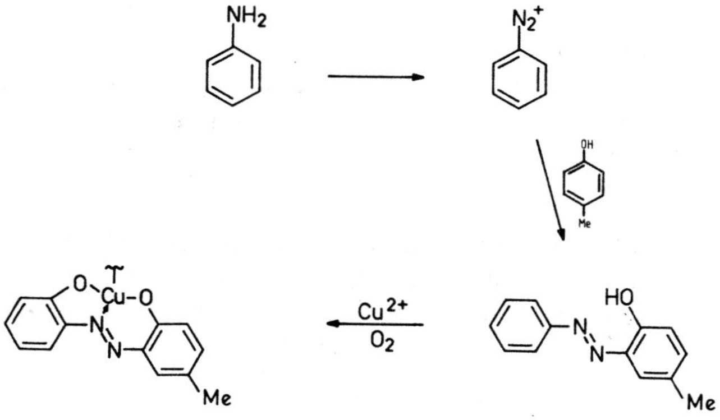

Scheme 2.28

2.4.2 Anthraquinone Dyes

In contrast to azo dyes, anthraquinone dyes are usually synthesised by preparing the anthraquinone skeleton and then introducing the desired substituents.

(i) Synthesis of the Anthraquinone Skeleton

The two industrial methods used for the synthesis of anthraquinone (4) are the oxidation of anthracene and the Friedel-Crafts reaction using benzene and phthalic anhydride. As mentioned earlier anthracene is available from coal tar and on oxidation gives 9,10-anthraquinone (Eq. 2.33) in good yield. The two most widely used oxidising agents in this process are nitric acid and chromic acid. When chromic acid is used the reduced chromium salts are isolated and sold to the Tanning industry. Indeed, the latter depends largely upon this process for its source of chromium.

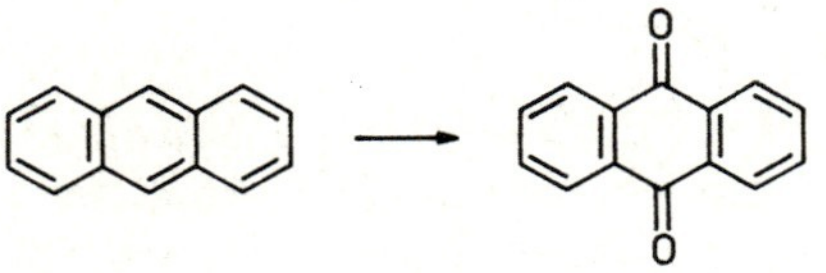

Eq. 2.33

(4)

In the second method phthalic anhydride gives anthraquinone when it is heated with benzene and a Lewis acid such as aluminium chloride; the reaction is therefore an example of a Friedel-Crafts acylation, Eq. 2.34. The final cyclisation to anthraquinone is carried out by heating the benzophenone (108) in concentrated sulphuric acid.

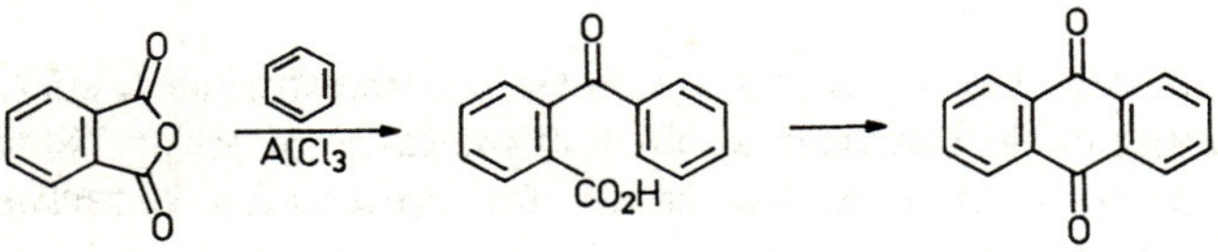

$AlCl_3$ CO_2H Eq. 2.34

(108) (4)

Once formed, anthraquinone may be converted into a variety of substituted derivatives by the type of chemistry outlined in Section 3. However, before these reactions are discussed further, some of the few direct routes to substituted anthraquinones will be highlighted. Most of these direct routes lead into hydroxyanthraquinones and are variations on the phthalic anhydride synthesis. For instance, if phthalic anhydride is heated with *p*-chlorophenol and sulphuric acid using boric acid catalyst, quinizarin, (109), a synthetically very useful anthraquinone, is obtained in good yield. (This route, because it uses such inexpensive reagents, also provides quinizarin relatively cheaply).

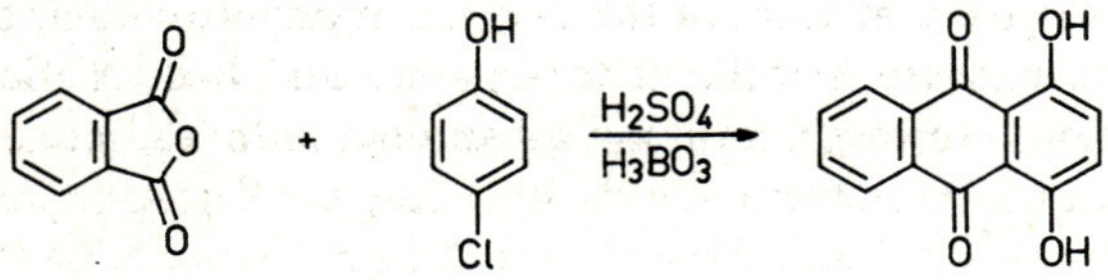

OH + Cl H_2SO_4 H_3BO_3 OH OH

(109)

Anthrarufin (110), an isomer of quinizarin, can be prepared by the dimerisation and dehydration of *m*-hydroxybenzoic acid. In a similar vein, anthragallol (111) may be synthesised from gallic acid (112) and benzoic acid, and the hexahydroxy anthraquinone (113) is synthesised by the dimerisation of gallic acid (112).

CO_2H OH → O OH HO O (110)

OH OH HO$_2$C OH (112) + CO_2H (benzoic acid), oleum 120° → O OH OH OH O (111)

→ HO HO HO O OH OH OH O (113)

(ii) The Introduction of Substituents

Electrophilic substitution (see Sect. 2.3.1) at the anthraquinone skeleton is a relatively difficult process because of the deactivating effect of the carbonyl groups. However, two electrophilic reactions are used extensively for synthesising anthraquinone derivatives: — nitration and sulphonation.

Nucleophilic attack upon the anthraquinone system is used often (see Sect. 2.3.1), particularly for the introduction of amino and hydroxy groups. Boric acid, which complexes with the carbonyl groups in anthraquinone, can markedly catalyse such nucleophilic attacks and is used in many reactions, as will be demonstrated throughout this section.

(a) Nitro Groups

As in the benzene series nitro groups are introduced via nitration using nitric acid or a mixture of nitric and sulphuric acids. The nitro group is almost always reduced to the amino group after it has been introduced.

Anthraquinone nitrates to give a mixture of the 1- and 2-mononitro compounds, Eq. 2.35. Unfortunately, the mixture is difficult to separate and this reaction has found only limited application. However, a donor substituted anthraquinone often yields, initially, a single product on nitration. For instance, the diphenoxyanthraquinones (114) and (115) give only the dinitrodiphenoxyanthraquinones (116) and (117) respectively, although the corresponding dihydroxy compounds still give mixtures.

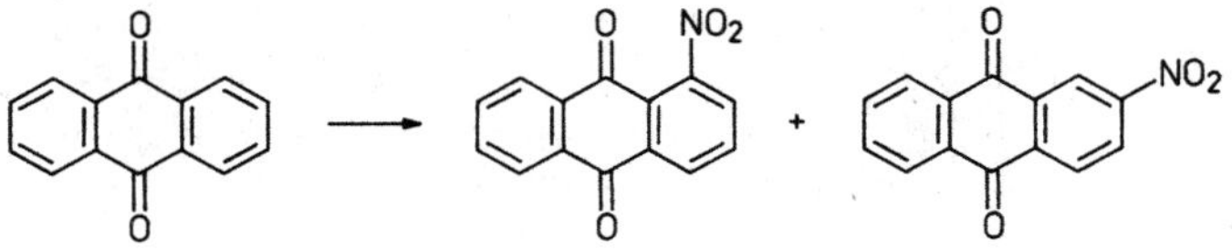

Eq. 2.35

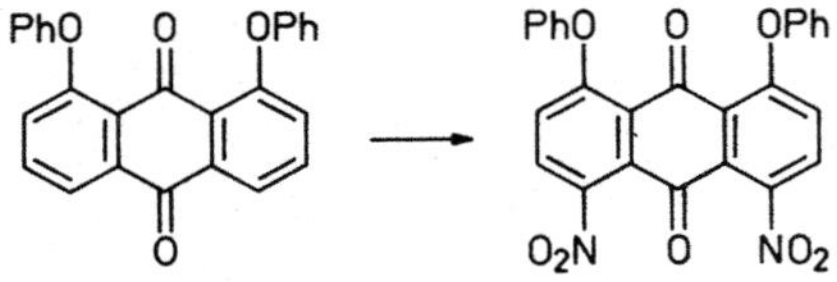

(114) (116)

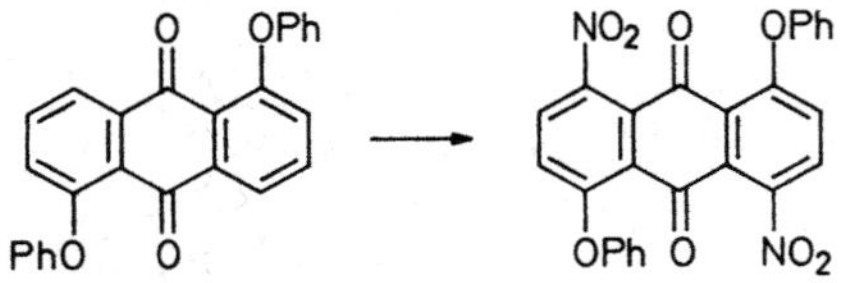

(115) (117)

Sometimes, by altering the reaction conditions and reagents, the position of electrophilic attack can be altered. This is the case for the nitration of alizarin (118). 4-Nitroalizarin is obtained by nitrating in oleum at —5 °C whilst 3-nitroalizarin is synthesised by nitrating alizarin in concentrated sulphuric acid containing boric acid. Here, the boric acid complexes with the oxygen atoms, *i.e.* the 1-hydroxy and 9-carbonyl oxygen atoms, to deactivate the 4-position towards electrophilic attack.

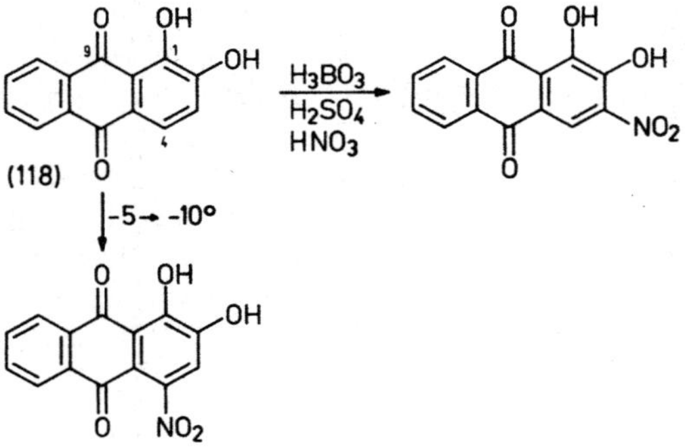

H_3BO_3, H_2SO_4, HNO_3

(118)

-5→ -10°

A useful general method for obtaining just one isomer is by blocking a position which, if free, would be attacked by the electrophile, Eq. 2.36.

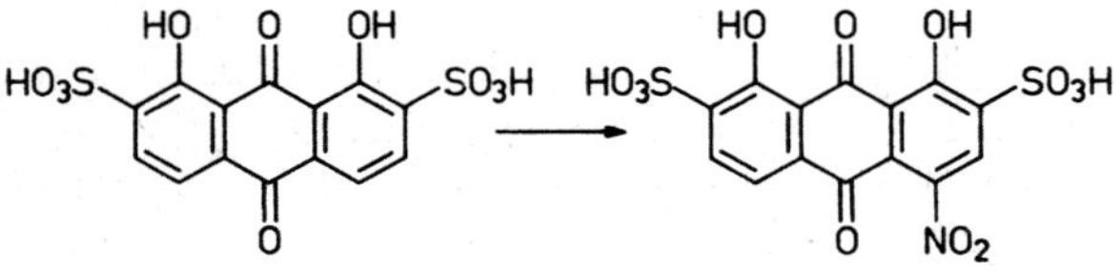

Eq. 2.36

Surprisingly, quinizarin, even though it contains two hydroxy groups, will not nitrate. This problem is circumvented by the unusual reaction sequence shown below, Eq. 2.37.

$$\text{quinizarin} \xrightarrow{SOCl_2} \text{1-chloro-4-hydroxyanthraquinone} \xrightarrow[\text{hydrolysis}]{HNO_3} \text{1,4-dihydroxy-5-nitroanthraquinone} \qquad \text{Eq. 2.37}$$

(b) Sulphonic Acid Groups

Sulphonic acid groups are introduced into anthraquinones for much the same reasons as they are introduced into benzenoid intermediates, *i.e.* to confer aqueous solubility on the anthraquinone, and as leaving groups in nucleophilic reactions.

Anthraquinone sulphonates to give the 2-sulphonic acid derivative ('silver salt'). However the 1-anthraquinone sulphonic acid is more useful synthetically and is prepared using mercury salts in oleum. The current view is that anthraquinone initially mercurates at the 1-position and that the sulphonating agent then displaces the mercury to give the desired 1-sulpho derivative. The only other metal which facilitates this reaction is thallium, a metal well known for its propensity to form arylthallium salts.

Disulphonation of anthraquinones in the presence of mercury salts leads to a mixture of the 1,5- and 1,8-disulphonated anthraquinones, Eq. 2.38. Fortunately, the mixture is easily separated.

$$\text{anthraquinone} \xrightarrow{Hg^{II}} \text{1,5-}(SO_3H)_2\text{-anthraquinone} + \text{1,8-}(SO_3H)_2\text{-anthraquinone} \qquad \text{Eq. 2.38}$$

Surprisingly, the introduction of a hydroxy group into anthraquinone does not appreciably activate the anthraquinone nucleus towards sulphonation. However, sulphonation is now directed *para* to the hydroxy group. Disulphonation results in sulphonic acid groups being introduced into different rings of the hydroxy anthraquinone. Thus, alizarin gives 3,6- and 3,7-disulphonated anthraquinones: in the presence of mercury salts, it gives 3,5- and 3,8-disulphonated anthraquinones (Scheme 2.29). Such mixtures are avoided in symmetrical dihydroxyanthraquinones, *e.g.* anthrarufin (110).

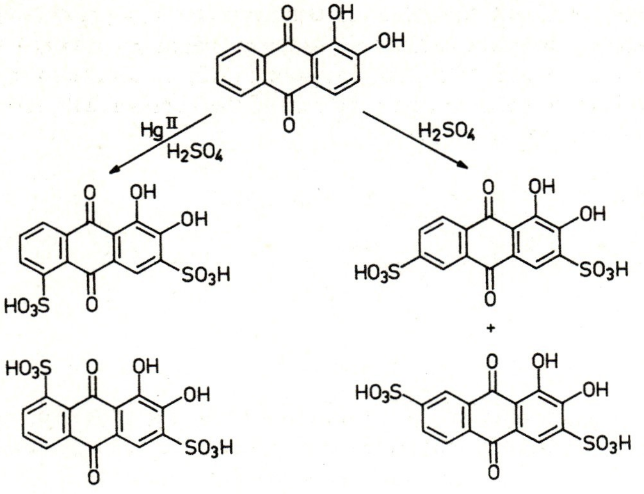

Scheme 2.29

In some cases, it is imperative to use mercury salts for sulphonation to occur at all. Thus, quinizarin will only sulphonate in the presence of boric acid and mercuric sulphate and then only at the 6-position, Eq. 2.39. Thus, quinizarin shows once again a surprising lack of reactivity (see nitration).

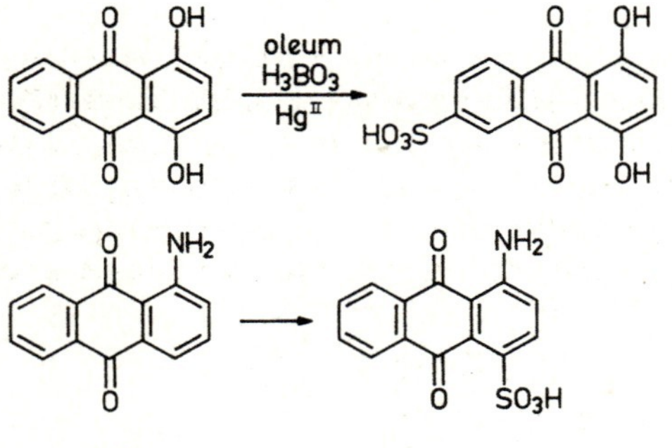

Eq. 2.39

(119)

Aminoanthraquinones generally sulphonate at the position *para* to the amino group as in the sulphonation of 1-aminoanthraquinone (119). As for the hydroxy-anthraquinones, further sulphonation leads to mixtures of disulphonated amino-anthraquinones.

(c) The Halogens

The halogens, particularly bromine and chlorine, are used chiefly for their synthetic value in anthraquinone dyes since they are fairly readily displaced by nucleophiles. Although bromine and chlorine are usually introduced into anthraquinones via the element, anthraquinone itself does not halogenate directly. Consequently the halogen has to be introduced into anthraquinone via the sulphonic acid group, *e.g.* Eq. 2.40. Predictably, the presence of strong electron-donor groups in the anthraquinone nucleus facilitate direct halogenation. Thus, 1-aminoanthraquinone (119) gives the dibromo derivative (120), and the aminosulphonic acid (121) gives the monobromo derivative (122). This latter compound (122), known as Bromamine acid, is a very useful intermediate in the synthesis of the commercially important 1,4-diaminoanthraquinone dyes (see Chap. 4).

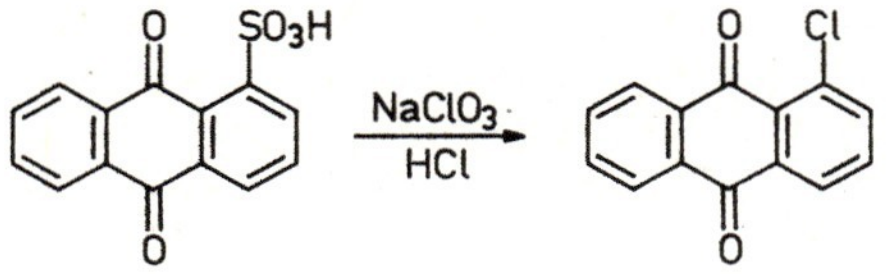

$NaClO_3$ / HCl

Eq. 2.40

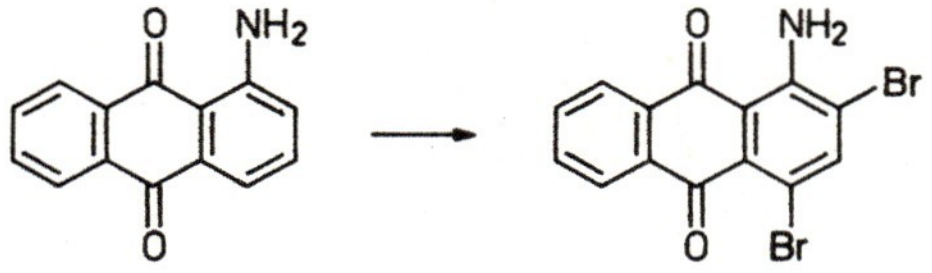

(119) (120)

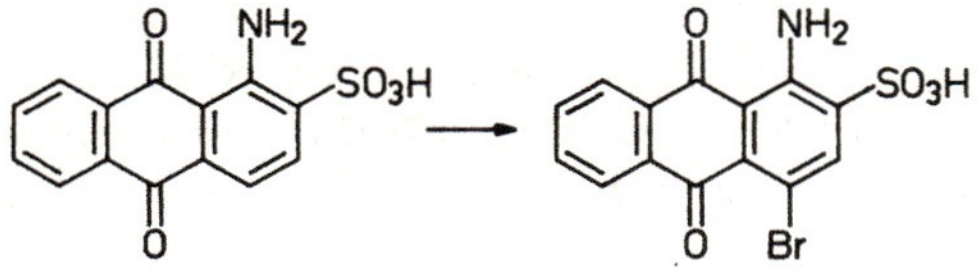

(121) (122)

Alizarin also brominates readily to give the 3-bromo derivative (123). Here, the 2-hydroxy group is the dominant directing group since the 1-hydroxy group is involved in intramolecular hydrogen-bonding with the carbonyl group. The position of substitution can be altered by using a 2-methoxy group, as in the methylated alizarin derivative (124). In this case bromination occurs at the 4-position.

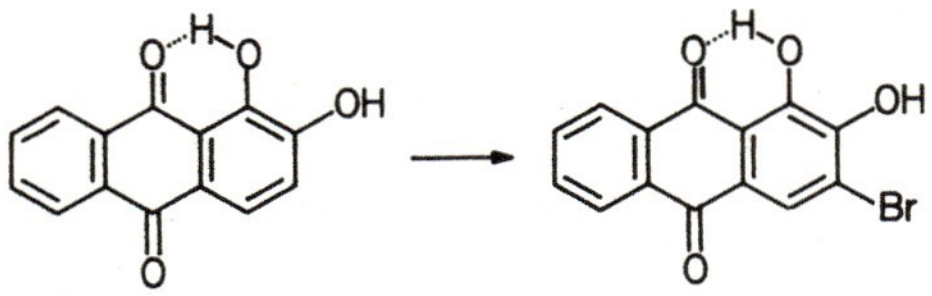

(123)

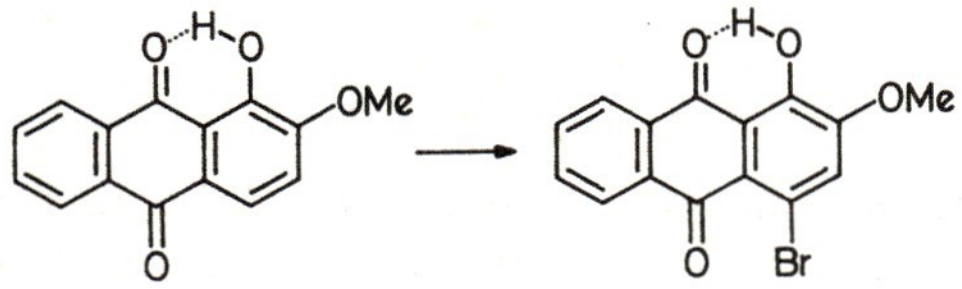

(124)

Interestingly, quinizarin can be chlorinated in the unsubstituted ring in preference to the ring containing the donor groups. However, the reaction requires boric acid which functions not as a catalyst but as an agent to block chlorination in the substituted ring via the complex (125).

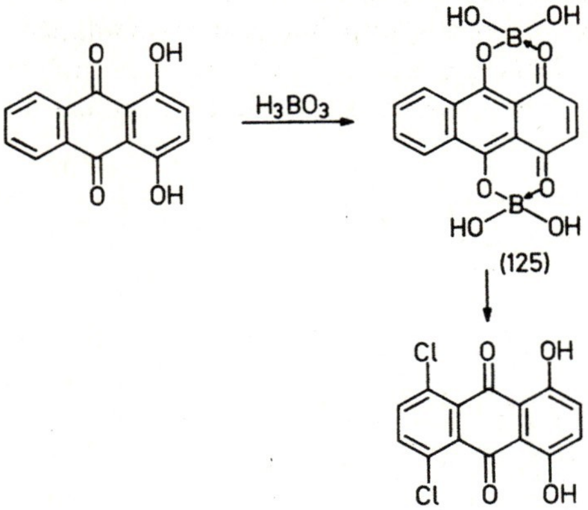

(125)

(d) Hydroxy Groups and their Derivatives

As was seen earlier in the synthesis of quinizarin, hydroxyanthraquinones can be synthesised directly. However, the hydroxy group is more often introduced into anthraquinones via nucleophilic attack at a carbon bearing a good leaving group, *e.g.* SO_3H, Br, Cl. For instance, 2,4-dibromo-1-aminoanthraquinone readily undergoes substitution at the 2 and 4 positions. Interestingly, this substitution can be induced to go stepwise so that two different groups may be introduced consecutively. The hydroxyalkoxyaminoanthraquinone (126) can therefore be obtained by regio-specific hydroxylation to give first the 4-hydroxy derivative (127) and then alkoxylated to give the 2-alkoxy-4-hydroxy derivative (126). In a similar manner many other hydroxyanthraquinones can be obtained from the corresponding chloro- and bromoanthraquinones.

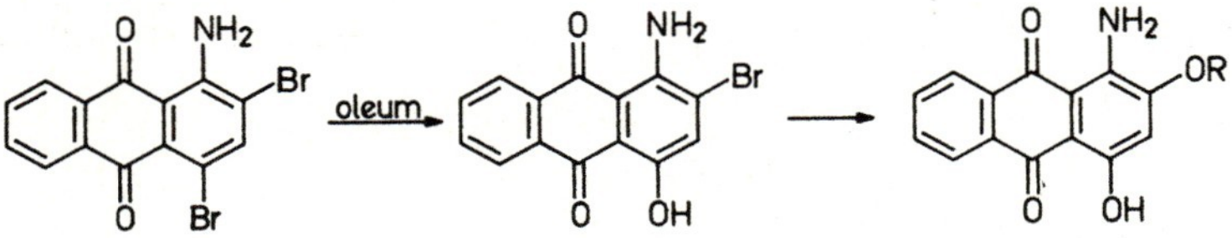

(127) (126)

Sulphonated anthraquinones are also useful intermediates for the introduction of hydroxy groups. Two important dihydroxyanthraquinones, chrysazin (128) and anthrarufin (110), are easily obtainable by heating 1,8- and 1,5-disulphonated anthraquinones respectively with lime under high pressure.

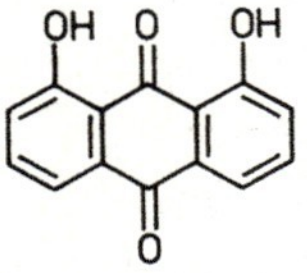

(128) 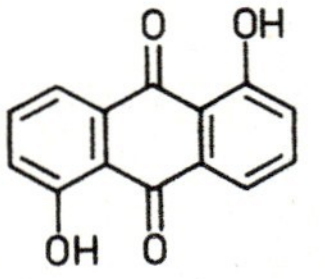 (110)

Sulphonic acid groups at the 2-position are also susceptible to nucleophilic attack and provide the corresponding 2-hydroxy or 2-alkoxy derivative, as in Eq. 2.41 a, b. If the 1-position is free, *e.g.* for 'silver salt', then not only is the sulphonic acid group replaced by a hydroxy group but also an oxidation of the molecule to a dihydroxyanthraquinone takes place (Eq. 2.42). This reaction was first discovered in the last century and formed the basis for the first industrial manufacture of alizarin (see Sect. 1.3.5).

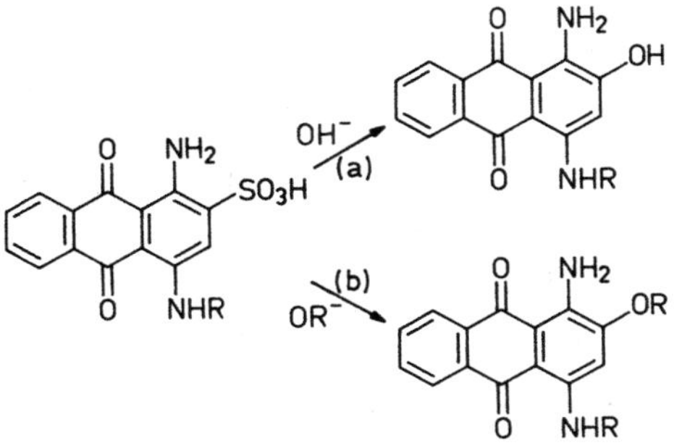

O NH2 SO3H NHR O; (a) OH⁻ → O NH2 OH NHR O; (b) OR⁻ → O NH2 OR NHR O

Eq. 2.41

O SO3H O; [O] OH⁻ → O OH OH O

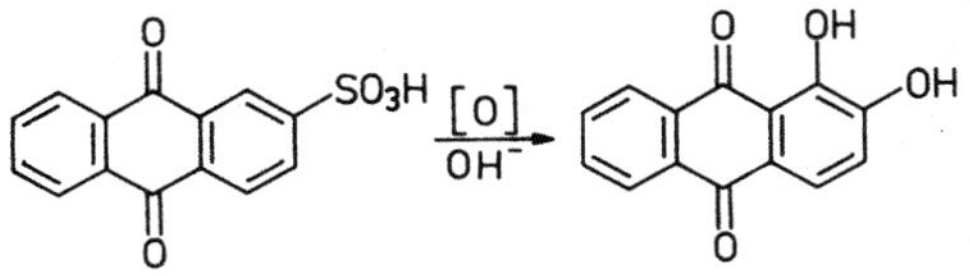

Eq. 2.42

As the last example shows hydroxy groups can be introduced by an oxidation reaction and further examples of this reaction are the oxidation of quinizarin to purpurin (129) and the boric acid catalysed oxidation of 1,4-diaminoanthraquinone to the dihydroxy derivative (130). This latter example once again demonstrates the use of the ubiquitous boric acid 'catalyst' and just as in the boric acid catalysed chlorination of quinizarin it prevents reaction in the substituted ring via a complex. However, in general the oxidation reaction has a lesser importance than the substitution reaction.

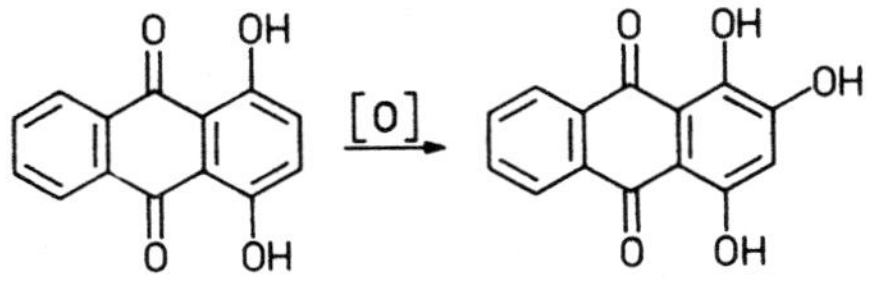

O OH OH O [O] → O OH OH OH O

(129)

O NH2 NH2 O → HO B OH O NH NH O HO B OH → OH O NH2 OH O NH2

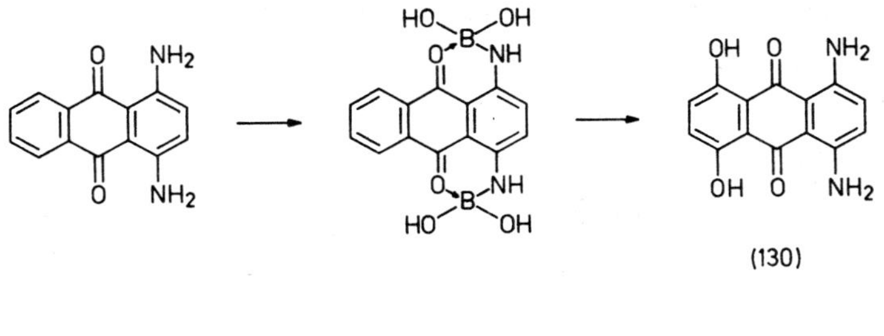

(130)

(e) Amino Groups

Aminoanthraquinones account for a large percentage of the commercial anthraquinone dyes (see Chap. 4). In the majority of cases the amino group is introduced via nucleophilic attack at a carbon atom bearing a good leaving group (*cf.* the synthesis of aminoazo dyes).

As for the hydroxy group, the amino group will displace a halogen from an anthraquinone to give the corresponding amino substituted anthraquinone. An example of this reaction is shown for the bromoanthraquinone (122), Bromamine acid. Bromamine acid, the synthesis of which has already been described, has achieved some considerable importance for it presents the industrial chemist with a flexible route to a variety of 1,4-diaminoanthraquinones in which two different amino groups can be introduced. Displacement of bromine is catalysed quite markedly by copper ions. Copper is known to complex with the anthraquinone oxygen atom and the amino group, and therefore, probably aids nucleophilic attack by bringing the reactants into close proximity as well as exerting an electronic effect on the anthraquinone ring. Thus, the displacement reaction occurs under relatively mild conditions. It even occurs with the less nucleophilic amines (*e.g.* aromatic amines). Indeed, the displacement of halogen provides a very general route into aminoanthraquinones, especially those containing more than one amino group.

Another very important procedure for the preparation of 1,4-diaminoanthraquinones is via *leuco*-quinizarin (131), a reduced form of quinizarin. Because quinizarin can be prepared relatively cheaply (see earlier), and can be converted easily to 1,4-diaminoanthraquinones, it has achieved great importance as an intermediate in the synthesis of anthraquinone dyestuffs. Alkylamines react with *leuco*-quinizarin to give the corresponding 1,4-diamino derivatives which are then oxidised (by air) to the corresponding 1,4-diaminoanthraquinones. However, arylamines, which are less nucleophilic than alkylamines, do not react with *leuco*-quinizarin unless boric acid is present as catalyst. Boric acid functions by complexing with the *leuco*-quinizarin (131) and this greatly facilitates nucleophilic attack by the arylamine. The 1,4-diamino compound (132) thus obtained is readily oxidised, even by air, to the 1,4-diaminoanthraquinone (133).

Unsymmetrical 1,4-diaminoanthraquinones can be prepared from quinizarin by using an equimolar mixture of an arylamine and an alkylamine or alternatively a stepwise reaction may be employed. The same strategy is also applicable to 1,4,5-trihydroxy and 1,4,5,8-tetrahydroxyanthraquinones.

If quinizarin is oxidised, then amino groups can be introduced into the 2-position under extremely mild conditions via the reactive quinone (134). This further underlines the synthetic versatility of quinizarin.

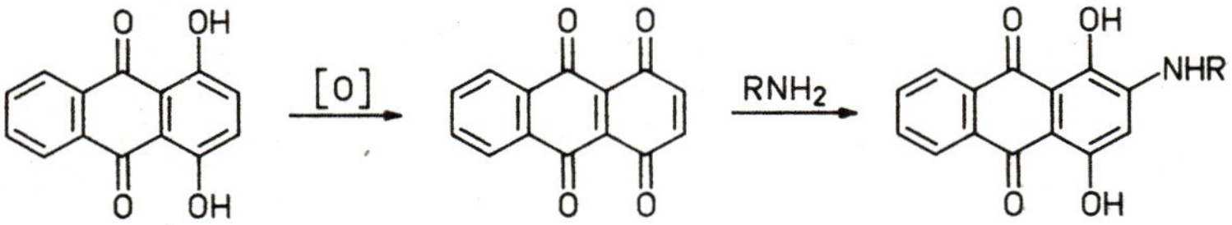

O OH [O] O O RNH_2 O OH NHR

O OH O O O OH

(134)

As well as the hydroxy group, other oxygen functionalities may be replaced in the anthraquinone nucleus. The most familiar is the methoxy group, which can be replaced by alkylamines. The example below also illustrates the reversibility of the amination reaction since the two amino groups are also displaced (Eq. 2.43).

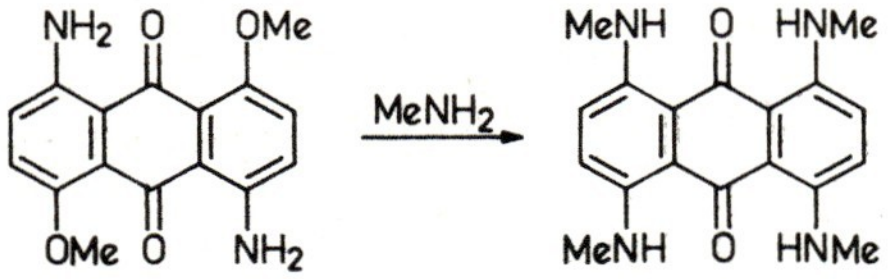

NH_2 O OMe $MeNH_2$ MeNH O HNMe

OMe O NH_2 MeNH O HNMe

Eq. 2.43

The nitro group, as in other aromatic systems, provides an easy access to aminoanthraquinones. The corresponding amino compound can be obtained either by reduction or by nucleophilic displacement of the nitro group by an amine. It is even possible to carry out these reactions in a stepwise fashion on a dinitrated anthraquinone, *e.g.* Scheme 2.30.

(f) Miscellaneous Reactions

The above discussion only outlines a minute proportion of the total number of reactions carried out on anthraquinone and substituted anthraquinones. There are many other groups which are used, *e.g.* CN, CO_2H, SR, *etc.*, but the strategy previously described is applicable. However, one reaction is worthy of mention for it describes a method by which heterocyclic rings may be annelated to the anthraquinone system and it also records the first example of a general synthesis for quinolines. In 1877, the anthraquinone (135) was obtained by chance from the reaction of alizarin, 3-nitroalizarin, glycerol and concentrated sulphuric acid. The yield was poor but was vastly improved by substituting 3-aminoalizarin (which was obtained in the original

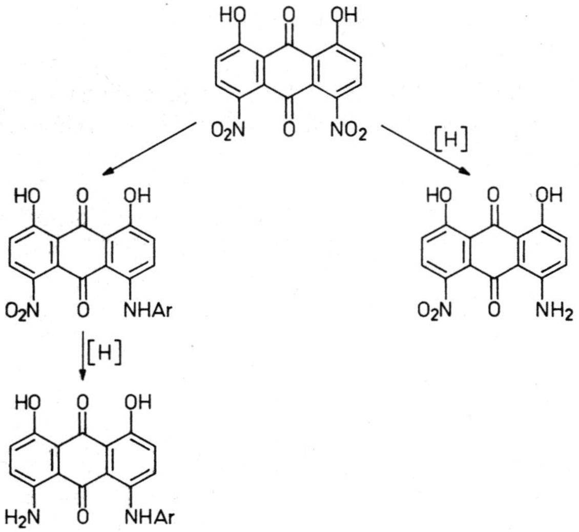

Scheme 2.30

reaction by reduction of 3-nitroalizarin with alizarin) for alizarin. The sequence is described in Scheme 2.31. This reaction was later generalised to the synthesis of quinolines from anilines and is now called the Skraup synthesis.

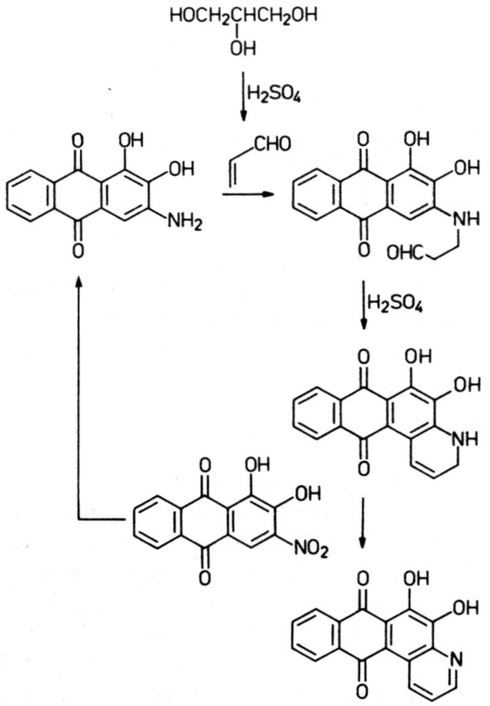

(135)

Scheme 2.31

2.4.3 Vat Dyes

Unlike the azo dyes, which are all structurally related, vat dyes can possess widely differing structures. Thus, the term vat dye does not refer to a distinct chemical class but rather to a class of dyes which is applied to a fibre by a specific process. Briefly, the process requires the dye to be in a reduced and solubilised form whereupon it is impregnated into a fibre and then oxidised. The vat dye, upon oxidation, precipitates in an aggregated form and therefore remains locked in the fibre. Despite the fact that no discrete chemical class can be applied to vat dyes, anthraquinone type dyes find themselves pre-eminent in the class. Of course, there are other types of vat dyes, *e.g.* indigoids, and these will be discussed in their respective chemical class. Thus, the vat dyes in this classification are those dyes which bear a structural resemblance to anthraquinone dyes, *i.e.* they are extensively conjugated aromatic sytems containing two or more carbonyl groups.

The simplest vat dyes are modified anthraquinones, *e.g.* (136), and their synthesis is therefore an extension of those given in the previous section. However, the majority of the important vat dyes are polycyclic condensed systems some of which are derived from anthraquinone dyes. For instance, indanthrone (5), the first major vat dye (see Sect. 5.2.2), is still prepared by the action of fused sodium hydroxide on 2-aminoanthraquinone. Similarly, the carbazole (137) is prepared via the Scholl reaction on the anthrimide (138); this involves heating in an aluminium chloride-sodium chloride flux. The mechanism for the oxidation step is not known although it is thought that a radical mechanism is possible.

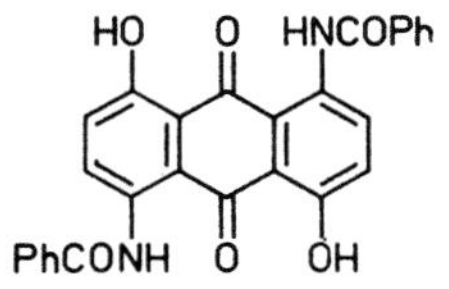

HO O HNCOPh

PhCONH O OH

(136)

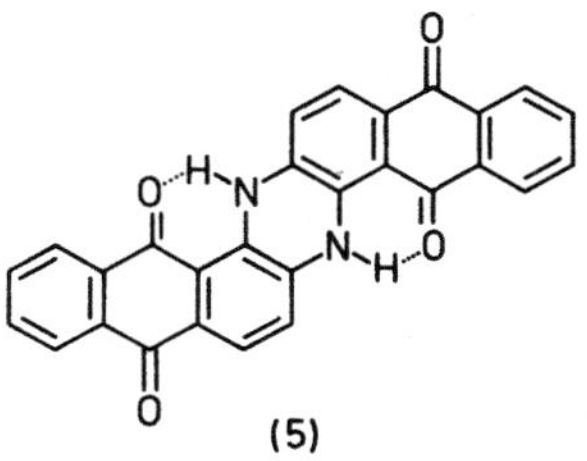

(5)

(138)

[O] $AlCl_3$ - NaCl

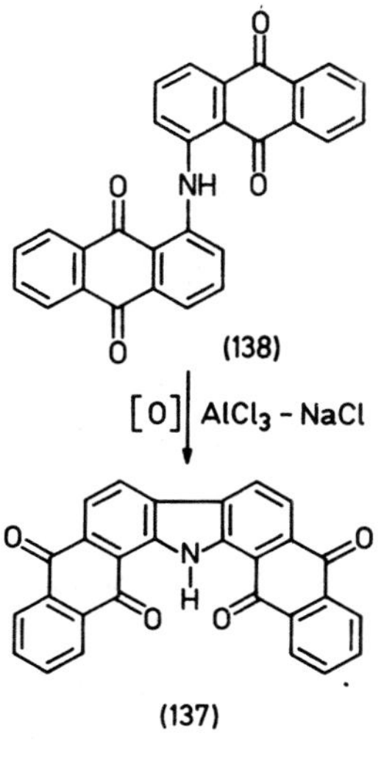

(137)

Lewis acid catalysts are used extensively as agents to induce the cyclisation reactions necessary to give vat dyes, as illustrated by the last reaction. A further example is the cyclisation of 1,5-dibenzoylnaphthalene by aluminium chloride to the dibenzopyrenequinone (139). A mechanism for the transformation, another example of the Scholl reaction, is outlined in Scheme 2.32, although radical mechanisms have not been rigorously excluded. Sulphuric acid is also a useful catalyst and facilitates the double cyclisation of the diacid (140) to the anthanthrone (141; X=H); this is then converted to the more important anthanthrones (141; X=Cl, Br) by chlorine or bromine respectively.

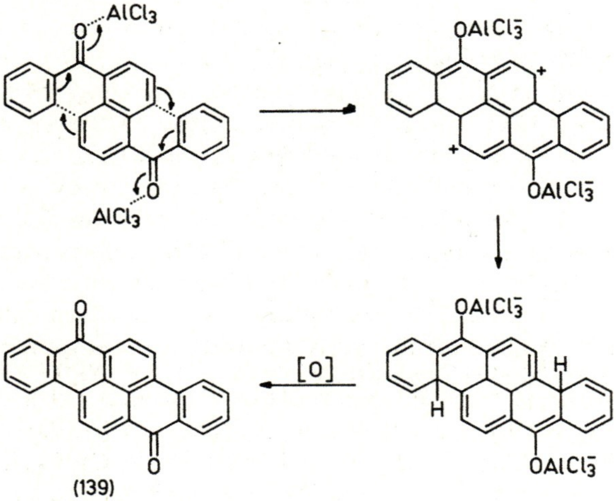

Scheme 2.32

(139)

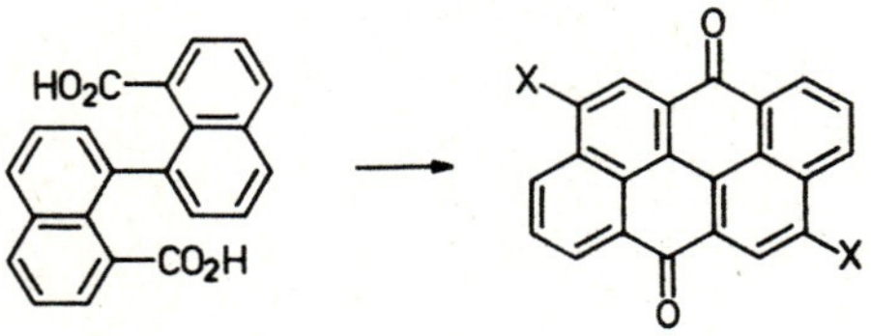

(140) (141)

The pyranthrone (142) is prepared by a base induced cyclisation of the bisanthraquinone (143). The methyl groups at the 2,2′-positions are sufficiently activated by the anthraquinone carbonyl groups to be ionised and subsequently condense in a reaction analogous to an aldol condensation.

Benzanthrone (6), an important intermediate for Caledon[11] Jade Green XBN, can be prepared by two methods. The first is by the Scholl reaction on 1-benzoylnaphthalene (see earlier) and the second synthesis starts from anthraquinone, iron powder, sulphuric acid and glycerol. The mechanism for this interesting reaction has not been fully substantiated but is thought to proceed as in Scheme 2.33. Thus,

[11] Caledon is the trademark of ICI.

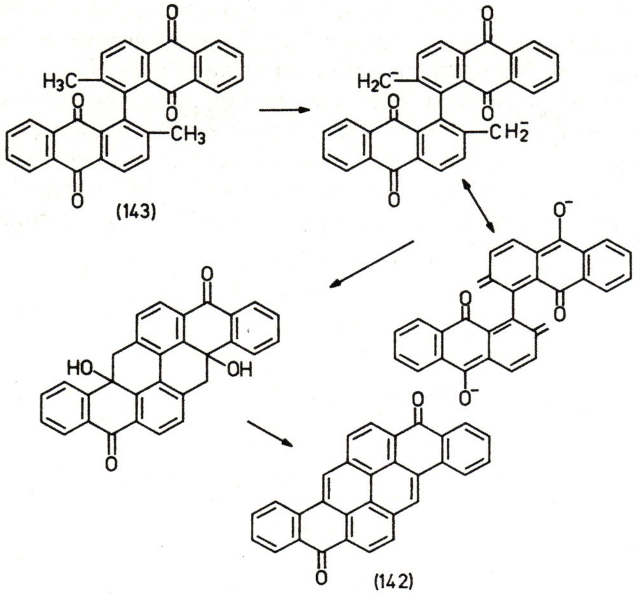

H3C
CH3
(143)
H2C
CH2
O
HO
OH
(142)

the iron and sulphuric acid reduce one carbonyl group of the anthraquinone completely and the molecule so produced undergoes a Michael reaction with acrolein, which is formed from the glycerol and concentrated sulphuric acid (*cf.* the Skraup synthesis). The intermediate (144) then undergoes an internal Friedel-Crafts reaction followed by dehydration to the dihydrobenzanthrone (145). The dihydrobenzanthrone

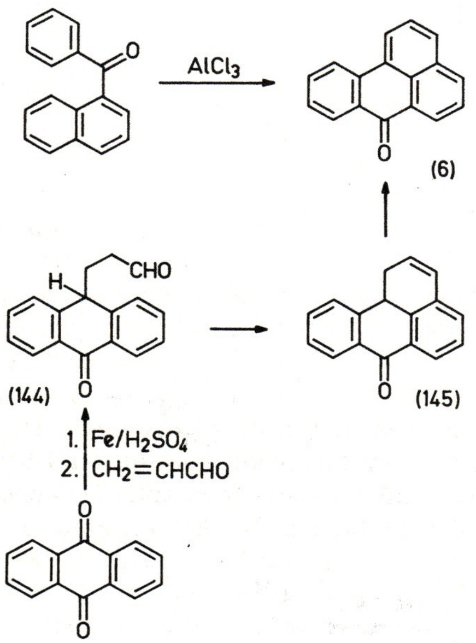

AlCl3
(6)
H
CHO
(144)
(145)
1. Fe/H2SO4
2. CH2=CHCHO

Scheme 2.33

(145) is then oxidised to benzanthrone (6); this latter reaction is usually carried out in the presence of copper II sulphate. Whether the mechanism is quite this simple is rather doubtful although the later steps do seem reasonable.

As mentioned previously benzanthrone, although itself a vat dye, is prepared primarily as an intermediate for the important vat dye, Caledon Jade Green XBN (146). Thus, benzanthrone is oxidised (by air) in the presence of potassium hydroxide to 4,4′-dibenzanthrone (147), possibly by a radical dimerisation. Further oxidation with manganese dioxide and sulphuric acid at 35 °C gives (148) which is then partially reduced and methylated to give Caledon Jade Green (146).

KOH
O_2
(147)
MnO_2
(148)
1. $NaHSO_3$
2. Me_2SO_4
(146)

Thus, although the major vat dyes are very complex systems their synthesis from smaller molecules tends to be rather simple, *e.g.* by condensations, oxidative cyclisations, *etc.*

2.4.4 Phthalocyanine Dyes

Copper phthalocyanine (23) and its halogenated derivatives are still the most important phthalocyanine colourants although many other metallised phthalocyanines, *e.g.* Cr, Mo, Fe, Co, Rh, have been made. Hence, the synthesis of copper phthalocyanine will be used to illustrate the principles involved in the synthesis of phthalocyanines in general.

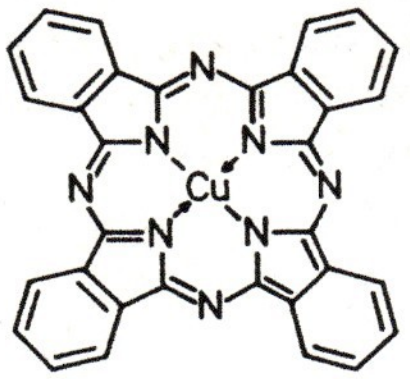

(23)

There are two important routes to copper phthalocyanine; the first is the phthalic anhydride-urea process and the second is the phthalonitrile process. In the first process phthalic anhydride is heated with urea, copper I chloride and a catalytic amount of ammonium molybdate in a high boiling solvent. The function of the urea is to act as a source of nitrogen since the carbonyl group of the urea is lost as carbon dioxide; this has been confirmed by labelling studies. The first step in the transformation is the conversion of phthalic anhydride to phthalimide by the ammonia liberated from the urea. More ammonia then converts phthalimide to isoindoline (149), and finally to the isoindoline (150); the latter has been isolated and identified. However, in the presence of copper chloride, (150) spontaneously tetramerises and is oxidised to phthalocyanine which then forms a complex with the copper.

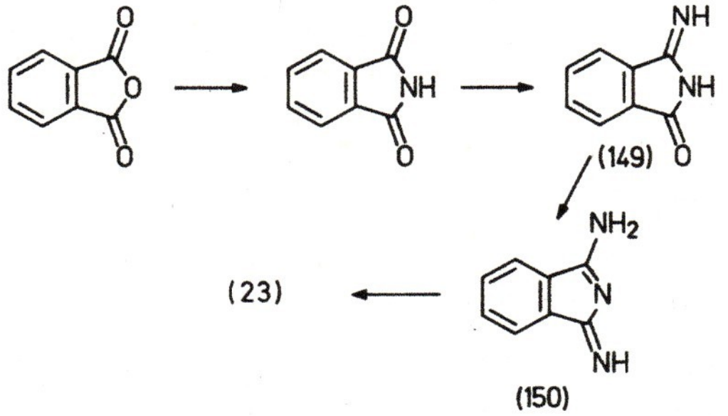

(149)

(23)

(150)

In the phthalonitrile process phthalonitrile (151) is heated in the presence of a base and cupric and ammonium acetates, with or without a solvent. The mechanism has been less studied than in the case of the phthalic anhydride-urea process. However, an intermediate such as (152) is postulated to form before tetramerisation to phthalocyanine occurs. If the reaction is carried out in the absence of copper acetate then metal free phthalocyanine is obtained.

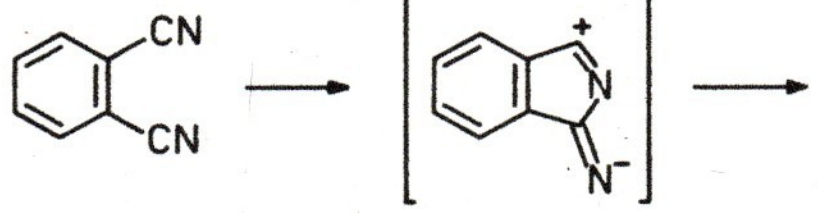

(151) (152)

The shade of a phthalocyanine may be altered slightly by incorporating groups into the aromatic carbocyclic rings. This may be done by reaction on the phthalocyanine; for instance, chlorination with SO_2Cl_2—$AlCl_3$—NaCl, and also chlorosulphonation with chlorosulphonic acid. Alternatively, a suitably modified phthalic acid derivative may be used in the phthalocyanine synthesis. Of course, in the latter case a symmetrical tetrasubstituted phthalocyanine is obtained.

2.4.5 Indigoid Dyes

The basic structure of the indigoids is (7).

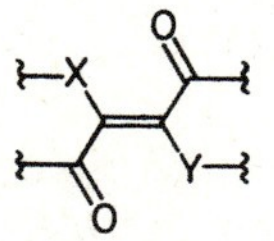

(7)

However, the vast majority of the indigoids have *X*/*Y* and the carbonyl group in a ring system because these are more accessible synthetically and also much more stable. By far the most important indigoids are the indigos (153) and thioindigos (154). Indigoids such as (155) and (156) are of less importance. For the sake of brevity only synthetic routes to (thio)indigo and their derivatives will be described.

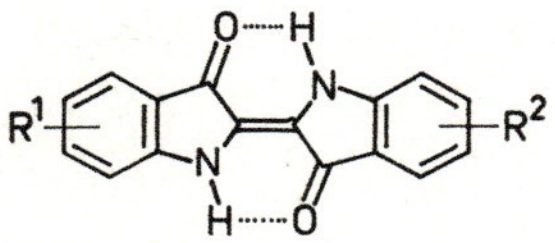

(153) 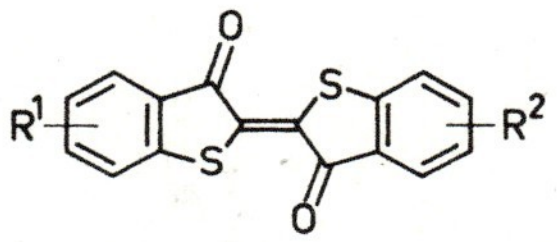 (154)

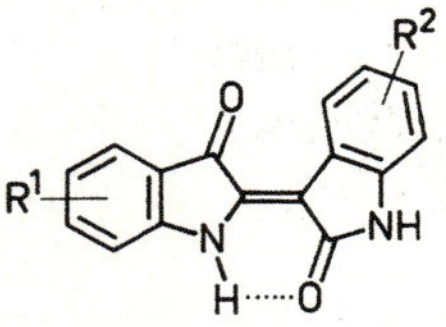

(155) 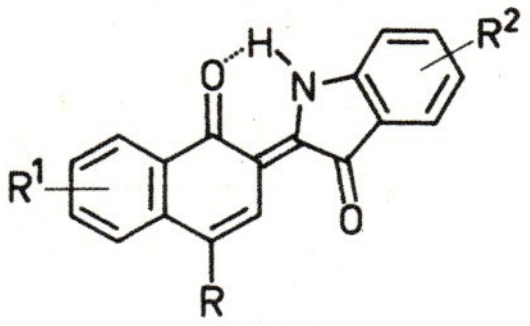 (156)

The first syntheses of indigo have already been discussed (Sect. 1.3.6), including the once important Heumann synthesis. However, the synthesis of indigo has been improved further and now starts from aniline. In the first step aniline is condensed with formaldehyde-sodium bisulphite and sodium cyanide to give N-cyanomethylaniline. This compound is then hydrolysed to sodium phenylglycinate which is cyclised in molten sodium and potassium hydroxide (containing sodamide) to give indoxyl. The latter is then oxidised in air to indigo (Scheme 2.34).

The above route fails for N-alkylglycines and for thioglycollic acid derivatives. Hence, N-alkylglycines (157) are converted first to the corresponding acid chloride (158) and then cyclised by an intramolecular Friedel-Crafts reaction with aluminium chloride. As before, the indoxyl derivative is oxidised to the indigo derivative.

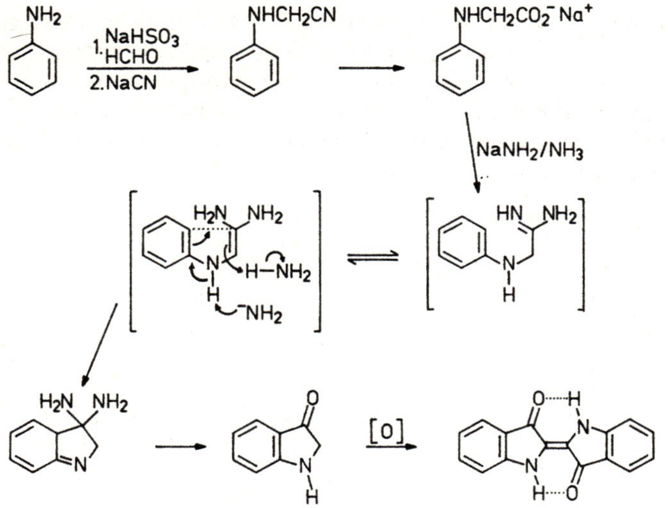

Scheme 2.34

Thioindigos are prepared by the condensation of a substituted benzenethiol (159) with chloroacetic acid to give the corresponding thioglycollic acid (160). The route thereafter is similar to that shown for NN′-dialkylindigo. The benzenethiol is prepared from aniline as shown.

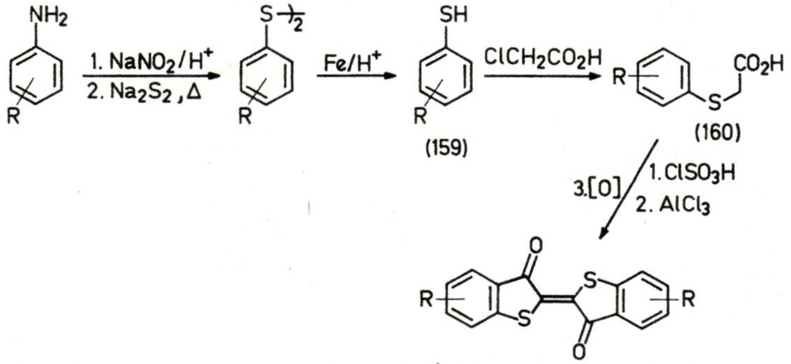

(159)

(160)

The route provides better yields of substituted thioindigos from the corresponding thiols than it does for thioindigo itself. Thioindigo is best prepared from *o*-carboxybenzenethioglycollic acid (161) by a reaction analogous to the Heumann synthesis of indigo from *o*-carboxyphenylglycine (see Sect. 1.3.6).

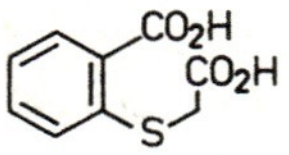

(161)

The final step in the preparation of the indigo or thioindigo is always an oxidation. On an industrial scale atmospheric oxygen is used as the oxidising agent in the synthesis of indigos, and sulphur in the synthesis of thioindigos. The mechanism for the oxidation has not been proved conclusively but the two mechanisms shown below (Eq. 2.44a, b) have been postulated.

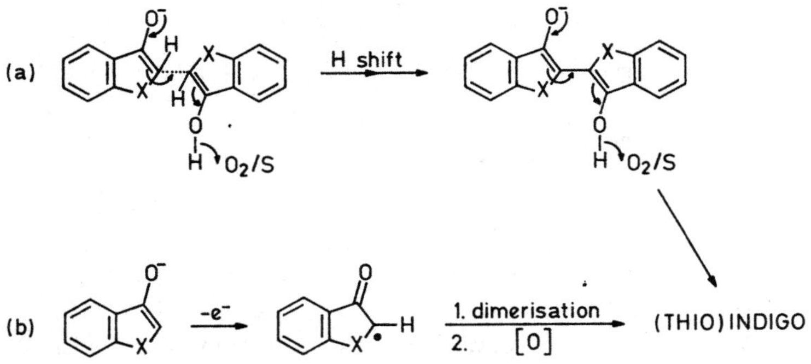

(a) H shift O_2/S O_2/S

(b) -e⁻ 1. dimerisation 2. [O] (THIO)INDIGO

Eq. 2.44

Derivatives of indigo and thioindigo containing a variety of substituents have been prepared. These substituents may be introduced either directly or via a suitably substituted aniline. For instance, 5,5′-dichloroindigo (162) may be prepared by chlorination of indigo or from *p*-chloroaniline (Scheme 2.35).

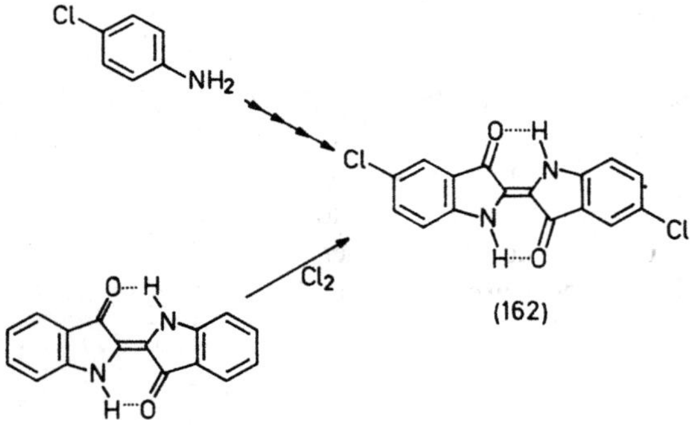

Cl_2

(162)

Scheme 2.35

However, in some cases the indigo can be prepared only by one method. An example is 6,6′-dichloroindigo (163), which must be prepared from *m*-chloroaniline since it cannot be obtained directly by chlorination.

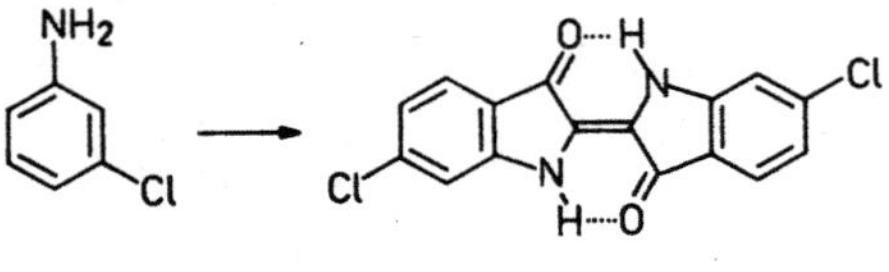

(163)

2.4.6 Polymethine Dyes

In general, polymethine dyes are of more importance as sensitisers in the photographic process than as dyes for textiles. The main reason for this is their inherently poor light fastness. Nevertheless, some polymethines have found use as dyes for textiles and these will therefore be used to illustrate the general synthetic strategy. In particular, the indolenines (60a) (see earlier) will be used to illustrate specific examples where possible.

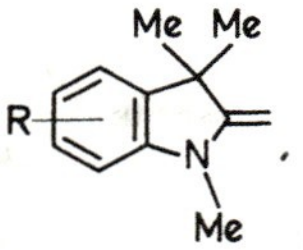

(60a)

Apocyanines (164; n = 0, m = 1), cyanines (164; n = 0, m = 0), carbocyanines (164; n = 1, m = 0) and polycarbocyanines (164; n ≧ 2, m = 0) are structurally very similar. The synthesis of carbocyanines is therefore fairly representative of the synthetic strategy normally employed.

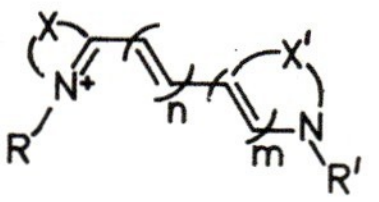

(164)

Symmetrical carbocyanines, *e.g.* (165) can be made by heating two moles of the indolenine (60a) with a trialkylorthoformate, *e.g.* triethylorthoformate, in the presence of a base such as pyridine. Upon acidification the carbocyanine (165) is obtained in a good overall yield. Unsymmetrical carbocyanines are also readily prepared utilising a very similar method. For instance, the Fischer's Base derivative (60b) when heated with the Fischer's aldehyde derivative (166) and a base, yields the corresponding unsymmetrical carbocyanine (167).

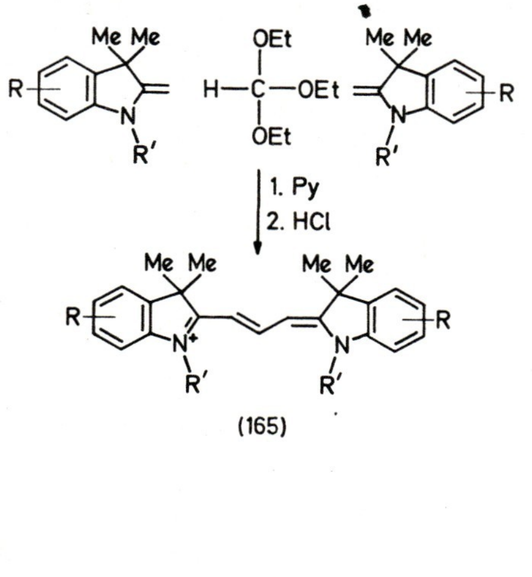

1. Py
2. HCl

(165)

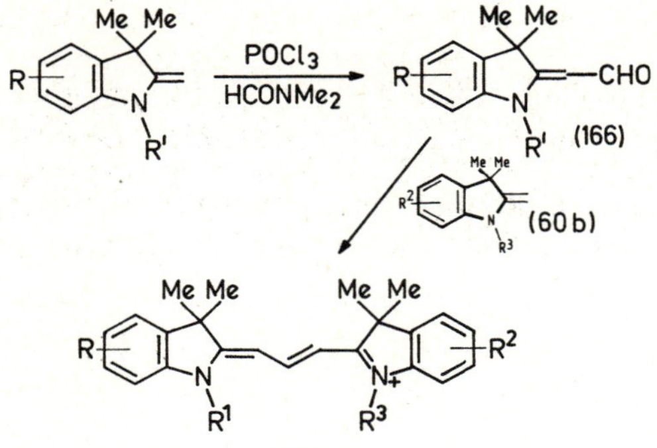

(166)

(60b)

(167)

In a similar manner hemicyanines such as (168) can be prepared by heating a Fischer's Base derivative with an aldehyde. Here the reaction is illustrated for CI Basic Violet 7.

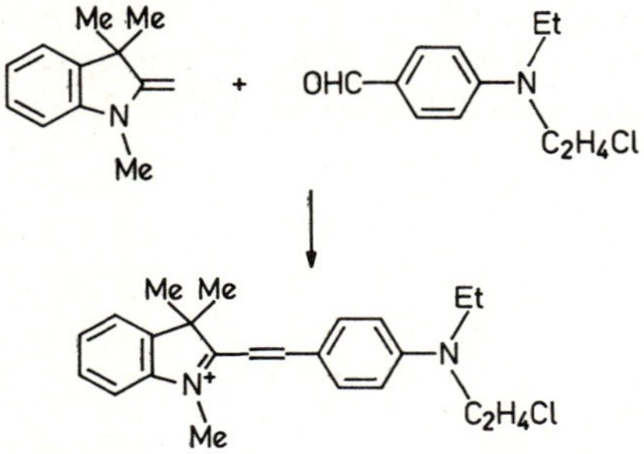

CI Basic Violet 7 (168)

The polymethine dyes just discussed have not achieved much importance as textile dyes. However, the aza analogues of polymethine dyes have proved useful, particularly for polyacrylonitrile fibres, because of their strong, bright shades and their affinity for these fibres (see Chaps. 5 and 6). In contrast to the polymethines, azapolymethine dyes are ineffective as photosensitisers.

The yellow azacarbocyanine dye (13) is a typical azapolymethine. It is synthesised by heating Fischer's aldehyde and 2,4-dimethoxyaniline in acetic acid; many simple variants of this structure are known.

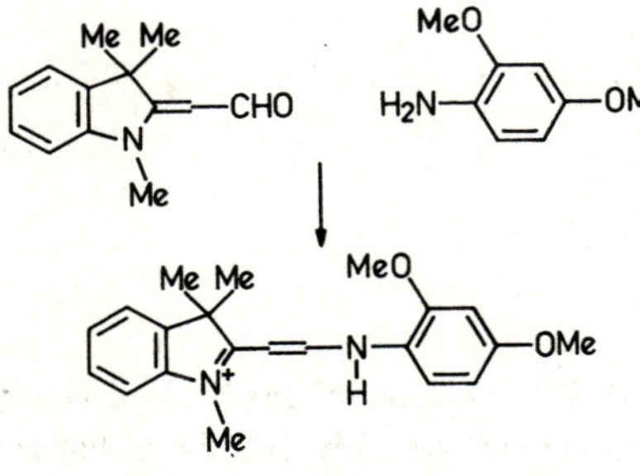

CI Basic Yellow 11 (13)

The diazacarbocyanine dye (14), which contains two heterocyclic atoms in the conjugated chain, is an extremely important yellow dye for polyacrylonitrile. It is tinctorially strong and gives a bright golden yellow shade on polyacrylonitrile. The dyed fibre exhibits excellent all round properties including good light fastness. The dye is prepared in excellent yield by coupling the diazonium salt (169) with Fischer's Base and then quaternising with dimethylsulphate.

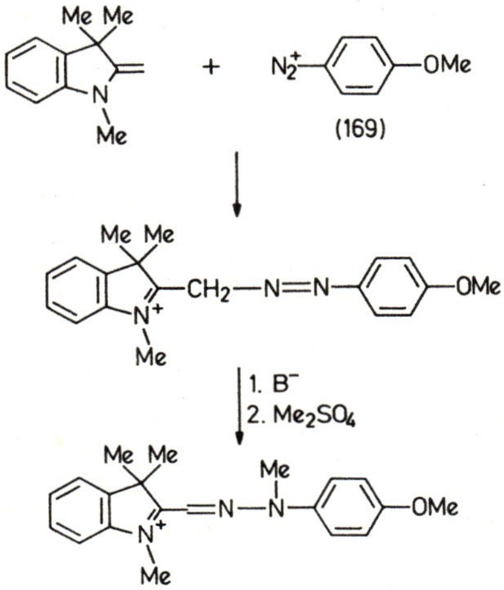

(169)

1. B^-
2. Me_2SO_4

CI Basic Yellow 28 (14)

The azapolymethines span a wide shade area and shades ranging from yellow to blue are readily obtained. In this respect diazahemicyanines are particularly important in the red to blue shade areas. Two important general routes exist for the preparation of these dyes; the first route is the oxidative coupling of a heterocyclic hydrazine or hydrazone to an aromatic amine and, if necessary, quaternisation: the second route is accomplished by the quaternisation of a heterocyclic azo dye prepared by coupling a heterocyclic diazonium salt with an aromatic amine.

The first route is employed in cases where the aromatic amine either will not diazotise or where the diazonium salt produced is so unstable that it decomposes in preference to coupling, Eq. 2.45.

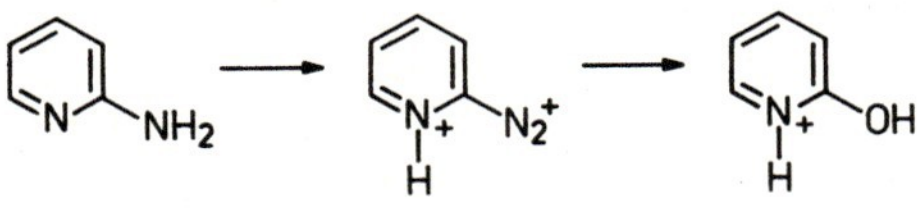

Eq. 2.45

The synthesis of CI Basic Red 30 (170) is a typical example of this type of synthesis. The hydrazone oxidises in preference to aniline, although the latter can give a troublesome side reaction.

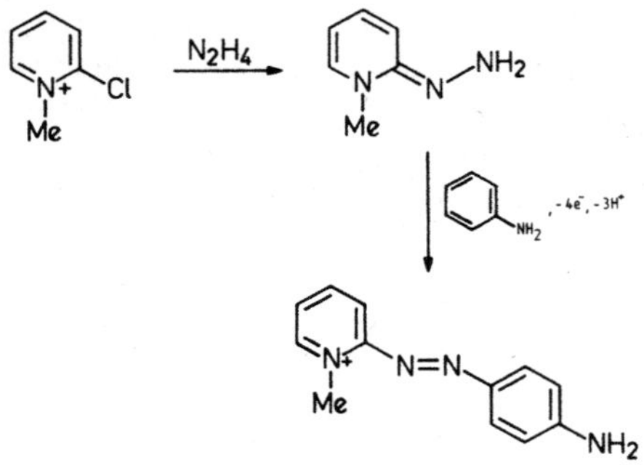

(170)

Where possible the second route is used in preference to the first. A typical synthesis is that of the benzothiazolium dye (12), CI Basic Blue 41. The heterocyclic amine (63) is diazotised with nitrosylsulphuric acid (from $NaNO_2$ and conc. H_2SO_4) to give the diazonium salt (171) and this is coupled with N-hydroxyethyl-N-ethylaniline (172) to give the heterocyclic azo dye (173). Quaternisation with dimethylsulphate gives the blue cationic dye (12) in excellent yield.

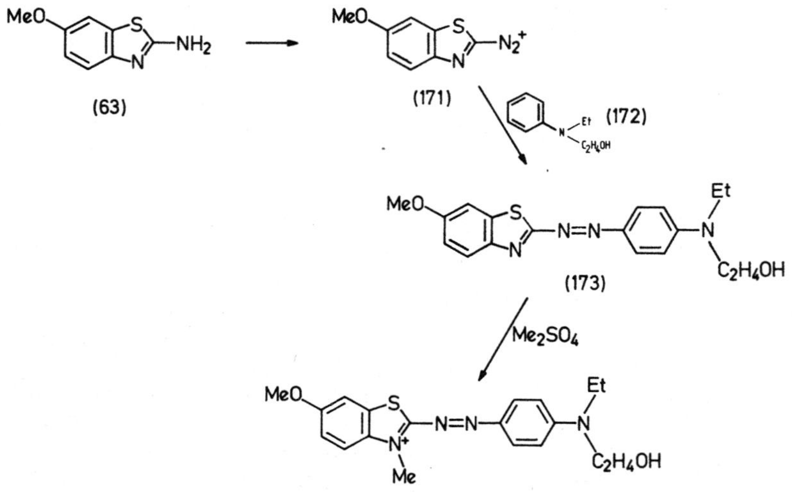

(63) (171) (172) (173) Me_2SO_4 (12)

However, as pointed out earlier, not all the polymethine dyes are cationic, and neutral and anionic types do exist. These latter dyes have little commercial importance and so only one synthesis of each type is shown below.

Reaction of the active methylene compound (174) with the benzothiazoliumthiol (175) gives the merocyanine (15). The dye, however, is not important commercially.

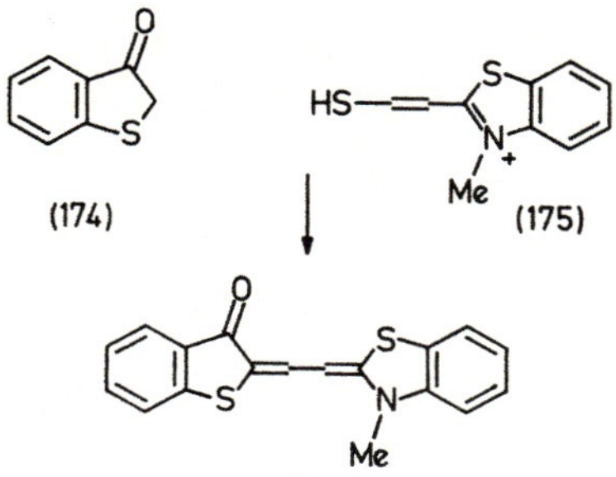

(174) (175)

(15)

Oxonols such as (16) can be used as dyes for basic fibres, *e.g.* nylon; a typical synthesis is the condensation of rhodanine (176) with triethylorthoformate to give the oxonol (16). (Compare the synthesis of the carbocyanine, Astraphloxine FF) [165].

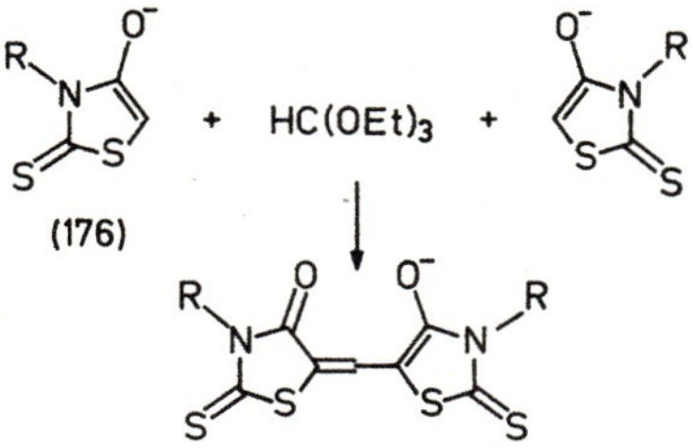

(176)

(16)

2.4.7 Di- and Tri-Arylcarbonium Dyes

In common with the polymethines, the arylcarbonium class of dyes covers a large number of different, though structurally related, dyestuffs. In general the dyes from this class are much less important than the azo and anthraquinone dyes and the following discussion will therefore be kept fairly brief.

Many of the early synthetic dyes belong to this class and some are even in use today, particularly on polyacrylonitrile fibres. One of the simplest arylcarbonium dyes is Auramine O (19), a diphenylmethane dye which has been used for nearly one hundred years. At one time this yellow dye was prepared from Michler's Ketone (177), obtained from dimethylaniline and phosgene. However, the modern synthesis of Auramine O starts from the diaryl compound (178) which is then heated with sulphur and ammonium chloride in a stream of ammonia at 200 °C. The sulphur participates in a *redox* reaction to give the thiobenzophenone (179) which subsequently reacts with ammonia.

The arylcarbonium class of dyes span a huge shade gamut. Thus, the triphenylmethane dye Malachite Green (22) is at the opposite end of the colour spectrum to Auramine O (yellow). Malachite Green is prepared using a similar strategy to that for Auramine O. Thus, dimethylaniline is condensed with benzaldehyde in the presence of acid to give the benzhydrol. This is then reacted with dimethylaniline to give the *leuco* base (180) which is oxidised with freshly prepared lead dioxide to the carbinol base (181). Malachite Green (22) is obtained upon acidification.

By using different aromatic aldehydes and amines a variety of triphenylmethanes may be similarly prepared. The introduction of more than one sulphonic acid group into a triphenylmethane, *e.g.* by direct sulphonation, gives dyes which may be used on polyamide fibres, *e.g.* nylon, wool. A second important method for preparing triphenylmethane dyes utilises the diaryl compound (178). The diaryl compound (178) is oxidised with lead dioxide to the benzhydrol (182) which is then reacted with an activated aromatic compound, *e.g.* a phenol, naphthol or aniline, to give the *leuco* base (183). Oxidation and acidification convert this to the triphenylmethane (184). (*cf.* Malachite Green synthesis).

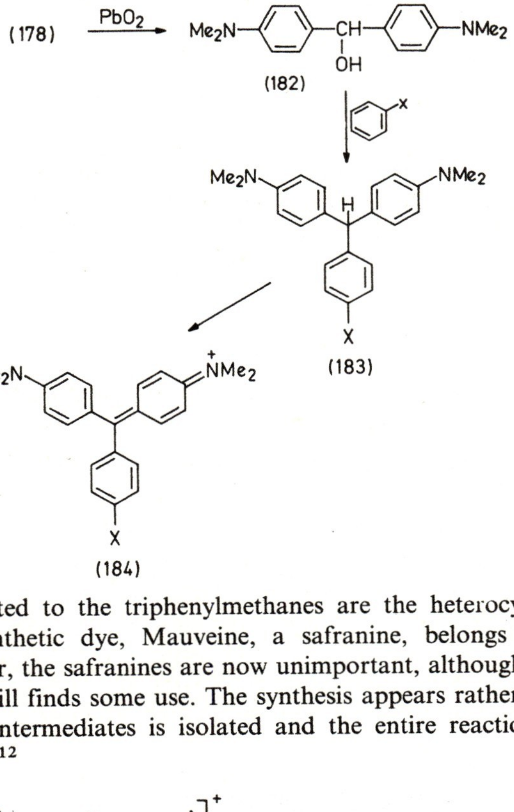

(178) $\xrightarrow{PbO_2}$ Me_2N–C$_6$H$_4$–CH(OH)–C$_6$H$_4$–NMe_2

(182)

(183)

(184)

Related to the triphenylmethanes are the heterocyclic dyes (17). In fact, the first synthetic dye, Mauveine, a safranine, belongs to this class (see earlier). However, the safranines are now unimportant, although the paper dye, Safranine T (185), still finds some use. The synthesis appears rather long but interestingly none of the intermediates is isolated and the entire reaction is essentially a 'one-pot' method.[12]

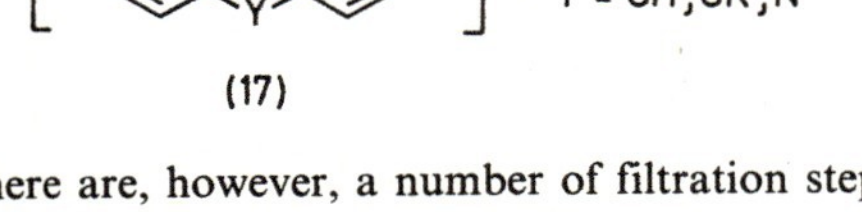

X = O, S, NH, NR
Y = CH, CR, N

(17)

[12] There are, however, a number of filtration steps to remove inorganic salts.

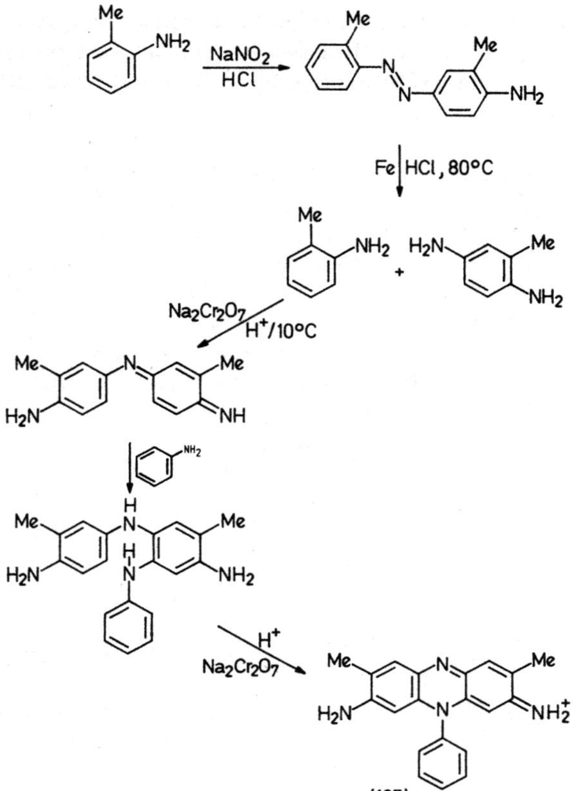

(185)

Rhodamine 6G (20) is representative of the oxygenated heterocyclic dyes of this class although it finds little use because of its poor fastness properties. However, the synthesis is interesting since it once again illustrates the similarity in the synthetic strategy.

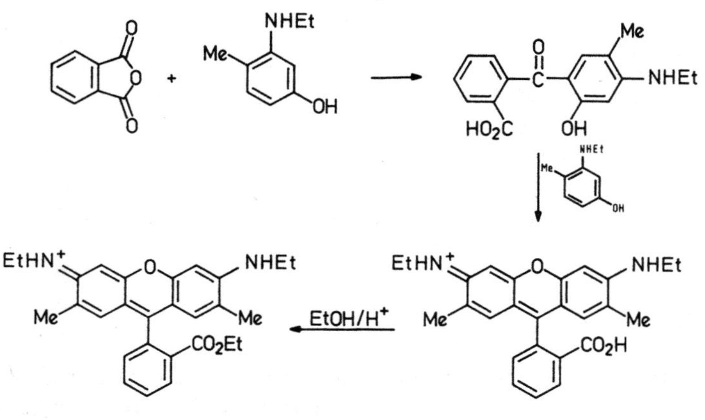

(20)

Methylene Blue, which is nowadays used as a pH indicator and a biological stain, is the most notable of the thiazine dyes. The synthesis once again uses simple chemical transformations and cheap, readily available chemicals to provide fairly good overall yields of product (Scheme 2.36).

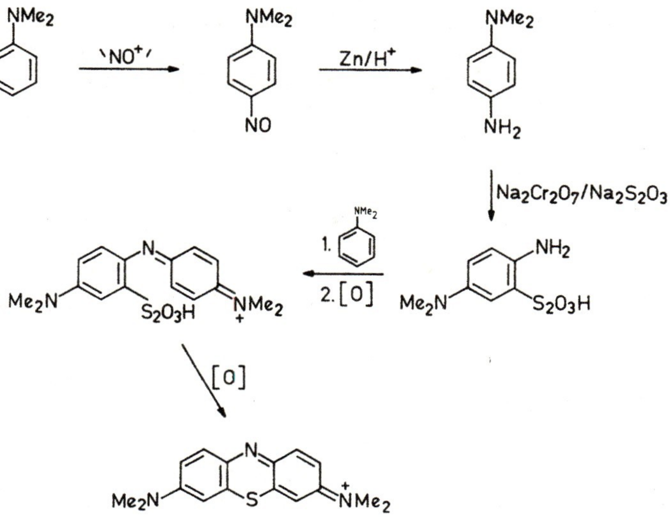

Scheme 2.36

Finally, the oxazine (186) is still used as a blue dye for polyacrylonitrile fibres. The dye is synthesised by nitrosation of NN-diethyl-*m*-anisidine followed by condensation with *m*-diethylaminophenol.

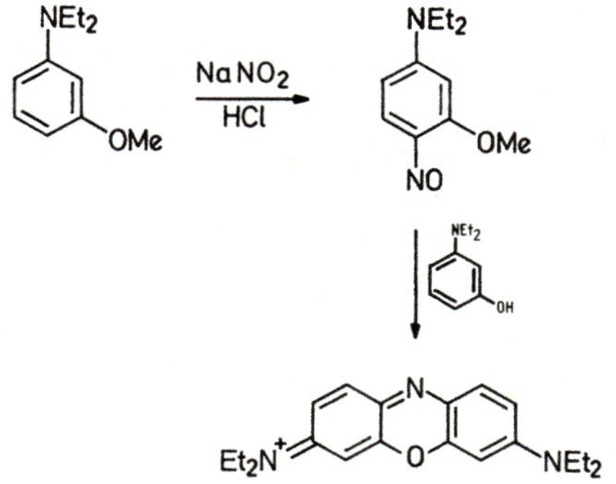

CI Basic Blue 4 (186)

2.5 Summary

The majority of the commercially important synthetic dyes can be classified under eight headings; azo, anthraquinone, vat, phthalocyanine, indigo, polymethine, carbonium and nitro dyes.

The synthesis of each separate class of dyes has been discussed and for the majority of the classes the chromogen is assembled from previously prepared intermediates. The exceptions to this rule are the anthraquinone dyes and the nitro dyes. Anthraquinone dyes are usually prepared by preforming the anthraquinone skeleton and introducing the substituents in subsequent steps. The nitro dyes are prepared in much the same way as the intermediates used for the preparation of the azo dyes.

For most dyes the intermediates can be broken down into two classes: aromatic carbocycles and aromatic heterocycles. The former are prepared from the parent hydrocarbon by a series of unit processes. In contrast, the heterocycles are frequently prepared from alicyclic intermediates and not from the parent heterocycle.

In more recent years great advances have been made in the synthesis of heterocyclic azo dyes and therefore azo dyes have increased further in their commercial importance.

2.6 Bibliography

Much useful information may be obtained on the synthesis of dyes and their intermediates from the invaluable series, 'The Chemistry of Synthetic Dyes' Vols. I–VIII by K. Venkataraman (ed.) and published by Academic Press (1952–1978) and references cited therein. Additional information may be obtained from the sources detailed below: —

A. *Azo Dyes* (including diazotisation and coupling).

1. Abrahart, E. N.: Dyes and their intermediates. London: E. Arnold 1977
2. Butler, R. N.: Chem. Revs. *75*, 241 (1975)
3. Ridd, J. H.: Quart. Revs. *15*, 418 (1961)
4. Zollinger, H.: Chem. Revs. *51*, 347 (1952)
5. Zollinger, H.: Azo and diazo chemistry. New York, London: Interscience Publishers 1961
6. Hegarty, A. F.: Kinetics and mechanisms of reactions involving diazonium and diazo groups. In: The chemistry of diazonium and diazo groups, Patai, S. (ed.), Part 2, pp. 511 to 593. New York: John Wiley & Sons 1978
7. Schenk, K.: Preparation of diazonium groups. In: The chemistry of diazonium and diazo groups, Patai, S. (ed.), Part 2, pp. 645–659. New York: John Wiley & Sons 1978
8. Wulfman, S.: Synthetic applications of diazonium ions. In: The chemistry of diazonium and diazo groups, Patai, S. (ed.), Part 1, pp. 241–341. New York: John Wiley & Sons 1978
9. Lewis, J. R.: Chem. and Ind. *1962*, 159
10. Lewis, J. R.: Chem. and Ind. *1964*, 1672

B. *Other Dye Classes*

1. See Ref. A1
2. Greenhalgh, C. W.: Endeavour *126*, 134 (1976)

Chapter 3

Azo Dyes

3.1 Introduction

Azo dyes are by far the most important class of dyes, comprising over 50% of total world dyestuffs production. They have achieved this prominence for the following reasons: they are tinctorially strong (azo dyes are about twice the strength of anthraquinone dyes, the next most important class of dyes); they are usually easy to prepare in a multi-purpose chemical plant from cheap, readily available starting materials; they cover the whole shade range; and they have good fastness properties. The first two features make azo dyes cost effective against most dye classes, but particularly anthraquinone dyes. Furthermore, their ease of preparation from a host of readily available intermediates allows azo dyes to be 'tailor-made' to meet almost any end use.

However, azo dyes do have some deficiencies. Compared to their main rivals, the anthraquinone dyes, they tend to be duller in shade and generally cannot equal the excellent light fastness of such dyes, especially in the blue shade area, although recent research suggests that both these deficiencies can be overcome. These defects are insufficient to outweigh the cost advantage which azo dyes enjoy.

The study of azo dyes has also been of great value in the development of theoretical organic chemistry. Thus, azo dyes have been widely used for developing and then testing theories of colour and constitution, tautomerism, indicator action, and acid-base equilibria.

3.2 Basic Structure of Azo Dyes

All azo dyes contain at least one, but more usually two, aromatic residues attached to the azo group. They exist in the more stable *trans* (1) rather than the *cis* form (2). Both nitrogen atoms are sp^2 hybridised so that the carbon-nitrogen bond angles are *ca.* 120°.

R–N=N–R' (trans) (1)

R–N=N–R' (cis) (2)

Trans-azobenzene (3), the basic azo chromogen, is essentially planar in both the solid state and in solution, though it is apparently non-planar in the vapour phase. Bond length determinations indicate some contribution from resonance forms such

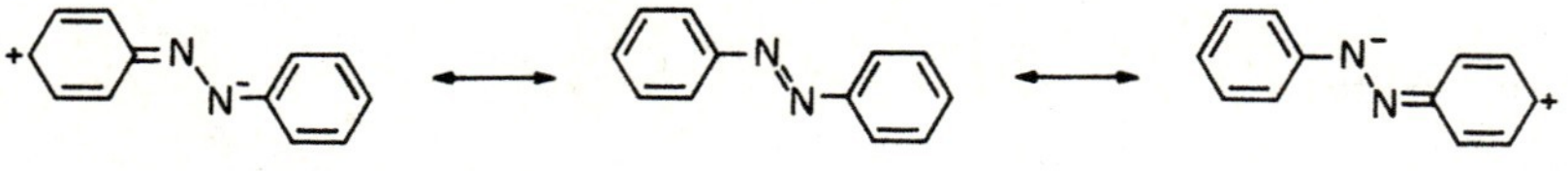

(3A) (3) (3B)

as (3A) and (3B), so that the carbon-nitrogen bond lengths are slightly shorter than expected and the nitrogen-nitrogen bond length slightly longer. Also, both phenyl rings show some quinonoid character. Electron donor groups in one ring and electron acceptor groups in the other, especially when they are conjugated to the azo linkage, enhance such resonance contributions. Crystallographic data for Methyl Orange (4) illustrate these effects quite clearly (Fig. 3.1).

	Bond length (in Å)		
	Actual	Expected	
a	1.244	1.22–1.24	(aliphatic azo compounds)
b	1.407	1.472	(aliphatic C—N)
b′	1.44		
c	1.392	1.39	(benzene C=C)
c′	1.406		
d	1.372		

Fig. 3.1. Crystallographic data for Methyl Orange

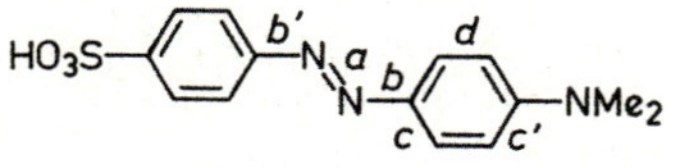

(4)

Steric effects are discussed later (Sect. 3.5.9).

3.3 Tautomerism

3.3.1 Tautomerism of Hydroxyazo Dyes — Azo-Hydrazone Tautomerism

In 1884, Zincke and Bindewald obtained the same product from coupling benzenediazonium chloride with 1-naphthol and from condensing phenylhydrazine with 1,4-naphthoquinone. Since the azo dye (5) and the hydrazone dye (6) were the expected products, the authors correctly suggested a mobile equilibrium between the two forms, *viz.* tautomerism. Extensive research followed this initial observation of azo-hydrazone tautomerism and, indeed, is still continuing today.

This intensive activity has continued because the phenomenon is not only of the utmost importance to the dyestuff manufacturer but also to other areas of chemistry. The azo and hydrazone tautomers not only have different colours (see Sect. 3.5.7), they

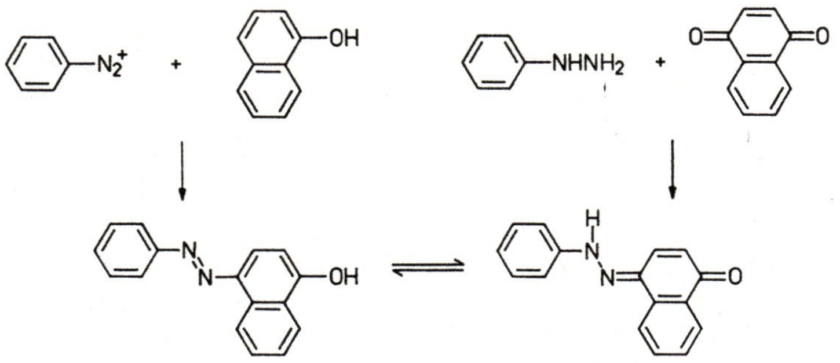

(5) (6)

also have different tinctorial strengths (and hence economics), and different properties, *e.g.* light fastness. Usually, the hydrazone form is bathochromic compared to the azo form and has higher tinctorial strength. For the 4-phenylazo-1-naphthol system, the azo tautomer (5) is yellow (λ_{max} ~ 410 nm, ε_{max} ~ 25,000), and the hydrazone tautomer (6) orange (λ_{max} ~ 480 nm, ε_{max} ~ 35,000). Hence commercial hydroxyazo dyes have been chosen which exist predominantly, if not exclusively, in the hydrazone form.

Hydroxyazo dyes capable of undergoing azo-hydrazone tautomerism are those in which the hydroxy group is conjugated to the azo group, *i.e. ortho*-hydroxyaryl- and *para*-hydroxyaryl-, but not *meta*-hydroxyarylazo dyes. Whether such tautomerism is actually observed depends on several factors but principally on the relative thermodynamic stabilities of the azo and hydrazone tautomers. If these are similar, tautomerism is generally observed. Where one tautomer is much more stable, then the dye will exist exclusively in that form. This concept is best illustrated by examples.

Consider 2-phenylazophenol: in the azo form (7), the sum of the bond energies of the azoenol system (7a) is some 108 kJ mol^{-1} less than the ketohydrazone system (8a). Thus, the hydrazone tautomer appears more stable than the azo tautomer. However, in aromatic systems, the resonance stabilisation energy (RSE) must also be considered. The azo form (7) contains two fully aromatic benzene rings, each with a RSE of 150.5 kJ mol^{-1}. In the hydrazone form (8), the aromaticity of one benzene ring has been lost completely and with it approximately 134 kJ mol^{-1} of RSE (Table 3.1). Thus, taking into account both effects, the azo tautomer (7) is some 26 kJ mol^{-1} (134–108) more stable than the hydrazone tautomer (8), an amount sufficient to ensure that most azophenol dyes exist exclusively in the azo form.

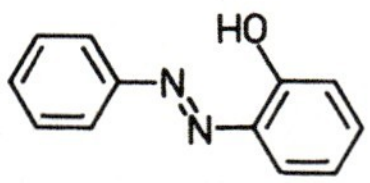

(7) 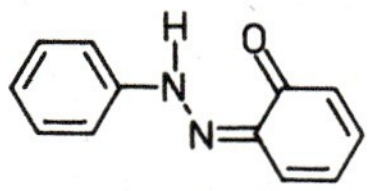 (8)

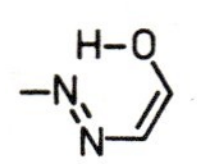

(7a) 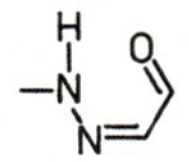 (8a)

Table 3.1. Comparative RSE of aromatic hydrocarbons and their quinone derivatives

Substance	RSE (kJ mol^{-1})	Difference
benzene	150.5	
1,4-benzoquinone	16.5	134
1,2-benzoquinone[a]	—	
naphthalene	255	
1,4-naphthoquinone	164.3	90.7
1,2-naphthoquinone[a]	—	
anthracene	349	
anthraquinone[b]	346	3

[a] Data is apparently not available for these compounds but the values would be expected to be similar to the corresponding 1,4-isomers.

[b] Throughout this book, anthraquinone refers to the 9,10-isomer.

The presence of the fused benzene ring in 1,2- and 1,4-naphthoquinones markedly stabilises these compounds relative to the corresponding benzoquinones. Consequently, the loss in RSE in going from a dihydroxynaphthalene to a naphthoquinone is less than that in going from a dihydroxybenzene to a benzoquinone (Table 3.1). Indeed, the value of 90.7 kJ mol^{-1} is insufficient to offset the 108 kJ mol^{-1} greater bond energy stability of the hydrazone tautomer. Therefore, in contrast to the azophenol dyes, it is the hydrazone tautomer which predominates in azonaphthol dyes. However, because of the smaller energy difference between the azo (5) and hydrazone (6) tautomers, tautomerism is more widespread in azonaphthol dyes than it is in azophenol dyes. An interesting exception is provided by 3-phenylazo-2-naphthol. Here, only the vastly more stable azo tautomer (9), with a fully aromatic naphthalene nucleus, is observed.

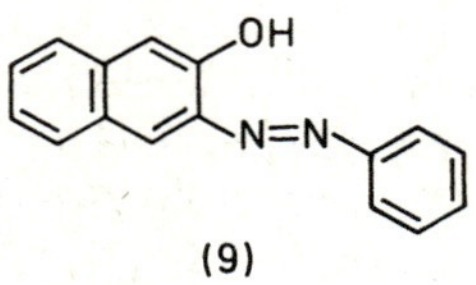

OH N=N

(9)

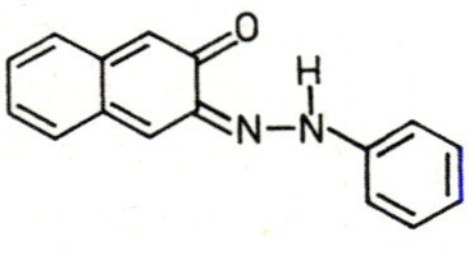

O H N–N

With azoanthrol dyes, the relative loss in RSE of the hydrazone form is even less than for the azonaphthol dyes since the RSE of anthraquinone approaches that of anthracene (Table 3.1). Consequently, the higher bond energy stability of the hydrazone tautomer predominates and the dyes exist totally in the hydrazone form (10).

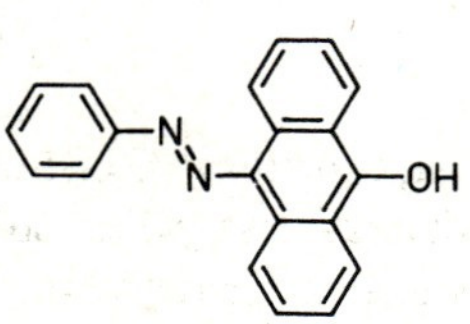

N N– OH

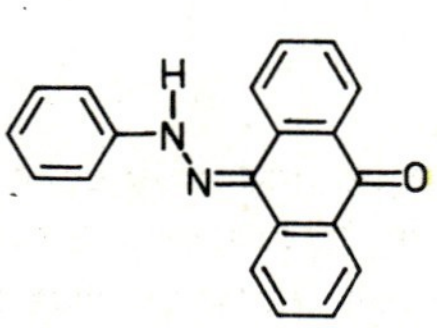

H N N= =O

(10)

The azo-hydrazone tautomerism of the two most commercially important classes, the azonaphthols and the azophenols, including heterocyclic analogues, is now discussed in more detail, showing how the equilibrium is affected by other factors such as solvent, hydrogen-bonding, and both electronic and steric effects.

3.3.2 Hydroxyazo Dyes of the Naphthalene Series

The initial proposition of Zincke and Bindewald that a mobile equilibrium existed between the azo (5) and hydrazone (6) tautomers of 4-phenylazo-1-naphthol sparked off a controversy that was not resolved until the classic work of Kuhn and Bar in 1935. These latter workers suggested that the electronic absorption spectrum of the azo tautomer (5) should be analogous to that of its O-methyl derivative (11) and that the spectrum of the hydrazone tautomer (6) should resemble that of its N-methyl derivative (12). (The O-methyl derivative is prepared by aqueous methylation of 4-phenylazo-1-naphthol under strongly alkaline conditions, and the N-methyl derivative by condensation of 1-methyl-1-phenylhydrazine with 1,4-naphthoquinone). Comparative spectra in various solvents supported the existence of azo-hydrazone tautomerism (Fig. 3.2). Thus, the spectrum of the O-methyl derivative shows a peak at λ_{max} ~ 410 nm (yellow) and that of the N-methyl derivative a peak at λ_{max} ~ 480 nm (orange). In the spectrum of 4-phenylazo-1-naphthol peaks due to the azo and hydrazone tautomers can be seen at these wavelengths. Because of the sensitivity of the 4-phenylazo-1-naphthol system to various effects it has been used extensively as a model to study azo-hydrazone tautomerism.

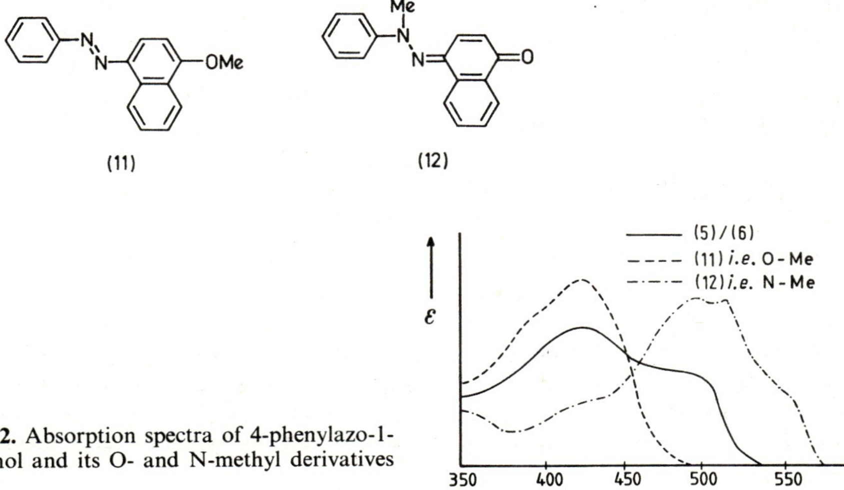

Fig. 3.2. Absorption spectra of 4-phenylazo-1-naphthol and its O- and N-methyl derivatives

The solvent can exert a profound influence on the equilibrium. Generally, a more polar solvent favours the hydrazone form (14) whereas less polar solvents favour the azo form (13), though there are exceptions to this generalisation. More recent work provides a better explanation for solvent effects. Briefly, it is not the

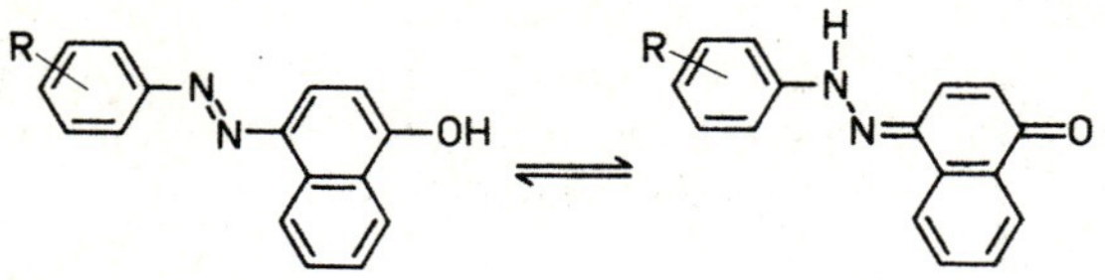

(13) ⇌ (14)

bulk solvent properties (as measured by dielectric constant, *etc.*), which determine the proportion of azo and hydrazone tautomers, but rather the solvent structure and the microscopic environment of the dye in the solvent matrix. For solvent sensitive dyes the hydrazone form is predominant in those solvents capable of forming a 3-dimensional structure, whereas the azo form is favoured by solvents that form a 2-dimensional structure, or are unstructured. Water, formamide and acetic acid are examples of the former type and pyridine, liquid alcohols and hydrocarbons are examples of the latter type. The effect of the solvent is shown in Fig. 3.3, the curves passing through an isosbestic point, behaviour typical of an equilibrium.

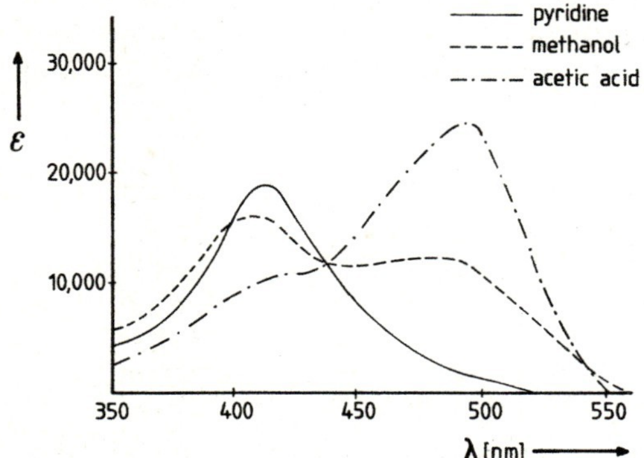

Fig. 3.3. Effect of solvent on 4-phenylazo-1-naphthol

The steric effect of *ortho* substituents *R* in the phenyl ring prevents effective solvent stabilisation of the hydrazone form (14) and this gives rise to an anomalous solvent effect: polar, strongly hydrogen-bonding 'hydrazone-favouring' solvents now favour the azo form (13), since solvation at the hydroxy group is unaffected by the *ortho* substituent. However, if the *ortho* substituent is capable of forming an intramolecular hydrogen-bond with the hydrogen of the hydrazone form, *e.g.* a carboxy group, then it is the hydrazone form (15) which predominates, irrespective of solvent.

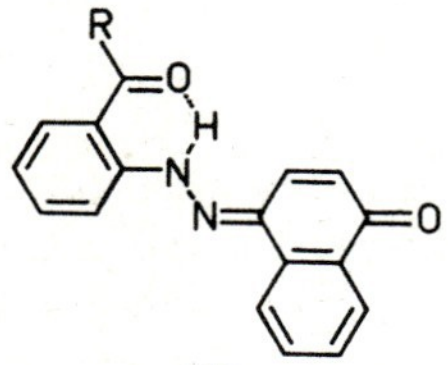

R = OH, OR'

(15)

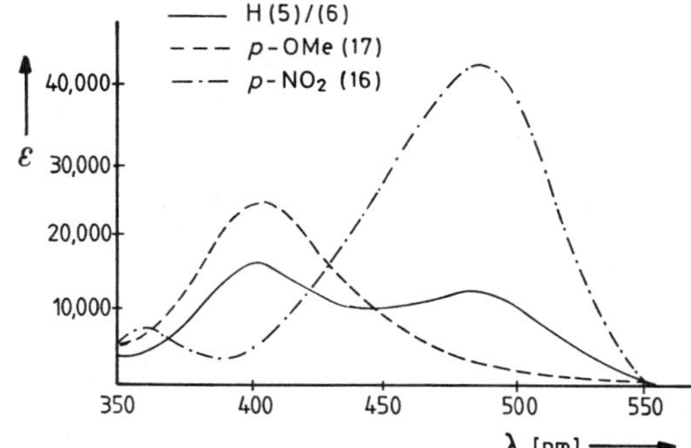

Fig. 3.4. Electronic effect of substituents in 4-phenylazo-1-naphthol

Electron-accepting groups, especially the nitro group, generally increase the proportion of the hydrazone form (14), whereas electron-donating groups, especially methoxy, generally increase the proportion of the azo form (13) — Fig. 3.4. These electronic effects are powerful, particularly for *para* substituted derivatives, and can override solvent effects. Thus, a *p*-NO_2 group ensures that the dye (16) exists exclusively in the hydrazone form in all solvents, whereas a *p*-OMe group causes the azo form (17) to predominate.

(16) (17)

The sensitivity of the tautomeric equilibrium of the 4-phenylazo-1-naphthol system, particularly to pH and solvent effects, make it unacceptable for providing commercial dyestuffs. For example, the dyes are readily ionised, even by weak bases, to the red, mesomeric anion (18 : R = H).

base

(18)

Therefore, washing garments dyed with these dyes in the presence of sodium carbonate or even soap would not only cause a colour change from yellow, or orange, to red, but may also result in the removal of the dye from the fabric because of the lower affinity of the ionised form. Dry cleaning of such garments could also cause a temporary colour change by altering the azo-hydrazone equilibrium.

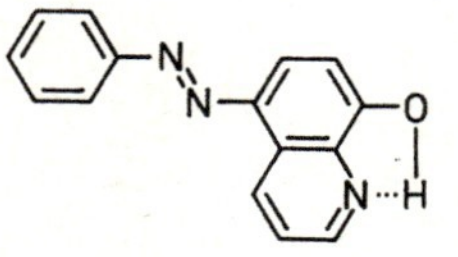

(19) 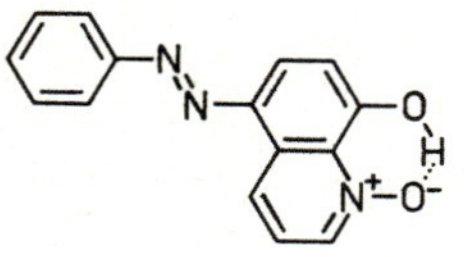 (20)

As well as intermolecular hydrogen-bonding between dye and solvent, intramolecular hydrogen-bonding is also important. It is not usually encountered in *p*-hydroxyazo dyes, although special cases such as (19) and (20) exist where the extra stabilisation it confers on the azo form makes this the predominant tautomer. However, intramolecular hydrogen-bonding is very important in *o*-hydroxyazo dyes such as 2-phenylazo-1-naphthol (21) and 1-phenylazo-2-naphthol (22). In both these systems, the predominance of the hydrazone form (21A and 22A respectively) is such that the effect of both solvents and substituents is much less marked than in the 4-phenylazo-1-naphthol system. One explanation for the dominance of the hydrazone form is that intramolecular hydrogen-bonding is stronger in that form than in the azo form. Whether this is the case or not, *o*-hydroxyazo dyes certainly do form strong intramolecular hydrogen bonds; an average value of 30 kJ mol^{-1} has been calculated, although a figure as high as 67 kJ mol^{-1} has been quoted.

(21) (21A)

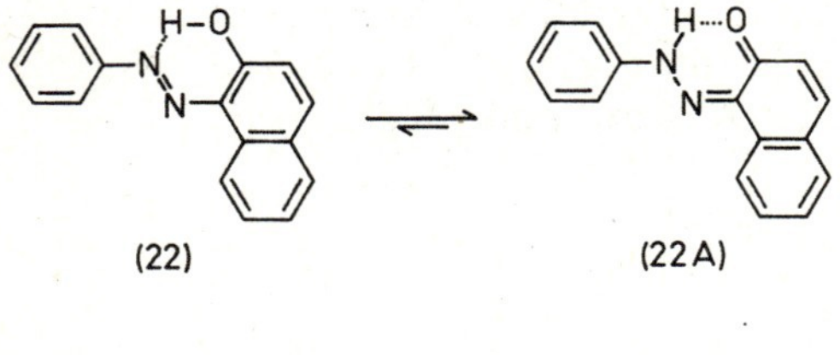

(22) (22A)

This high strength is reflected by the stability of the dyes to alkali (they are difficult to ionise), and by their reluctance to undergo deuterium exchange. It is not surprising, therefore, that both these systems, in contrast to the 4-phenylazo-1-naphthol system, have found widespread commercial use. Their existence in the tinctorially stronger hydrazone form, plus their pH and solvent insensitivity, make them ideal colourants. As seen from Fig. 3.5, they are orange → red in colour (λ_{max} 475 → 500 nm).

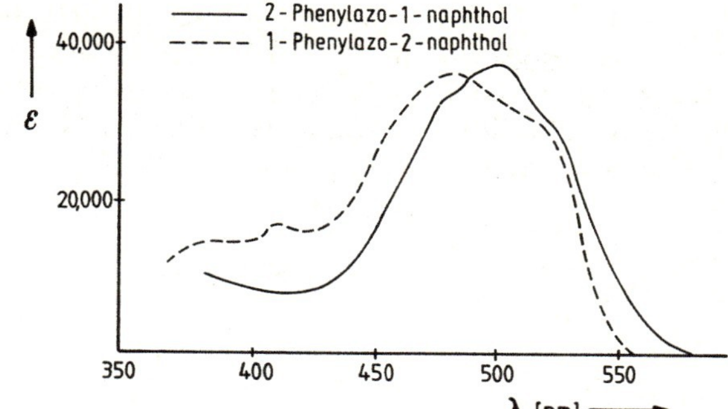

Fig. 3.5. Spectra of 1-phenylazo-2-naphthol and 2-phenylazo-1-naphthol (in CH_3CO_2H)

Sulphonated derivatives of 2-phenylazo-1-naphthol have found extensive use in most dyestuff classes, *e.g.* as direct dyes for cotton (CI Direct Orange 26) (23) and acid dyes for nylon and wool (CI Acid Red 138) (24).

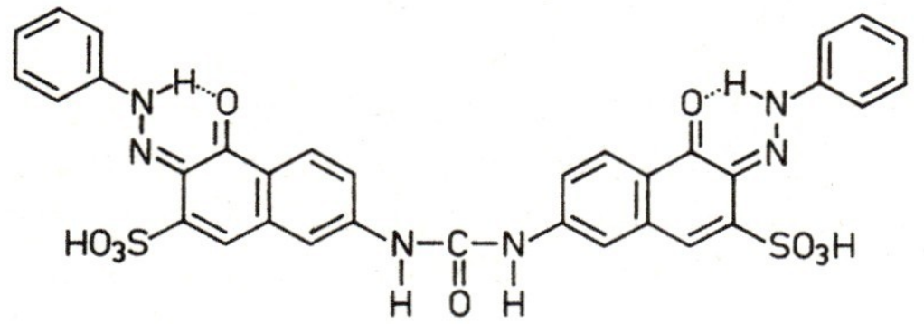

(23)

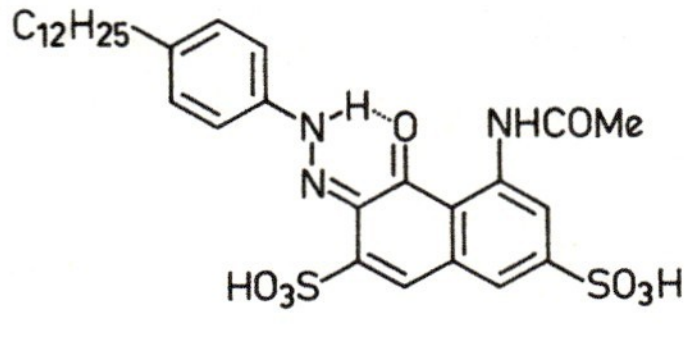

(24)

They also provide the majority of bright red and orange reactive dyes for cotton. Typical examples are CI Reactive Red 1 (25), derived from H-acid (26), and Procion[1] Orange MX-G (27), derived from J-acid (28).

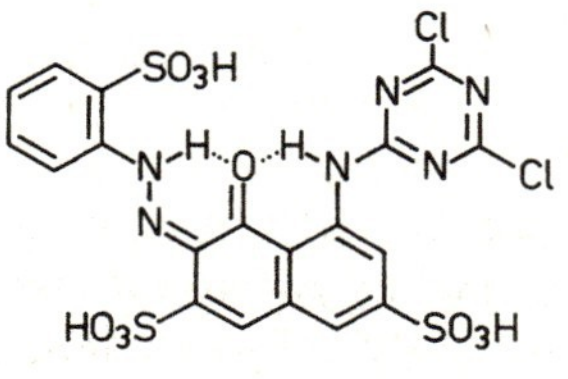

(25)

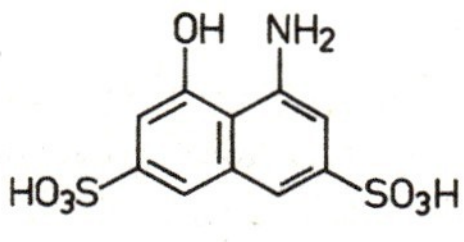

(26)

[1] 'Procion' is an ICI trademark.

(27)

(28)

In contrast, 1-phenylazo-2-naphthol derivatives have been used primarily as pigments: those derived from 2-hydroxy-3-naphthoic acid arylamides (29) provide the majority of orange and red pigments, and azoic dyes (see Sect. 6.3.4). A typical example is CI Pigment Red 2 (30).

(29)

(30)

3.3.3 Hydroxyazo Dyes of the Benzene Series

As mentioned previously, both *ortho*- and *para*-hydroxyazobenzene exist exclusively in the azo form because of the much greater thermodynamic stability of that form compared to the hydrazone form. The hydrazone tautomer is only observed in those special cases when combinations of electronic, steric, and hydrogen-bonding effects provide sufficient stability to offset the loss in aromaticity. These combinations may be summarised as follows:—

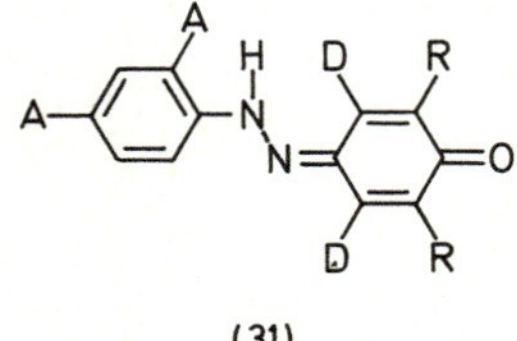

(31)

(i) (31) — where *R* is a bulky alkyl group,
(ii) (31) — where *D* is an electron-donating group, especially one capable of intramolecular hydrogen-bond formation,

and

(iii) (31) — where *A* is an electron-accepting group, especially one capable of forming an intramolecular hydrogen-bond with the hydrogen atom.

Derivatives of 2,6-di-*t*-butylphenol provide examples of the first type. In these dyes, the driving force for the formation of the hydrazone tautomer (33) is the relief of steric interactions present in the azo tautomer (32) between the bulky *t*-butyl groups and the hydroxy group (the steric requirements of a quinone carbonyl

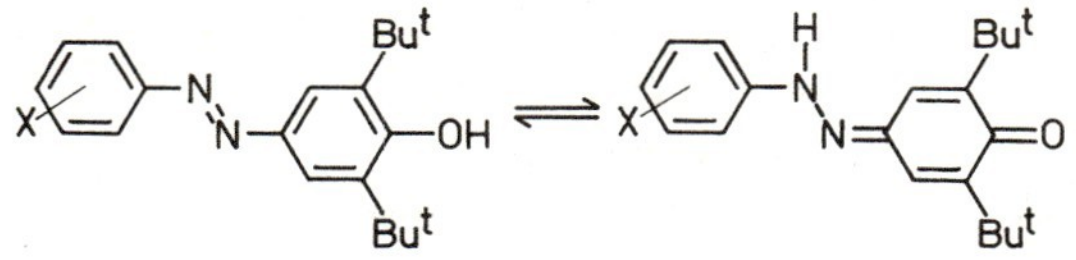

(32) (33)

group are less than for a hydroxy group). As in the azonaphthol series, electron-accepting groups X increase the proportion of the hydrazone tautomer, whereas electron-donating groups decrease it.

The second case is exemplified by the dimethoxy dye (34). Here, the methoxy groups stabilise the hydrazone tautomer by both intramolecular hydrogen-bonding and by electronic effects, as shown by (34A). The magnitude of these two effects is such that the dye exists solely in the hydrazone form.

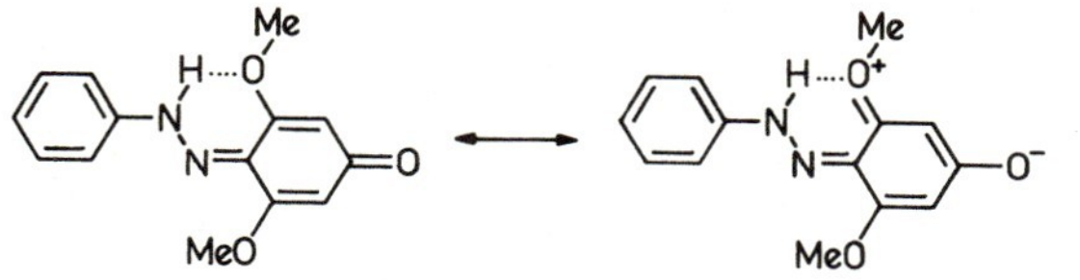

(34) (34 A)

The dinitroazophenol dyes (35) provide an example of the third type, especially when the hydroxy group is flanked by two *ortho* alkyl groups (Table 3.2). The importance of the *ortho* nitro group is shown by comparing the two dyes (36) and (37): only 4% of the former exists as the hydrazone tautomer whereas the latter is totally in that form. The most likely explanation for this difference lies in the ability of the *ortho* nitro group to stabilise the hydrazone tautomer by forming an intra-

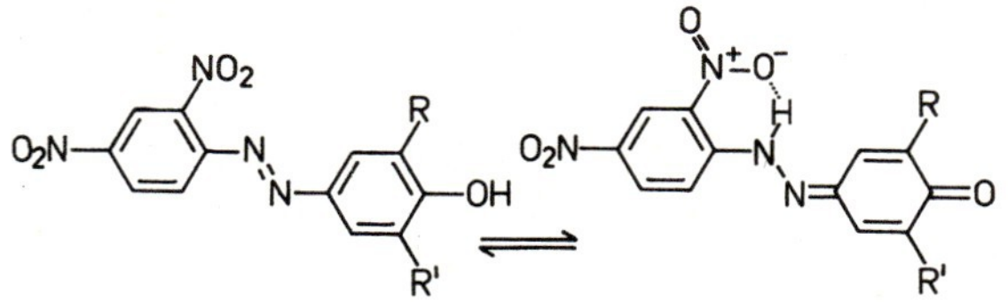

(35) (35A)

Table 3.2. Spectral data for the azo and hydrazone forms (in tetrachloroethylene)

R	R_1	Azo form		Hydrazone form		Percentage hydrazone
		λ_{max} (nm)	ε_{max}	λ_{max} (nm)	ε_{max}	
H	H	383	—	—	—	8
H	Me	393	26,100	429	25,900	60
H	*n*-Bu	394	—	435	—	84
n-Bu	*n*-Bu	401	28,600	440	33,100	100

molecular hydrogen-bond with the hydrogen atom, a stable 6-membered chelate ring system being formed.

(36) (37)

As seen from Table 3.2, alkyl groups *ortho* to the hydroxy group increase the proportion of the hydrazone tautomer. However, when the *ortho* substituent can form an intramolecular hydrogen-bond, it is the azo form which predominates. Thus, (38) exists totally in the azo form.

(38)

Because of the scarcity of tautomerism in the azophenol dyes, a direct comparison of the relative colour and tinctorial strength of the azo and hydrazone tautomers is difficult to achieve. However, comparing the spectra of the azo dye (38) with that of the isomeric hydrazone dye (34) shows the differences quite clearly (Fig. 3.6). Not only has the azo dye a lower ε_{max} than the hydrazone dye, it also absorbs at shorter wavelengths and thus the principal absorption is in the ultraviolet rather than the visible region of the spectrum. For this reason, azophenol dyes have not been used much commercially. However, sulphonated derivatives (coloured ones of course) are used as acid dyes for nylon, since on this fibre the azo form has superior light fastness to the hydrazone form (see Sect. 6.4.4). CI Acid Yellow 135 (39) is a typical example.

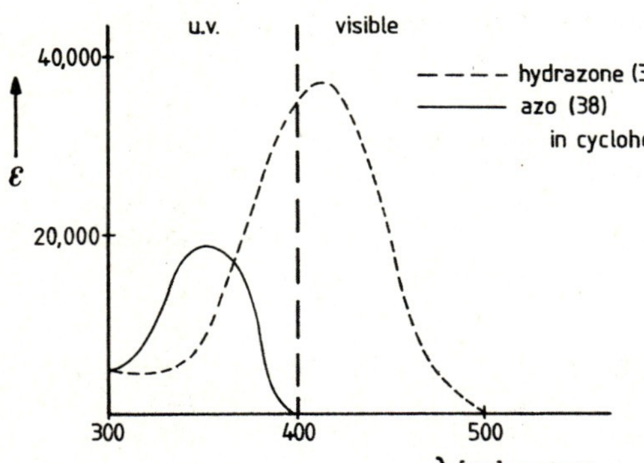

Fig. 3.6. Azophenol dyes — absorption spectra of azo and hydrazone tautomers

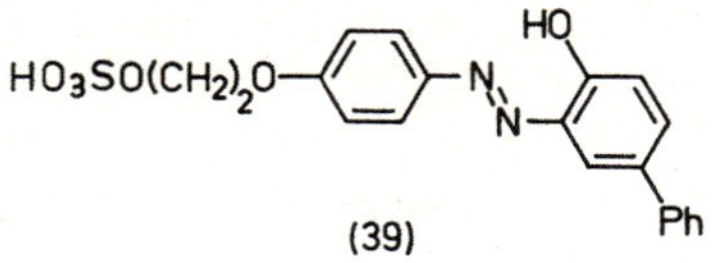

$HO_3SO(CH_2)_2O$–C₆H₄–N=N–(2-hydroxy-5-phenylphenyl)

(39)

In contrast, several 5- and 6-membered heterocyclic analogues of azophenol dyes are used extensively as commercial dyestuffs: the more important are the azopyrazolone[2] and, more recently, the azohydroxypyridone dyes.

The structure of the azopyrazolone dyes was in dispute for many years with all four possible tautomers (40, 41, 42 and 43) having been proposed. Only recently have nmr and ir spectroscopy provided the crucial evidence to establish which of these four possibilities is correct. Thus, nmr spectra clearly indicated signals due to a highly deshielded proton whose chemical shift was independent of concentration, and three other equivalent protons attributable to a vinylic methyl group. From the ir spectra, the presence of a hydrogen-bonded carbonyl group was deduced. Only the hydrazone (43) is consistent with these observations, and it was concluded that pyrazolone dyes exist in the hydrazone form.

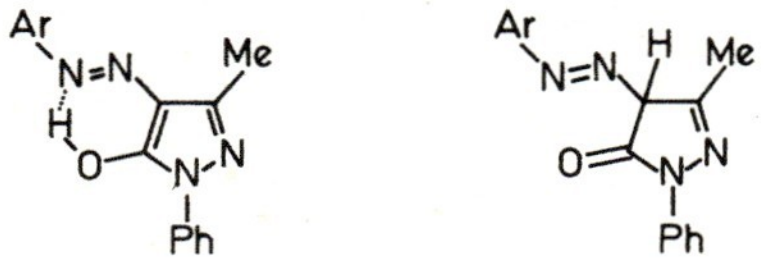

(40) (41)

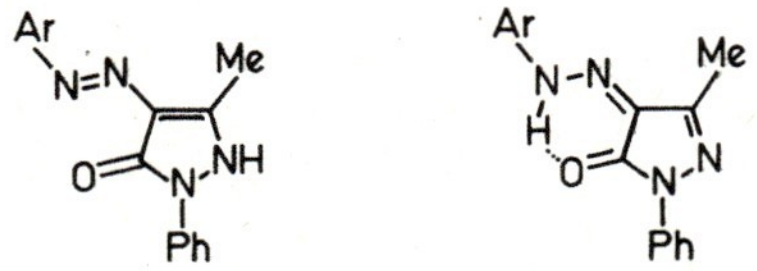

(42) (43)

Electronic absorption spectra confirmed the above findings. Indeed, one would not expect (41) to be coloured owing to the lack of conjugation between the arylazo and pyrazolone moieties.

CI Reactive Yellow 1 (44) is an example of a commercial pyrazolone dye.

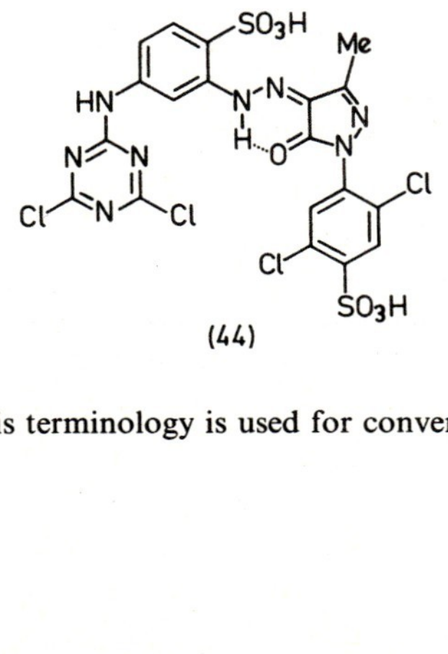

(44)

[2] This terminology is used for conventional reasons.

The predominance of the hydrazone form (43) over the azo (40) and other forms is again rationalised by considering their relative thermodynamic stabilities. The crucial factor is the much lower RSE of the pyrazolone ring compared to benzene; it is insufficient to offset the 108 kJ mol^{-1} greater bond energy stability of the hydrazone grouping.

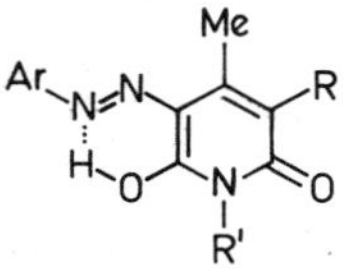

(45) $R = H, CN, CONH_2$; $R' = H$, alkyl, aryl (46)

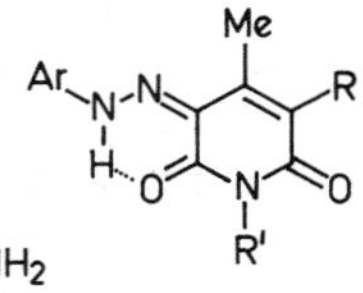

For a similar reason, hydroxypyridone dyes also exist predominantly in the hydrazone (46) rather than the azo form (45). Commercially, these dyes have become increasingly important since they provide greenish-yellow to orange dyes of significantly higher tinctorial strength than azopyrazolone dyes. Because of the intense competition between dyestuff manufacturers, no commercial azohydroxypyridone dye structures have been disclosed, but (47)[3] is typical of such dyes.

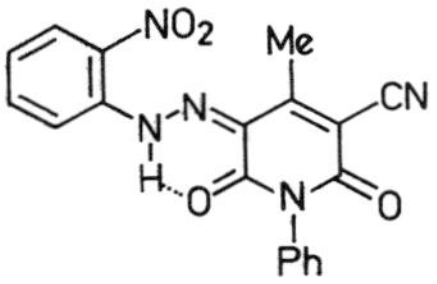

(47)

Commercial yellow dyes are also obtained from acetoacetarylamide coupling components (48), especially azoic yellows. Since there is a negligible difference in RSE between the azo (49) and hydrazone (50) forms (no loss in aromaticity is involved), only the hydrazone tautomer, with its greater bond energy stabilisation, is observed.

$MeCOCH_2CONHAr$

(48) (49) 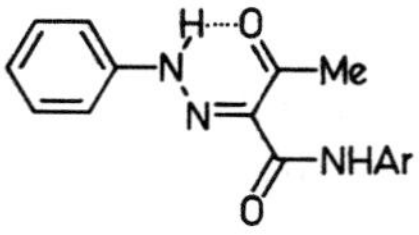 (50)

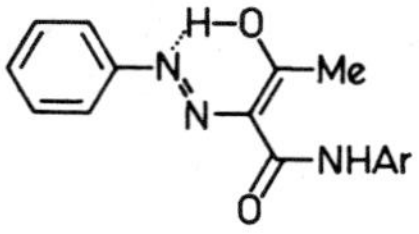

3.3.4. Tautomerism of Aminoazo Dyes — Amino-imino Tautomerism

Unlike their hydroxy analogues, 4-phenylazo-1-naphthylamine dyes exist totally in the aminoazo form (51): there is no evidence for the iminohydrazone form (52). Similarly, the aminoazobenzene dyes (53) and (54) also exist exclusively in the aminoazo form. These conclusions have been reached by several workers from a

[3] Example 2 of GB 1,256,093 (ICI).

comparison of the absorption spectra of the primary amino dyes with those of the corresponding NN-dialkyl- and N-acetylamino derivatives.

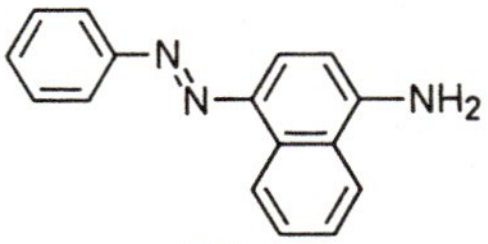

(51)

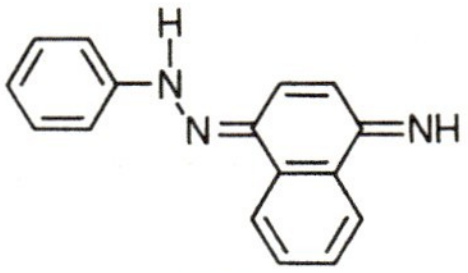

(52)

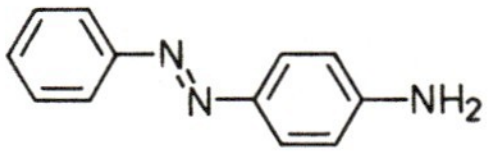

(53)

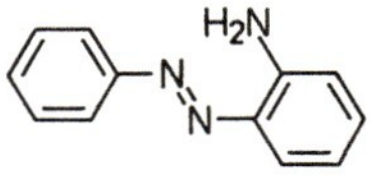

(54)

It has not been firmly established whether tautomerism exists in *o*-phenylazonaphthylamine dyes, though it is probable that these dyes also exist in the aminoazo forms (55) and (56).

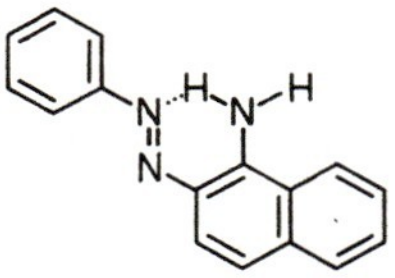

(55)

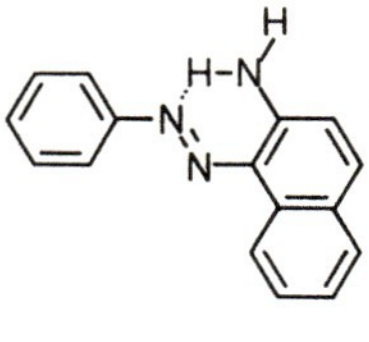

(56)

By analogy with the azoanthrol dyes, aminoazo derivatives (57) of anthracene are more likely to form the iminohydrazone tautomer (58). Unfortunately, these dyes have not been studied in detail.

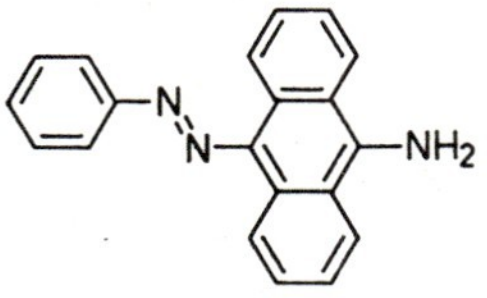

(57)

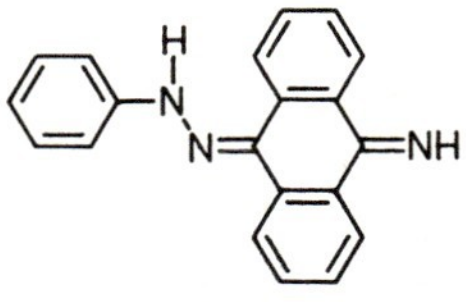

(58)

The reason for the dominance of the azo form in aminoazo dyes is not obvious, especially since the thermodynamic stability of that tautomer is less than that of the iminohydrazone tautomer. However, the relative instability of the imino group (C=NH, BE 614.5 kJ mol^{-1}) must be significant.[4]

In total contrast to the hydroxyazo dyes, the commercially important aminoazo dyes are those derived from *p*-aminoazobenzene (53), not those from the *o*-phenylazonaphthylamine dyes (55) and (56). There are several reasons for this, the two principal ones being ease of preparation and tinctorial strength. As discussed in

[4] See: — Dudek, G. O., Holm, R. H.: J. Amer. Chem. Soc. *84*, 2695 (1962)

Sect. 2.4.1, diazo coupling to aniline derivatives occurs either on nitrogen, to give a diazoamino compound (59), or on the *para* carbon atom to give an aminoazo compound (53).

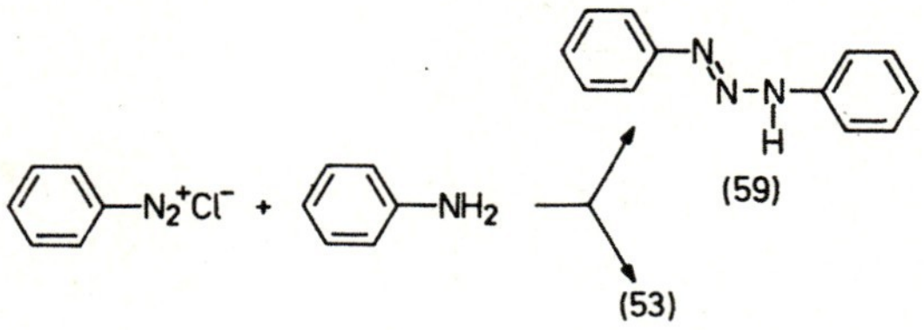

(59)

(53)

Use of NN-dialkylaniline derivatives eliminates diazoamino formation and the sole product is the 4-phenylazo dye (60). These dyes are tinctorially stronger and more bathochromic than the aniline dyes (53) and allow greater scope for 'fine-tuning' their properties by variation of the alkyl groups. Neutral dyes of this type have been widely employed commercially as 'disperse' dyes for polyester fibres — CI Disperse Red 1 (61) is a typical example. Dyes of low molecular weight are required for dyeing polyester fibres (see Sect. 6.3.3) and azo dyes of this type ranging in shade from yellow to blue can easily be obtained by varying the substituents.

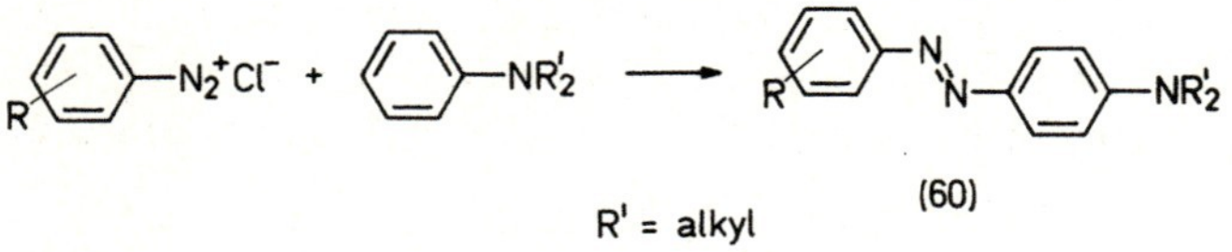

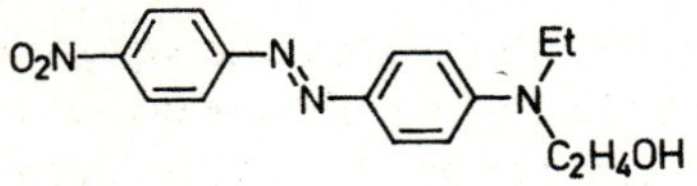

R' = alkyl

(60)

(61)

In contrast, *ortho* coupled phenylazo aniline dyes are difficult to obtain. Blocking the *para* position to force coupling *ortho* to the amino group is not normally successful: N-coupling predominates to give the diazoamino compound (62).

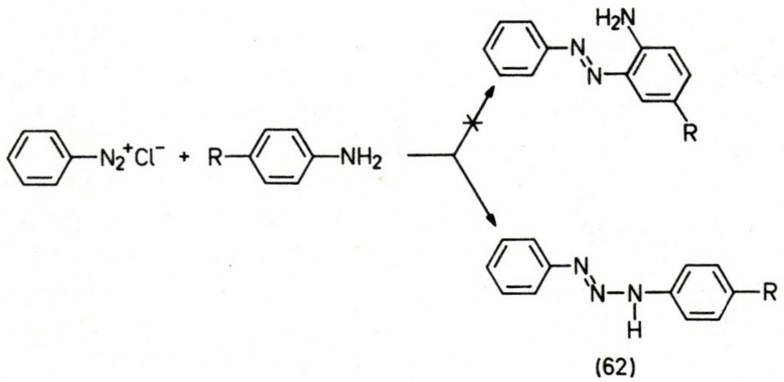

(62)

Analogous dyes (63) from the NN-dialkyl homologues can be prepared, with difficulty, but, because of steric interactions which cause non-planarity of the *o*-dialkylamino group, they are tinctorially weaker than their *para* counterparts (60) (see Sect. 3.5.9). They are also duller.

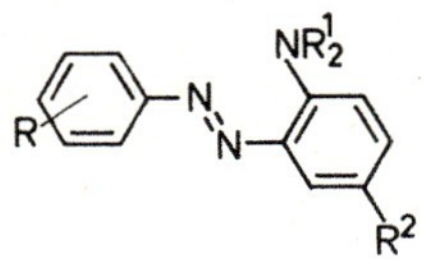

R^1 = alkyl

(63)

Coupling *ortho* to an amino group is readily achieved by using *meta*-phenylenediamine as a coupling component but dyes of this type have found only limited commercial use. However, judging from current patent activity, dyestuff companies are actively engaged in evaluating the use of polyamino heterocyclic compounds, particularly 2,6-diaminopyridine dyes such as (64). Again, the dyes probably exist in the azo form.

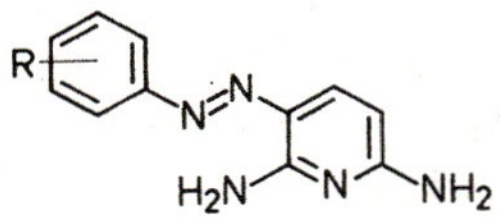

(64)

In contrast to the azo dyes derived from aniline coupling components there are many commercial dyes from sulphonated aminonaphthols where the arylazo group is *ortho* to the amino group. The reason for this difference is mainly ease of preparation since the naphthylamines tend to undergo predominantly C-coupling to give the aminoazo dye. Typical examples are the red monoazo dye (65) for nylon, and the disazo dye, CI Acid Black 1 (66) for cotton.

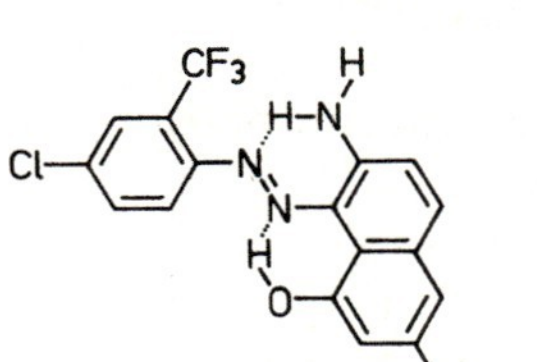

(65)

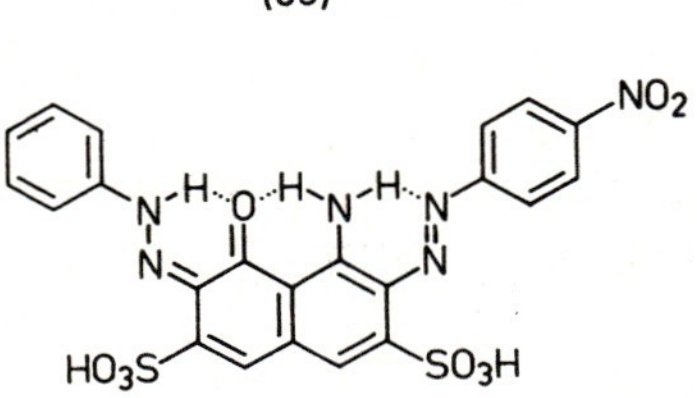

(66)

3.3.5 Protonated Azo Dyes — Azonium-Ammonium Tautomerism

Protonation of a 4-aminoazobenzene derivative can occur either at the terminal amino group to give the ammonium tautomer (67), or at the β-nitrogen atom of the azo group to give the azonium tautomer (68). The ammonium tautomer, in which the amino group no longer possesses a free lone pair of electrons, resembles azobenzene substituted in the *para* position with an electron-withdrawing group and thus absorbs in the ultra-violet at *ca.* 325 nm with a lower ε_{max} than the neutral dye.

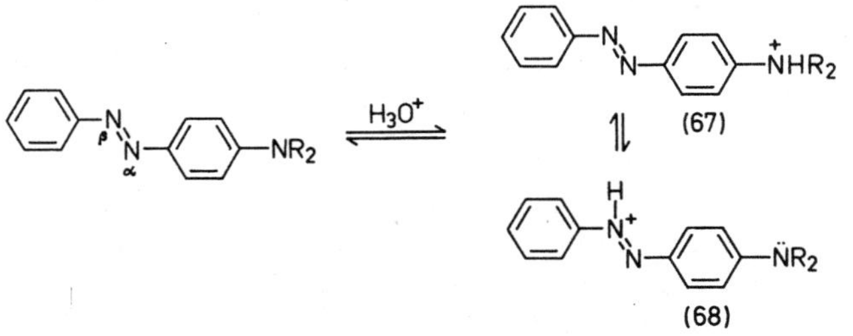

In contrast, the azonium tautomer contains a delocalised positive charge (68↔69) and hence bears some resemblance to cyanine dyes (see Sect. 5.5.3). It is more bathochromic (usually bluish-red, λ_{max} *ca.* 520 nm) and tinctorially stronger than the neutral dye.

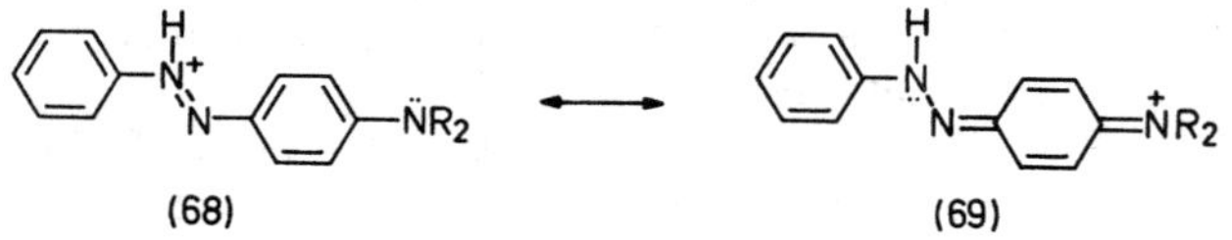

In the past, the nature of the species formed on adding acid to an aminoazo dye provoked a considerable amount of controversy. However, the evidence in favour of the above tautomeric equilibrium is overwhelming. Thus:—

(i) the electronic absorption spectra of the ammonium tautomer (67:R=Me) is very similar to that of the methylated compound (70), as is the electronic spectrum of 4-NN-dimethylaminostilbene in acidic solution to that of stilbene in neutral solution, as expected from proton addition to the dimethylamino group (71);

(ii) 4,4′-dimethoxyazobenzene exhibits only one band (λ_{max} 500 nm, ε_{max} 43,000) in acidic solution, typical of the azonium tautomer;

(iii) the effect of substituents *ortho* to the azo linkage on λ_{max} and ε_{max} (see Sect. 3.5.6) and finally;

(iv) the similarity of the electronic spectra of the N-phenylazo dyes (72),[5] which are locked in the azonium form, to the corresponding azonium tautomers (68). Analogous N-methyl compounds have not been reported, but N-methylation of the azo group to yield stable dyes has been achieved when the azo group is locked in the *cis* configuration as in benz-[c]-cinnoline (73)[6]: these dyes are also similar in colour and intensity to the azonium tautomers (68).

[5] GB 947,107 (CMC)

[6] GB 1326,323 (Bayer).

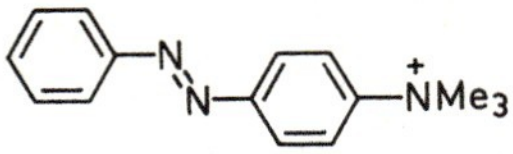

(70)

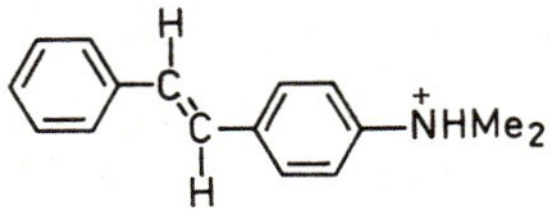

(71)

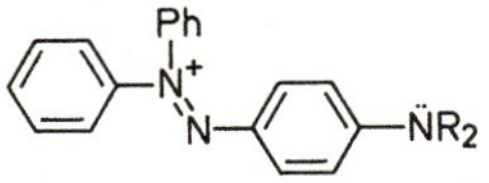

(72)

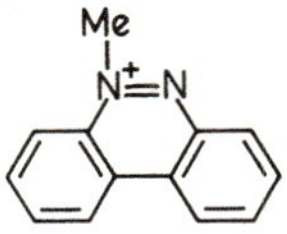

(73)

The major influences on the position of the azonium/ammonium tautomeric equilibrium are the strength of the acidic medium, and electronic, steric and hydrogen-bonding effects within the dye itself. Increasing the acidity increases the proportion of azonium tautomer and this effect has been ascribed to the different responses of the tautomers to the acidity function, H_o. For dyes derived from primary or secondary aniline coupling components, the difference stems from the extent of solvation by hydrogen-bonding. Thus, primary or secondary anilinium ions have 3 and 2 N—H bonds respectively to form hydrogen-bonds with the solvent, usually water or an alcohol, whereas the azonium tautomer has only 1 such N—H bond. Therefore, the azonium tautomer is more sensitive to a decrease in the solvent concentration than is the ammonium tautomer.

Substituent effects in derivatives of 4-aminoazobenzene, the most widely studied dyes, can be profound. Thus, the introduction of a bulky substituent *ortho* to the terminal dialkylamino group twists the latter out of the molecular plane. This increases the basicity of the terminal nitrogen atom and decreases the basicity of the azo nitrogen atom since delocalisation of the lone pair of electrons into the π-system, as depicted in (74), is diminished due to the loss in overlap. Therefore, protonation occurs to a much greater extent on the amino nitrogen atom and the equilibrium is shifted greatly in favour of the ammonium tautomer (75) — Table 3.3.

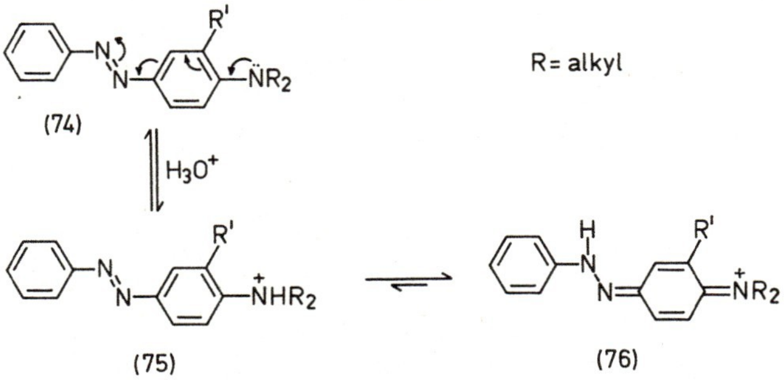

R= alkyl

(74)

H_3O^+

(75)

(76)

Electron-donating groups *meta* to the terminal amino group generally stabilise the azonium tautomer. The increase in stability is exceptional when the substituent also forms an intramolecular hydrogen-bond with the proton on the azo group.

Table 3.3. Effect of substituents *ortho* to the NN-dialkylamino group

R′	Ammonium tautomer (75) R = Me		Azonium tautomer (76) R = Me	
	λ_{max} (nm)	ε_{max}	λ_{max} (nm)	ε_{max}
H	320	9,800	512–520	35,500
Me	319	19,800	500	500

Prime examples of such groups are methoxy and amino and indeed dyes of this type are used commercially; for example, CI Basic Orange 2 (77), which is used primarily for dyeing paper. In contrast, electron-withdrawing groups tend to increase the proportion of the ammonium tautomer (Table 3.4).

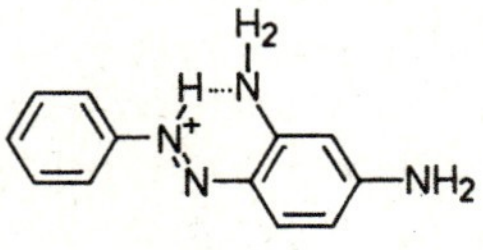

(77)

Table 3.4. Spectra of O_2N–C_6H_4–N=N–C_6H_3(X)–NEt_2 in EtOH: conc. HCl 2:1

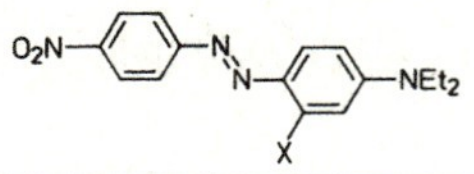

X	Azonium tautomer		Ammonium tautomer	
	λ_{max} (nm)	ε_{max}	λ_{max} (nm)	ε_{max}
NH_2	482	63,800	—	—
H	515	43,500	326	16,000
CF_3	530	21,000	324	22,800

For dyes of formula (78), electron-withdrawing groups *R*, *meta* or *para* to the azo linkage, stabilise the azonium tautomer (78A). Conversely, electron-donating groups stabilise the ammonium tautomer (78B). *Ortho* substituents *R* sterically hinder protonation at the β-azo nitrogen atom and hence the equilibrium is shifted in favour of the ammonium tautomer. One of the reasons why most commercial 4-aminoazobenzene dyes contain at least one substituent *ortho* to the β-nitrogen atom of the azo group is to suppress the formation of the azonium tautomer and hence the accompanying undesirable shade change. (Other important reasons are that an *ortho* substituent affects the colour and the light fastness of the dye). However, *ortho* substituents that are capable of intramolecular hydrogen-bonding with the azo proton form a stable 6-membered chelate ring: this greatly stabilises the azonium tautomer and displaces the equilibrium in its favour.

The ease with which certain azo dyes protonate has been exploited: they are widely used as pH indicators. Typical examples are Methyl Orange (79) and Methyl Red (80).

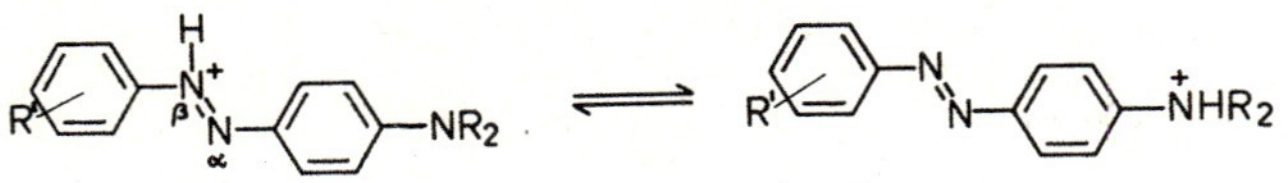

(78A) ⇌ (78B)

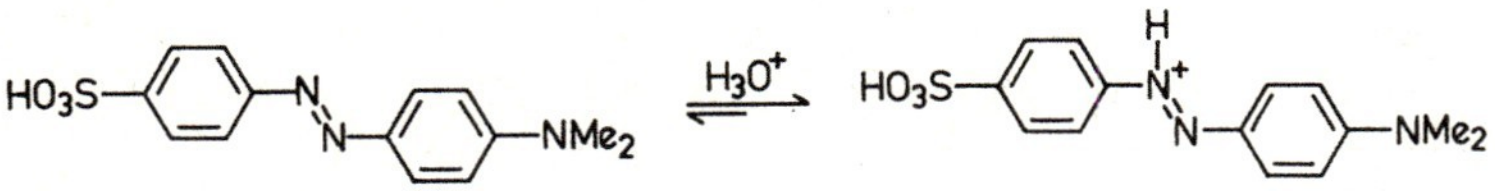

(79) $\xrightleftharpoons{H_3O^+}$ (79A)

yellow (λ_{max}^{EtOH} 442 nm) red ($\lambda_{max}^{H_2O}$ 506 nm) pH=1

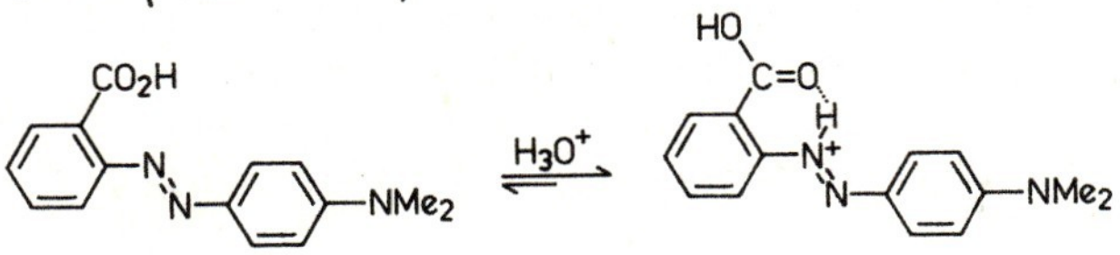

(80) $\xrightleftharpoons{H_3O^+}$ (80A)

orange (λ_{max} 435 nm; ε_{max} 19,000) red (λ_{max} 520 nm; ε_{max} 45,400)

In the case of disazo dyes (81) the first proton adds to the β-, and the second to the δ-azo nitrogen atom. This results in a progressive bathochromic shift of λ_{max} and an increase in ε_{max} — Table 3.5.

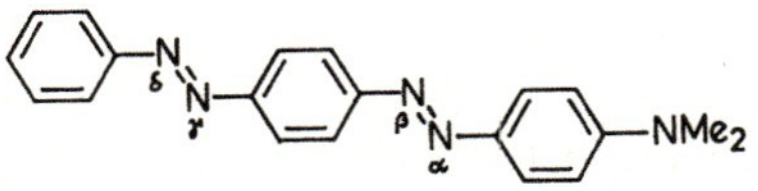

(81)

Table 3.5. The effect of acid on the disazo dye (81)

Solvent	λ_{max} (ε_{max})	Species	Colour
95% EtOH	476 (35,800), 330 (16,000)	Neutral dye	Scarlet
50% alcoholic 1.2 N HCl	548 (59,600), 355 (17,000)	β-protonated	Violet
50% alcohol, 50% H_2SO_4	635 (106,300) —	β-δ-diprotonated	Green

In contrast, bis-protonation of monoazo dyes results in a hypsochromic shift to give the yellow compound (82) — λ_{max} *ca.* 420 nm.

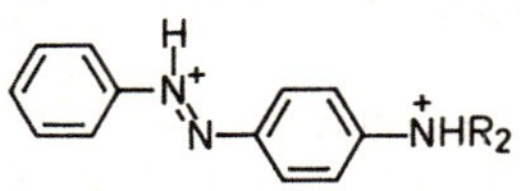

(82)

3.4 Metal Complex Azo Dyes

3.4.1 Introduction

The formation of metal complexes has figured prominently in dyestuffs chemistry from very early times. Indeed, mordant dyeing was based on this property. Thus, the cloth to be dyed was first impregnated with a soluble metal salt and then a soluble alkali was added to precipitate the insoluble metal hydroxide within the fibre pores. When this mordanted cloth was immersed in a solution of a suitable natural dye, *e.g.* Cochineal or Alizarin, the latter complexed with the metal hydroxide and thus became trapped within the fibre. Typical is Turkey Red, the aluminium/calcium complex of Alizarin (see Sect. 1.2.3). Polyvalent non-transition metal ions such as Al III, Ca II and Sn II were the usual mordants.

The natural dyes did not contain representatives of the azo class: the more important were bidentate anthraquinones, usually *ortho*-dihydroxy- or *ortho*-hydroxy-carboxyanthraquinone dyes. When the all important azo dyes were being pioneered metallisation was found to offer solutions to certain technical problems but the pre-eminent metals were now the transition metals, copper, chromium and cobalt.

Complexes of azo compounds can be either pre-formed or formed *in situ* in the fibre by after metallisation processes. However, it was not until 1919 that both Ciba and IG introduced the first premetallised dyes, 1:1 Cr III complex azo dyes for wool.

Metallised azo dyes are of two types: those in which the azo group is a coordinating ligand to the metal (medially metallised type) and those in which it isn't (terminally metallised type). The former are by far the most important commercially and consequently most attention will be devoted to them.

3.4.2 Medially Metallised Azo Dyes — Nature of the Bonding by the Azo Group

That the azo group participates in bonding to the metal ion was inferred from the observation that azobenzenes having *o*-hydroxy or *o*-amino groups form metal complexes whereas those having hydroxy or amino groups *meta* or *para* to the azo linkage do not. Whether the bonding involved the sp^2 lone pair electrons of the azo nitrogen atom or the π-electrons of the azo bond was not resolved until X-ray data on several azo compounds established that only one nitrogen atom was involved in bonding to the metal. In the copper complex (83), it is the β-nitrogen atom which bonds to the copper II ion with its sp^2 lone pair electrons to give a 6-membered chelate ring.

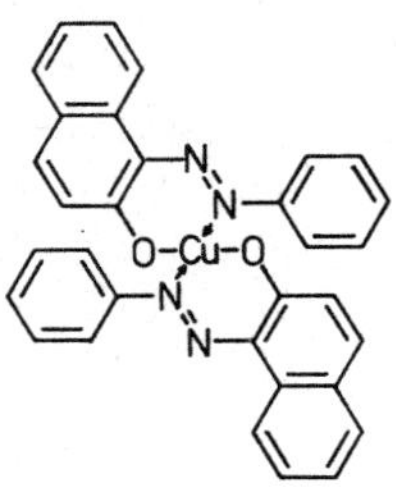

N N O Cu O N N

(83)

3.4.3 Types of Dyes and their Stability

The types of azo dyes capable of undergoing metal complex formation are illustrated by the general formula (84): —

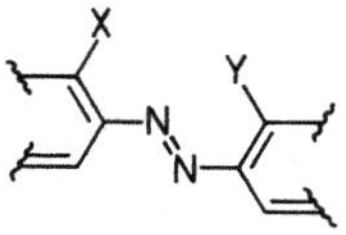

(84)

X	Y	
OH	H	bi-dentate
NH_2	H	
OH	OH	tri-dentate
OH	CO_2H	
OH	NH_2	
OH	OCH_2CO_2H	tetra-dentate

For obvious reasons only those metal complex dyes with a high degree of stability[7] and which are economic to manufacture are used commercially. The stability of the complex depends on the following factors:
1. the size of the chelate ring,
2. the number of chelate rings per molecule,
3. the basicity of the ligand, and
4. the nature of the metal.

Planar 5- and 6-membered rings are much more stable than non-planar 7-membered rings, and the more chelate rings there are per molecule, the greater the stability. As a rule, the most basic ligand forms the most stable complex and the most acidic the least stable complex. In accordance with these principles, the relative stabilities of the copper II complexes of bi- and tridentate azo dyes follow the sequence: —

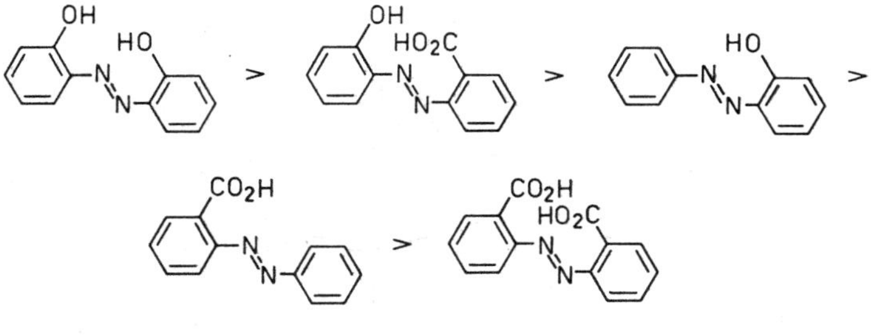

Complexes of the last dye necessarily have one 7-membered ring.

Chromium III, cobalt III and copper II are the three metal ions used in commercial metal complex azo dyes. Chromium III and cobalt III form octahedral complexes and in such complexes the Crystal Field Stabilisation Energy (CFSE) — and hence the stability of the complex — is at a maximum for the d^3 and d^6 electronic configurations: chromium III and cobalt III ions respectively possess these configurations. Copper II, which has a d^9 configuration, forms square planar complexes, which are significantly more stable than complexes formed from other divalent metal ions (Table 3.6).

[7] Stability refers to chemical stability in the dyebath and not thermodynamic stability.

Table 3.6. Stability constants of azo complexes with divalent cations

Azo compounds	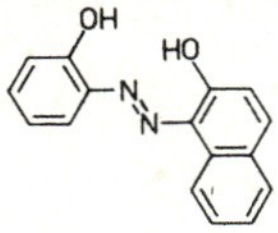 (o,o'-dihydroxyazo compound: 1-(2-hydroxyphenylazo)-2-naphthol)	
pK_a values of the two hydroxy groups	11.00	13.75
Metal ion	log K_1 in 75% dioxane at 30°	
Ba	5.74	
Sr	6.81	
Ca	8.61	
Mg	10.93	
Pb	14.65	
Zn	16.35	
Ni	19.62	
Cu	23.30	

$$K_1 = \frac{[ML]}{[M] \times [L]}$$

3.4.4 Structure and Stereochemistry

The 1:1 copper II dye complexes are essentially planar with the three coordination sites of the dye occupying three corners of a square planar arrangement (85): the fourth corner is occupied by a monodentate ligand, usually water.

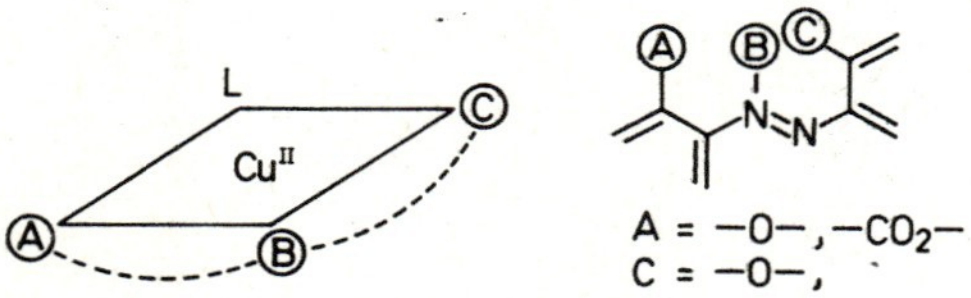

A = –O–, $-CO_2-$
C = –O–,

(85)

For the 1:2 chromium III and cobalt III dye complexes, two geometrical isomers are possible: the Drew-Pfitzner or meridial (*mer*) type (86) in which the two dyestuff molecules are mutually perpendicular and octahedrally coordinated to the central metal ion, and the Pfeiffer-Schetty or facial (*fac*) type (87) in which they are parallel. (Enantiomers of these isomers exist. However, since the enantiomers do not affect the colouration properties of the *mer* or *fac* isomers, they need not be discussed). The former are formed from *o,o'*-dihydroxyazo dyes, which have a 5:6 chelate ring system (86A), and the latter from *o*-carboxy-*o'*-hydroxyazo dyes, which have a 6:6 chelate ring system (87A).

Chromium III complexes having the *fac* configuration are less soluble in water, and, when dyed on wool, are brighter in hue and display higher light fastness but lower wet fastness than the *mer* isomer.

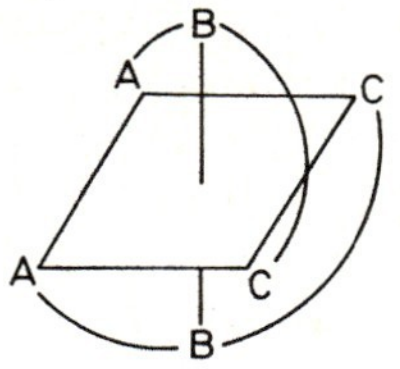

A = —O—
C = —O—

(86)

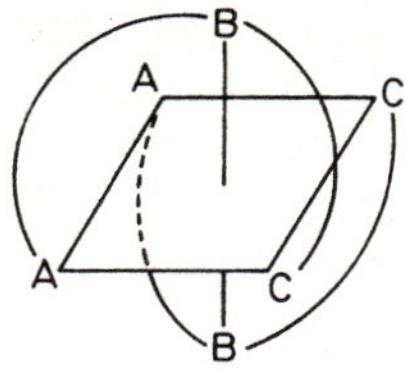

A = —CO_2—
C = —O—

(87)

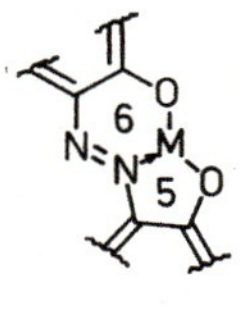

(86A)

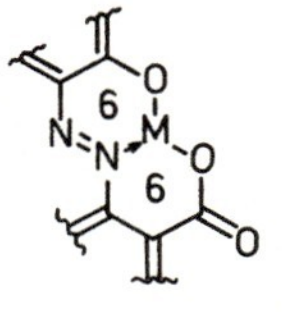

(87A)

3.4.5 Commercial Uses of Metal Complex Azo Dyes

The 1:1 copper II complexes are used widely in both reactive and direct dyes for cotton. CI Reactive Red 6, a dull rubine dye (88), is typical. Because of their instability under dyebath conditions the 1:2 copper II complexes have not found commercial use.

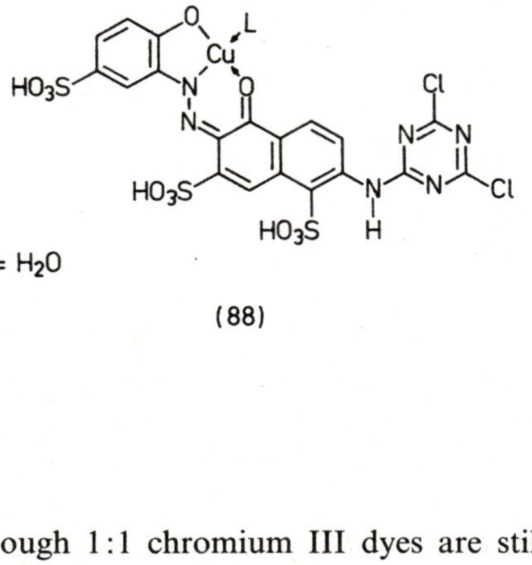

L = H_2O

(88)

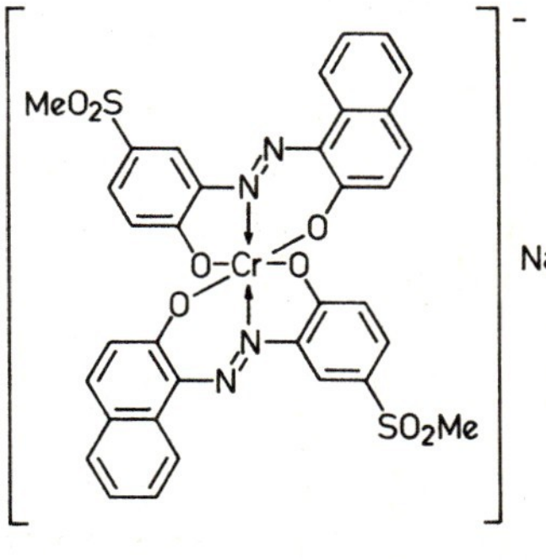

(89)

Although 1:1 chromium III dyes are still used, they are much less important than 1:2 chromium III dyes because the former are stable only at low pH values and hence have to be applied to the fibre, usually wool, from an acidic dyebath. This adversely affects the properties of the wool, particularly its softness. In contrast, the 1:2 chromium III complexes are stable over a wide pH range and can be applied to both wool and nylon from a neutral dyebath: indeed, so stable are 1:2 chromium III complexes that they tolerate the alkaline application conditions of reactive dyes. An example of a 1:2 chromium III dye for wool is CI Acid Violet 78 (89).

The relative stability of the 1:1 chromium III complexes allows the preparation of unsymmetrical (or mixed) 1:2 chromium III complexes by reaction with a second, different, tridentate azo molecule. This technique enables the properties of the final dye to be tailored. In contrast, cobalt III does not offer such a possibility due to the instability of the 1:1 cobalt III complexes. Hence, only symmetrical 1:2 cobalt III dyes can be made and this lack of versatility has rendered cobalt III dyes less important than chromium III dyes.

3.4.6 Properties of the Metallised Dyes

The metal ion alters the properties of the dye. Generally, a bathochromic shift results and the dye becomes duller. Usually the light fastness is improved, as is the resistance to oxidation, whilst aqueous solubility is decreased. These effects are shown for (90), a constituent of CI Reactive Violet 2, and its non-metallised precursor (91).

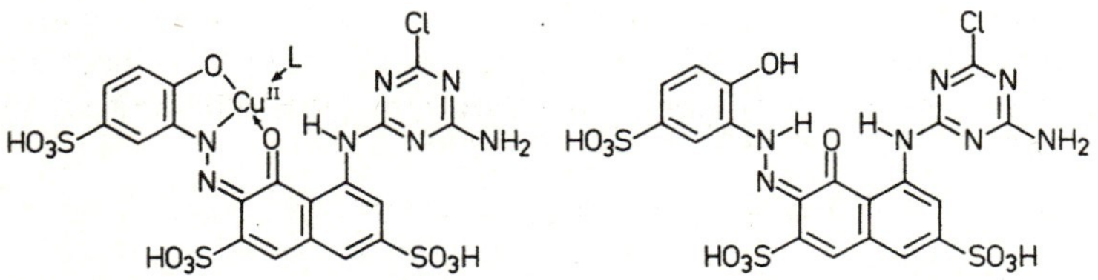

L = Ligand (90) (91)

Dye	Colour	λ_{max} (N/10 NaOH)	Light fastness	Resistance to hypochlorite	Aqueous solubility
90	violet	553	very good	very good	moderate
91	red	517, 539	poor	poor	good

The effect of a transition metal on the colour of a dye does not arise from the forbidden *d*→*d* transitions of the metal itself; these are so weak ($\varepsilon_{max} \ll 10^3$) that they are completely swamped by the allowed, intense $\pi \rightarrow \pi^*$ transitions of the chromogen (ε_{max} *ca.* 25,000). The colour change is caused by chelation of the metal to the major donor and acceptor groups, hydroxy and azo respectively, which results in a perturbation of the π-electron distribution.

It is this perturbation which improves the light fastness of the metallised azo dye. The electron deficient metal ion attracts electrons from the oxido ($O^{\ominus}$) and azo groups so that the electron density at the latter groups is reduced. Since photochemical oxidation at the azo group is an important pathway in the fading of azo dyes by light (see Sect. 6.4.4), oxidative attack at that position is retarded in the metallised dyes.

To explain the low aqueous solubility of copper II dyes, the formation of sheet like aggregates such as (92) have been proposed. Addition of an energetic ligand such as an aliphatic amine or urea displaces the second azo nitrogen from the fourth coordination site and improves the solubility markedly.

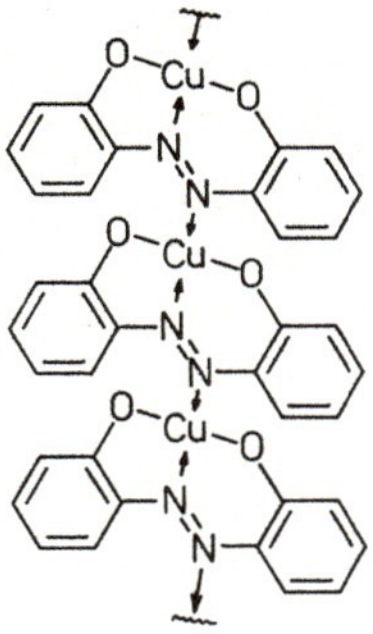

(92)

3.4.7 Terminally Metallised Dyes

As mentioned earlier, this type of dye is of little importance commercially. A typical example is the complex (93), derived from salicylic acid. In general, metal complexes derived from dyes of this type are brighter in hue than those in which the azo group is a ligand. In contrast to the latter, terminally metallised dyes normally undergo little shade change on metal complex formation and little enhancement of light fastness occurs.

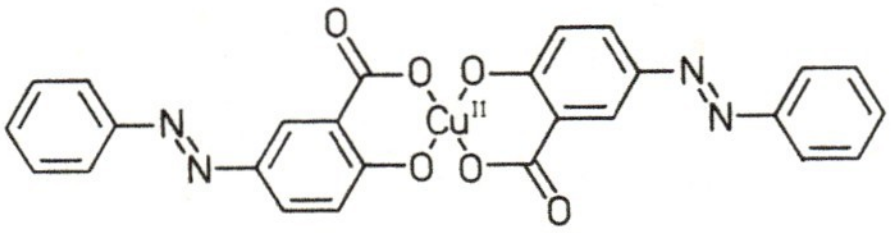

(93)

3.5 Colour and Constitution

3.5.1 Introduction

Before discussing the application of colour and constitution to azo dyes a brief review of the development of the subject is presented, beginning with the early theories and ending with the modern theories.

3.5.2 Early Theories

Chemists have long been intrigued by the relationship between the colour of a dye and its chemical constitution: to trace the unravelling of this relationship we have to go back over 100 years to the period when the dyestuffs industry was in its infancy. At that time little was known about the constitution of any of the coloured organic substances; indeed, Kekulé's structure for benzene had only just been published! (1865) — see Sect. 1.3.4. Although investigations into the structure of the natural dyes, Alizarin and Indigo, had begun (see Sect. 1.3.5), the only coloured compound whose structure was known was *p*-benzoquinone.

It was during this period that Graebe and Liebermann, in 1867, undertook the first study of the relation of colour to chemical structure. They found that reduction of the known dyes destroyed the colour instantly and from this observation they inferred that dyes were chemically unsaturated compounds.

It was from a study of compounds like azobenzene and *p*-benzoquinone that led Witt, in 1876, to formulate his celebrated theory which proposed that dyes consist of unsaturated groups, the chromophores, and salt forming groups, the auxochromes. Typical chromophores are ethylenic, carbonyl, azo and nitro groups, whereas hydroxy and (alkyl)-amino groups are typical auxochromes. Witt termed the complete coloured unit a 'chromogen'. Although Witt's theory has to some extent been superseded, his terminology hasn't — it is now the accepted terminology in dyestuffs chemistry. Witt also claimed that the auxochrome conferred dyeing properties on the molecule but it is now established that colour and dyeing properties are not directly related.

Three years later Nietzki stated that increasing the molecular weight of a dye by the introduction of substituents such as methyl, ethyl, phenyl, ethoxy or bromo produced a bathochromic shift. Though 'Nietzki's Rule' initially proved useful, its utility decreased as many exceptions were subsequently discovered.

The next major development was made by Armstrong who, in 1887, propounded the 'quinonoid theory' of colour. He argued that only compounds which can be written in a quinonoid form are coloured. Since at that time the majority of dyes fell into this category the quinonoid theory proved popular for many years.

In 1900 Gomberg discovered the first coloured free radical, the triphenylmethyl radical. Although its structure was not appreciated at the time this new type of coloured molecule — it was devoid of carbonyl and azo chromophores, and also auxochromes — aroused a flurry of activity amongst the contemporary organic chemists. Foremost amongst these was Baeyer. Research around the triphenylmethyl radical led Baeyer to propose his theory of halochromy, whereby a colourless compound is rendered coloured on salt formation. The term 'halochromism' is still used today to denote a colour change of a dye on the addition of acid or alkali.

From his researches into the halochromism of triphenylmethane dyes, Baeyer, in 1907, proposed the possibility of tautomerism to account for the colour of dyes such as Doebner's Violet (94). He suggested a rapid oscillation between the two 'tautomeric' forms (94)⇌(94A), the 'chlorine atom' flipping rapidly from one amino group to the other. It should be noted that at that time the distinction between ionic and covalent bonds was not appreciated. Hantzsch, from his work on the colourless nitrophenols and their coloured salts, also reached a similar conclusion to that of Baeyer, which he stated in his theory of 'chromotropy'.

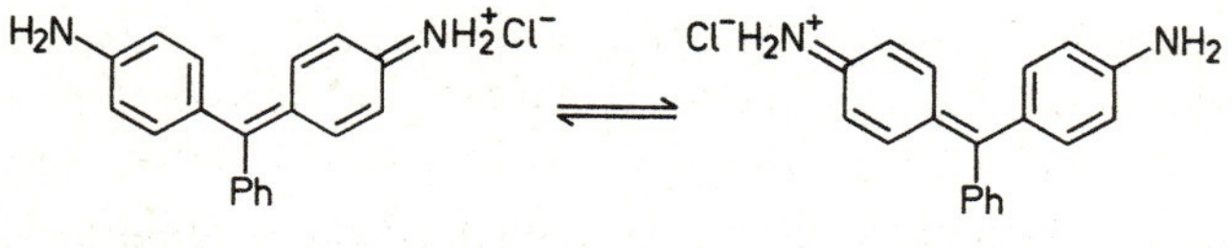

H_2N … $NH_2^+Cl^-$ … Ph ⇌ $Cl^-H_2N^+$ … NH_2 … Ph

(94) (94A)

In a series of papers published from 1904 to 1907 Baly proposed that colour arises from the rapid breaking and reforming of bonds as a molecule oscillates from one tautomeric form to another; he called this process, 'isorropesis'. In 1914, Watson put forward his tautomeric theory of the colour of dyes. This stated that to obtain intensely coloured compounds (dyes), tautomerism must occur, but that the conjugated chain must exist in a quinonoid form in all the possible tautomers. However, when dyes were discovered which couldn't exist in a quinonoid form, the theories of Armstrong and Watson became untenable.

The importance of conjugation was first realised by Hewitt and Mitchell in 1907. From a study of azo dyes these workers observed that the bathochromicity of a dye was directly related to the length of the conjugated chain of atoms in the chromogen — the longer the conjugated chain, the more bathochromic the dye — and this important observation became known as 'Hewitt's Rule'. Although there are exceptions to this rule, as Brooker and others later discovered, it remains one of the most valuable 'rules of thumb' in colour chemistry.

In 1928, Dilthey and Wizinger expanded and refined Witt's chromophore/auxochrome theory. They proposed that a dye consists of an electron-releasing, basic group, the auxochrome, connected to an electron-withdrawing, acidic group by a system of conjugated double bonds. They also found that the greater the nucleophilic and electrophilic character respectively of the two groups, and/or the longer the unsaturated chain joining them, then the greater was the resulting bathochromic shift.

The first chemists to show that it is the oscillation of *electrons* and not the oscillation, or migration, of *atoms* that produces colour were Adams and Rosenstein in 1914. They proposed, correctly, that atomic vibrations gave rise to absorption of infra-red radiation, whereas the 'oscillation' of electrons caused the absorption of ultraviolet or visible radiation — the latter would manifest itself as colour. Hence, they suggested that the colour of a dye such as the dibasic ion of phenolphthalein arises from the continuous oscillation of an electron between the two quinone oxygen atoms, *i.e.* (95) to (95A).

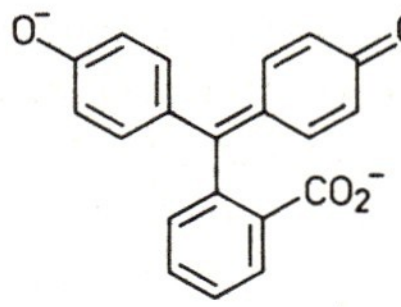

O^- O CO_2^-

(95)

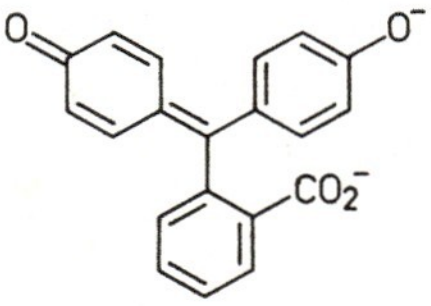

O O^- CO_2^-

(95A)

$H_2\ddot{N}$ NH_2^+ Ph

(96)

H_2N^+ $\ddot{N}H_2$ Ph

(96A)

However, it was not until 1935 that Bury highlighted the relationship between resonance and the colour of a dye. He refuted Baeyer's explanation that the colour of triphenylmethane dyes such as Doebner's Violet arose from the oscillation of atoms, *e.g.* (94) to (94A). Now that the distinction between covalent and ionic bonds was known, Bury realised that it was only the electrons which moved and not the atoms,[8] and that such a movement of electrons was merely an example of resonance, *i.e.* (96)↔(96A). He proposed that 'the intense absorption of light which characterises dyes is due to an intimate association of a chromogen and of resonance in the molecule'. He also realised that his theory explained the function of an auxochrome: it introduced the possibility of resonance. Bury's resonance theory explanation of the colour of dyes was generally accepted and further research by Swarzenbach and Brooker led to the conclusion that the more bathochromic dyes were those for which a greater number of limiting structures of similar energy could be written.

Of all the older theories, that proposed by Lewis in 1916 is perhaps the one which is most similar to the modern theories. Lewis argued that 'colour results from the selective absorption of light by valence electrons whose frequencies synchronise with light of a definite frequency'. By referring to the references in Appendix I the reader can ascertain for himself how close this concept comes to the modern ideas on the origins of colour.

3.5.3 Modern Theories

All the early theories had one thing in common — they were qualitative. A completely new approach was required which would not only provide a new insight into colour-structure relationships but would also allow quantitative assessments to be made. The Einstein-Planck quantum theory, which states that energy is not continuous but can only have certain discrete (quantised) values, constituted such a breakthrough. It was from this fundamental new theory that the two modern theories of colour and constitution evolved: Valence Bond (VB) theory and Molecular Orbital (MO) theory. The major difference between the two theories is that the former is based on the concept of bonding valence electron pairs being localised between specific atoms in a molecule, whereas the latter pictures electrons as being distributed amongst a set of molecular orbitals of discrete energies.

Using quantum theory, Schrödinger, in 1926, derived his now famous wave equation. Theoretically, solution of this equation enables the behaviour of atoms and molecules to be calculated without recourse to experimentation. Unfortunately, the equation can only be solved *exactly* for one-electron systems such as a hydrogen atom or the hydrogen molecule ion, $H_2^{\oplus}$. For polyelectronic systems the mathematical difficulties, due primarily to electron-electron interactions, are insurmountable at present.

VB theory was pioneered by Heitler, London, Pauling and Slater shortly after publication of the Schrödinger equation. Since it originated from a chemical point of view *qualitative* VB theory is used widely by chemists because it employs familiar chemical structures and is quick and easy to apply. In its qualitative form VB theory is similar to the resonance theory proposed by Bury. In contrast, *quantitative* VB theory is rarely used nowadays because it is unwieldy and, for large molecules, it is virtually impossible to apply. Indeed, quantitative VB theory has never been applied to dyestuffs.

At the same time that VB theory was being developed, Hund, Lennard-Jones and Mulliken were proposing the LCAO-MO theory. However, it was Hückel who

[8] Apparently, Bury was unaware that Adams and Rosenstein had already stated this: he did not mention or refer to their paper.

made the major breakthrough in utilising this theory for calculating the various properties of unsaturated molecules of interest to organic chemists. As already stated, the Schrödinger equation cannot be solved exactly for molecules (or atoms) which contain more than one electron. To obtain solutions for such molecules various approximations and assumptions have to be made. It was Hückel who made perhaps the most drastic assumption when he considered σ and π electrons as separable. Using Hückel MO (HMO) theory it was possible to calculate the properties of unsaturated molecules such as ethylene, butadiene and benzene. Perhaps it should be mentioned that the reasonable results obtained were somewhat fortuitous since it is now known that the erors incurred in such a simple treatment tended to cancel each other out.

In 1948, Kühn developed the Free Electron MO (FEMO) model, which is the simplest of the MO models and one with which the reader is probably familiar (electron-in-a-box model). It is suitable only to conjugated systems in which all the bonds have a similar π-bond order, such as the cyanine dyes, and discussion of the FEMO model is therefore deferred to Chap. 5.

The first general MO theory for dyes was proposed by Dewar in 1950. Using Perturbational MO (PMO) theory, he formulated a set of rules for predicting qualitatively the effect of substituents on the colour of a dye. These rules are stated and discussed later (Sect. 3.5.5) after the experimental facts have been presented.

In 1951, Roothaan[9] comprehensively formulated a set of equations based on LCAO-MO theory to describe the behaviour of molecules. If these equations could be solved for dye molecules it would enable both the colour and the tinctorial strength to be calculated precisely. However, the equations were theoretical in nature and were unsuitable for tackling the complex molecules of interest to organic chemists. Consequently, it was left to Pariser and Parr, and Pople, in 1953, to abstract from the Roothaan equations a semi-empirical theory designed for predicting the wavelengths and intensities of the main visible and ultra-violet bands of unsaturated organic molecules. The Pariser-Parr-Pople (PPP) model incorporated the π-electron only assumption of Hückel plus a self-consistent-field (SCF) treatment to take account of electron-electron interactions, and a configuration interaction (CI) treatment to include the effect of other transitions on the main highest occupied to lowest unoccupied MO (HOMO → LUMO) transition. Both these treatments lead to a better (more real) solution. In addition, the more complex integrals in the equations were neglected whilst values for others were obtained from experimental data.

It is this PPP model which has been used widely and successfully for predicting the colour and tinctorial strength of dyestuffs. With the advent of powerful computers, PPP MO calculations on azo dyes can take only a matter of seconds. However, the PPP method has its limitations. It deals only with planar molecules (since most dyes are essentially planar anyway, this limitation is not a great problem): it does not take into account the effects due to σ-electrons, *e.g.* hydrogen-bonding and steric effects; and finally, the results refer to the gas phase so that solvent effects, which are often pronounced, are not included. Nonetheless, in spite of these limita-

[9] The equations were also formulated independently by Hall [Proc. Royal Soc. *205*, 541 (1951)].

tions, the PPP model, as will be seen later, has been remarkably successful in calculating the colour and strength of dyes, particularly azo dyes.

How the VB and MO theories rationalise the observed effects of colour-structure relationships in azo dyes will be demonstrated after the experimental facts have been presented.

Azo Dyes

The colour and constitution of azo dyes will now be discussed. For convenience, these have been sub-divided into: — monoazo dyes — derivatives of 4-aminoazobenzene; protonated azo dyes; azo-hydrazone tautomers; and polyazo dyes. Though no rigid format is adhered to an attempt has been made to begin with the experimental observations in each case and then to rationalise these in terms of the VB and MO theories. Electronic and steric effects are discussed separately.

3.5.4 Experimental Observations

Monoazo Dyes — Derivatives of 4-Aminoazobenzene

Azobenzene, the simplest aromatic azo chromogen, is pale yellow in colour. The introduction of electron-accepting groups has only a minor effect on the colour of azobenzene. However, electron-donating groups produce a bathochromic shift. Powerful electron-donating groups such as diethylamino generate a totally new $\pi \rightarrow \pi^*$ transition which lies in the visible region of the spectrum (*i.e.* 400–700 nm). It is called a charge-transfer (CT) transition because of the relatively large redistribution of electron density that occurs on excitation. Consequently, in contrast to azobenzene, 4-NN-diethylaminoazobenzene has an intense yellow colour (Table 3.7).

Addition of a second group to give a symmetrical azobenzene results in little change, except when both the groups are dialkylamino groups. Thus, the intro-

Table 3.7. Effect of electron-accepting and electron-donating groups on azobenzene

R	R^1	λ_{max}^{EtOH} (nm)	ε_{max}
H	H	320	21,000
H	CO_2Me	325	21,900
H	NO_2	332	24,000
NO_2	NO_2	338	22,900
H	Me	333	23,400
H	MeO	349	26,000
H	NH_2	385	24,500
H	NMe_2	407	30,900
H	NEt_2	415	29,500
Me	NMe_2	407	—
MeO	NMe_2	407	—
NH_2	NMe_2	410	—
NH_2	NH_2	399, 435 s	33,900
NMe_2	NMe_2	460	33,100

s = shoulder

duction of a 4′-methoxy group into 4-NN-dimethylaminoazobenzene causes no change in λ_{max}, whereas a 4′-dimethylamino group causes a bathochromic shift of 53 nm (Table 3.7).

The introduction of an electron-accepting substituent into the diazo component ring *D* of an azobenzene which contains a donor group in the coupling component ring *C* produces a bathochromic shift. Indeed, the majority of commercial azo chromogens are of this type. Generally, the bathochromic shift is greater when both groups are conjugated to each other via the azo linkage: this is achieved by locating the substituents in positions *ortho* or *para* to the azo group.

The trend of increasing bathochromicity with increasing electron-donating power of the donor substituent observed with monosubstituted azobenzene compounds (Table 3.7) is maintained when electron-withdrawing groups are present in the second phenyl ring. For a given donor group the largest bathochromic shifts are produced by unsaturated electron-withdrawing groups such as nitro and cyano, rather than by saturated groups such as trifluoromethyl, or atoms such as halogens. These combined donor/acceptor effects are shown in Table 3.8.

Table 3.8. Combined effect of electron-donating and accepting groups in azobenzene

meta ortho
para→ (D) –N=N– (C) –D̈
A

D	A	λ_{max}^{EtOH} (nm)	ε_{max}
OH	*p*-NO_2	386	29,500
NH_2	*p*-NO_2	439	27,400
NEt_2	*p*-NO_2	486	34,000
NEt_2	*o*-Cl	427	29,500
NEt_2	*m*-Cl	423	29,800
NEt_2	*p*-Cl	422	26,700
NEt_2	*o*-CN	462	30,000
NEt_2	*m*-CN	446	28,100
NEt_2	*p*-CN	466	32,700

Table 3.9. Effect of electron-accepting groups in the coupling component ring *C*

O_2N–(D)–N=N–(C)–NEt_2 (Y on ring C) O_2N–(D)–N=N–(C)–NH_2 (X on ring C) [a]

Y	λ_{max}^{EtOH} (nm)	$\Delta\lambda$[b]	ε_{max}	X	λ_{max}^{EtOH} (nm)	$\Delta\lambda$[b]	ε_{max}
H	486		34,000	H	439		27,400
Cl	472	−14	33,800	Cl	427	−12	25,800
CN	472	−14	32,400	NO_2	394	−45	24,900
NO_2	470	−16	31,800				
CF_3	467	−19	31,400				

[a] These dyes have been chosen to minimise steric effects (see 4.5.9).
[b] Relative to X or Y = H.

Electron-withdrawing groups in ring *C* generally cause a relatively small hypsochromic shift. This is shown in Table 3.9.

There is a dearth of information as to the effect of electron-donating substituents in ring *D*. However, the introduction of a methoxy group into the dye (97; X=H) to give the dye (97; X=OMe) results in a small bathochromic shift.

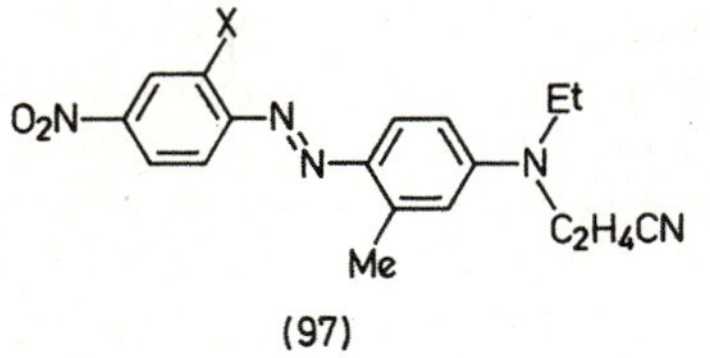

X, O_2N, N, N, Et, N, C_2H_4CN, Me

(97)

	$\lambda_{max}^{Me_2CO}$
X = H	470
X = OMe	474

The largest bathochromic shifts are obtained by adding more electron-withdrawing groups to ring *D* and more electron-donating groups to ring *C*. Electron-withdrawing groups exert a maximum effect when they are located *ortho* or *para* to the azo linkage, whereas electron-donating substituents are most effective when they are situated *meta*, and particularly *ortho* to the azo group. These effects are depicted in Tables 3.10 and 3.11 respectively.

These substituent effects on the colour of azo dyes have been used by dyestuffs chemists to provide commercial dyes ranging in shade from yellow to blue. To achieve the requisite fastness properties (see Chap. 6), ring *D* normally contains an electron-accepting group, usually nitro, located *ortho* or particularly *para* to the

Table 3.10. Effect of several electron-withdrawing groups in ring *D*

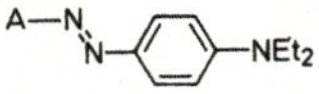

A—N=N—C_6H_4—NEt_2

A	λ_{max}^{EtOH} (nm)	ε_{max}
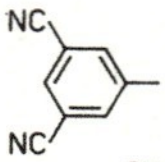 NC, NC	478	33,600
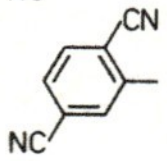 CN, NC	495	36,000
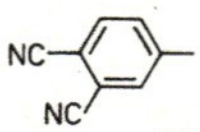 NC, NC	500	38,800
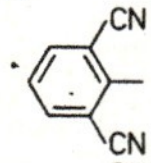 CN, CN	503	33,100
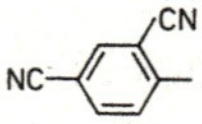 CN, NC	515	39,800
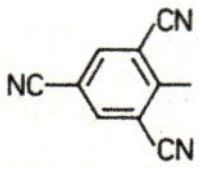 CN, NC, CN	562	46,600

Table 3.11. Effect of electron-donating groups in ring *C*

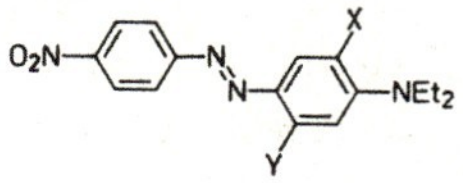

X	Y	λ_{max}^{EtOH} (nm)	$\Delta\lambda^{a}$	ε_{max}
H	H	486		34,000
H	OMe	501	+15	32,800
OMe	H	488	+ 2	22,600[b]
OMe	OMe	516	+30	29,600[b]
OMe	NHAc	530	+44	32,600[b]

[a] Relative to X=Y=H.
[b] Low because of steric hindrance (see 3.5.9).

azo linkage, and ring *C* is usually an aniline derivative. The basic chromogen of this type (98; X=H) is red. Many commercial red dyes for polyester fibres are based on the *meta* acylamino derivatives (98; X=NHAc) since they are brighter in hue and have a higher tinctorial strength than (98; X=H). Recent evidence suggests that intramolecular hydrogen-bonding is responsible for these effects, as depicted in (99): the homologous N-ethylacetylamino derivative, in which intramolecular hydrogen-bonding cannot occur, is similar in brightness and tinctorial strength to (98; X=H). Indeed, it is slightly hypsochromic relative to (98; X=H) because the bulky —N(Et)(Ac) group cannot lie in the plane of the molecule so that the —I inductive effect of the group predominates (Table 3.12).

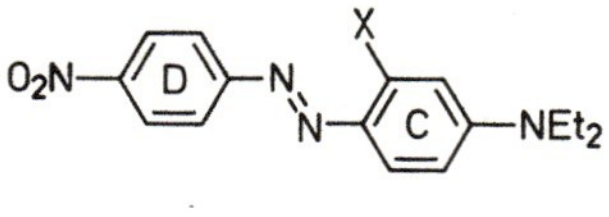

(98)

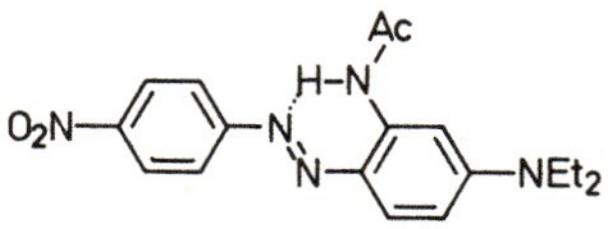

(99)

Table 3.12. Effect of an acylamino group in ring *C*

X in (98)	λ_{max}^{EtOH} (nm)	ε_{max}	$\Delta\nu_{1/2}$ (cm^{-1})
H	486	34,000	5,000
$NHCOCH_3$	511	47,000	3,900
N(Et)($COCH_3$)	482	36,000	4,900

As mentioned earlier, dyes more bathochromic than red, such as violet and blue, are obtained by adding further electron-accepting groups to ring *D* and electron-donating groups to ring *C*. Typical violet and blue dyes are (100) and (101) respectively. However, the increasing number of substituents tends to make such dyes rather dull in hue when compared to anthraquinone dyes, especially the blue dyes. Nitro groups in particular tend to cause excessive dullness. Presumably, the additional groups cause a divergence of the excited state and ground state geometries so that the absorption curve becomes broader.

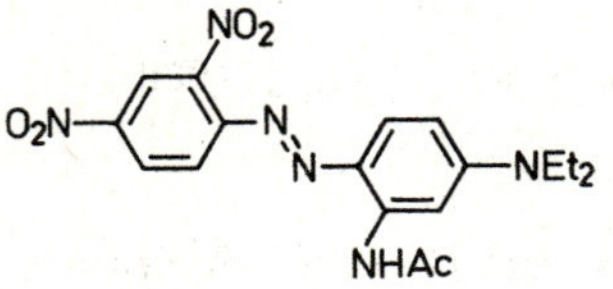

(100)
λ_{max}^{EtOH} 543 nm

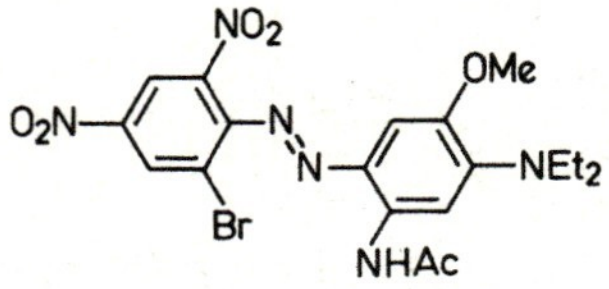

(101)
λ_{max}^{EtOH} 608 nm

To overcome this problem, heterocyclic diazo components have been employed since fewer electron-accepting groups are then required to produce the desired bathochromic shift. Thus, one nitro group is sufficient to produce blue dyes in the thiazole (102)[10] or benzoisothiazole (103)[11] systems. The dinitrothiophene dye (104)[12], which is analogous to the nitrothiazole dye (102), is greenish blue. Evidently, a nitro group exerts a more powerful electron-withdrawing effect than does a ring nitrogen atom. For comparison, the corresponding dinitroaniline dye (105) is red!

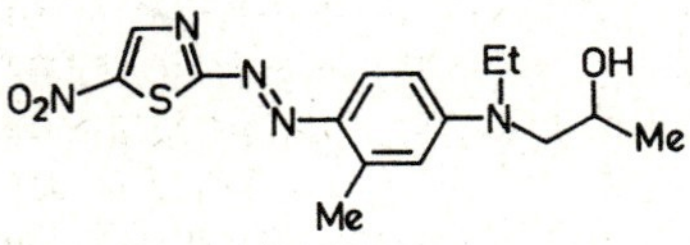

(102)
λ_{max}^{MeOH} 593 nm

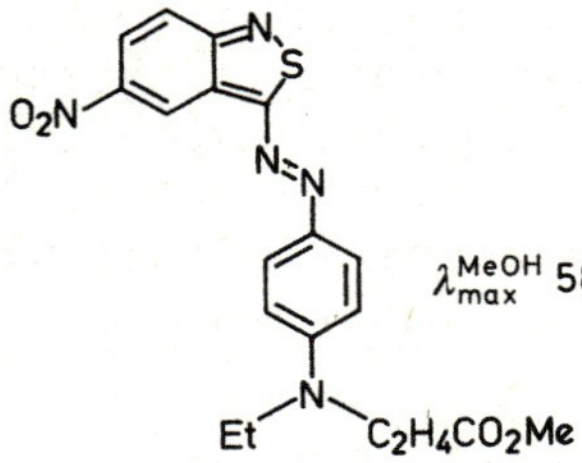

(103)
λ_{max}^{MeOH} 587 nm

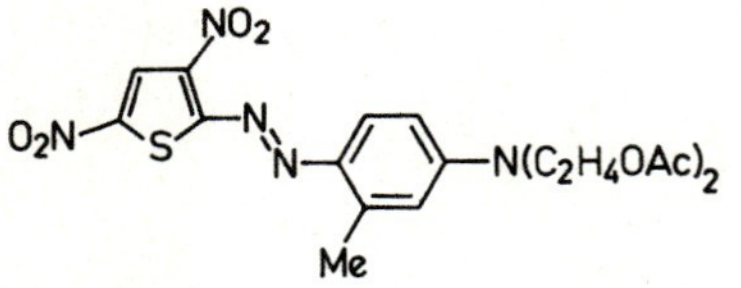

(104)
λ_{max}^{EtOH} 614 nm

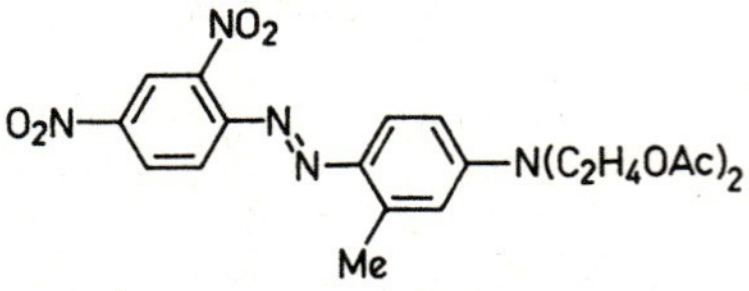

(105)
λ_{max}^{EtOH} 513 nm

Whilst heterocycles are used extensively as diazo components they are not normally encountered as coupling components. However, from the limited data available, azo dyes (106) from furan and thiophene coupling components are considerably more bathochromic than their benzenoid counterparts (107). Dyes in which both the diazo component and the coupling component are heterocyclic do not appear to have been studied.

[10] Example 3 of US 2659,719 (Eastman Kodak).
[11] Claim 5 of GB 1112,146 (BASF).
[12] Example 1 of GB 1394,367 (ICI).

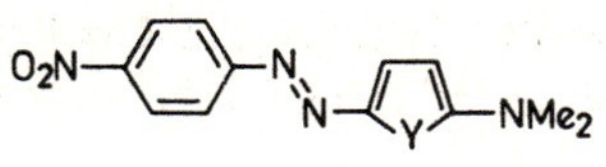

(106)

Y	$\lambda_{max}^{CHCl_3}$	ε_{max}
O	538	—
S	546	50,000

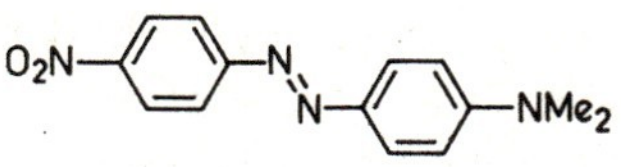

(107)

λ_{max}^{EtOH} 480 nm ε_{max} 33,000

Yellow dyes are obtained when the diazo component is devoid of electron-withdrawing groups, *e.g.* 4-NN-diethylaminoazobenzene is yellow (λ_{max} 415 nm). Dyes of this type are of little commercial value since other important properties are adversely affected, *e.g.* such dyes are usually photochromic (see Sect. 6.5). Most commercial yellow azo dyes therefore contain a *para* electron-withdrawing substituent in the diazo component and use modified coupling components to achieve the desired hue. One approach is to use a secondary rather than a tertiary aniline coupling component, or even aniline itself. Another ploy is to incorporate electron-accepting substituents into the N-alkyl chains so that their —I inductive effect will tend to reduce the +I inductive effect of the alkyl group. Obviously, the closer the substituent to the nitrogen atom the greater the hypsochromic shift. As seen from Table 3.13, the effect can be quite profound. Thus, the NN-bis-cyanoethyl dye absorbs at shorter wavelengths than the aniline dye! Incorporation of an electron-withdrawing group *ortho* to the amino group enhances further the hypsochromic shift. To avoid undesirable steric hindrance such substituents, *e.g.* Cl, are only normally used in primary or secondary aniline coupling components. CI Disperse Orange 1 (108), and the yellow dye (109) are typical examples which illustrates these principles.

Table 3.13. Substituent effect at the terminal amino group

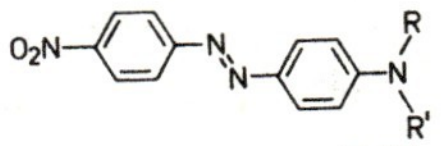

R	R′	λ_{max}^{EtOH} (nm)	ε_{max}
Et	Et	486	34,000
Et	C_2H_4CN	452	30,900
Me[a]	CH_2CN	422	26,000
C_2H_4CN	C_2H_4CN	432	30,000
H	H	439	27,400

[a] Data for Et compound is not available.

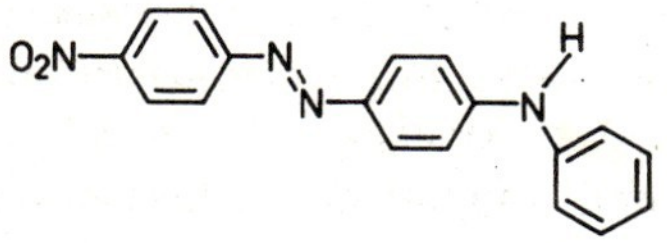

(108)

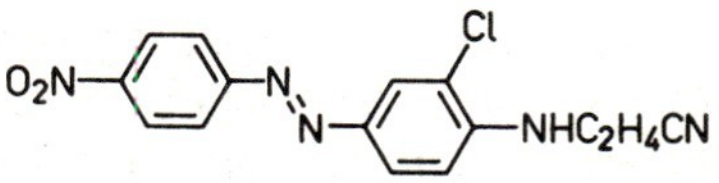

(109)

3.5.5 Application of VB and MO Theories

How are the observed effects of substituents on the colour of azobenzene rationalised by the VB and MO theories? In this section we will attempt to answer not only this question, but also how the theories fare in tackling that other important parameter, tinctorial strength.

Rationalisation of Colour

VB Theory (qualitative)

In qualitative VB theory the ground state of the molecule is assumed to resemble that of the most stable valence structure, whereas the first excited state is crudely represented by a less stable, usually charge separated, structure. The bathochromicity increases as the energy difference between the ground and first excited state decreases, since $\Delta E = hc/\lambda$. For azobenzene, the ground state is represented by the uncharged structure (110), and the first excited state by the much less stable charge separated structure (110 A), in which the positively charged carbon atom has only a sextet of electrons.

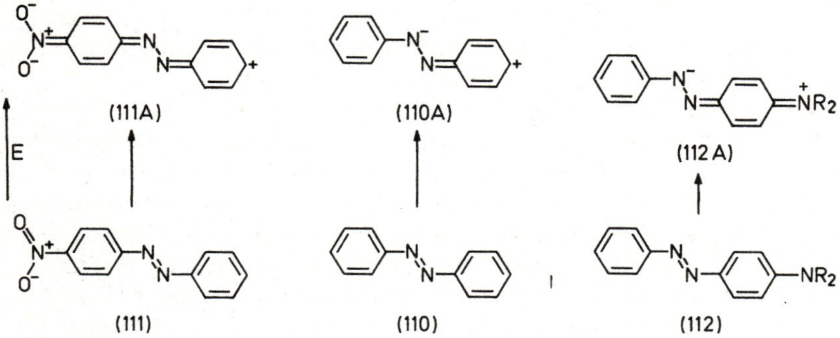

The introduction of an electron-accepting group such as nitro has only a minor effect on the colour. This is for two reasons: — the first, and principal reason, is that the structure representing the excited state (111 A) still contains a carbon atom having only six valence electrons, *e.g.* $C^{\oplus}$; second, the azo group itself is a good electron acceptor so there is no great advantage in this case in delocalising the electrons to some other group.

In contrast, a donor substituent *ortho* or *para* to the azo linkage markedly stabilises the charge separated excited state structure by donation of its lone pair of electrons, so that a resonance form can be written (112A) in which the outer shell of each atom has 8 electrons. Consequently, the energy difference between (112) and (112A) is much smaller than those between (110) and (110 A), and (111) and (111 A). Therefore *p*-aminoazobenzene is predicted to be more bathochromic than either azobenzene or its *p*-nitro derivative, as is the case experimentally (Table 3.7). Obviously, the same argument applies to *ortho* substituted azobenzenes but not to *meta* substituted azobenzene compounds. Here the substituent is not conjugated to the azo group and hence no stabilising resonance effect is possible between the substituent and the azo group. Consequently, electron-donating substituents are predicted to cause a much smaller bathochromic shift when they are in the *meta* position compared to when they are in the *ortho* or *para* positions.

According to VB theory a donor group introduced into the unsubstituted phenyl ring of a donor substituted azobenzene should cause a hypsochromic shift because it would tend to increase the electron density in the vicinity of an already electron rich site — the β-azo nitrogen atom. This clearly unfavourable situation is illustrated in (113A). In this case VB theory fails to rationalise the observed effects for if the reader refers back to Table 3.7 he will see that a second donor group has either no effect or produces a *bathochromic* shift.

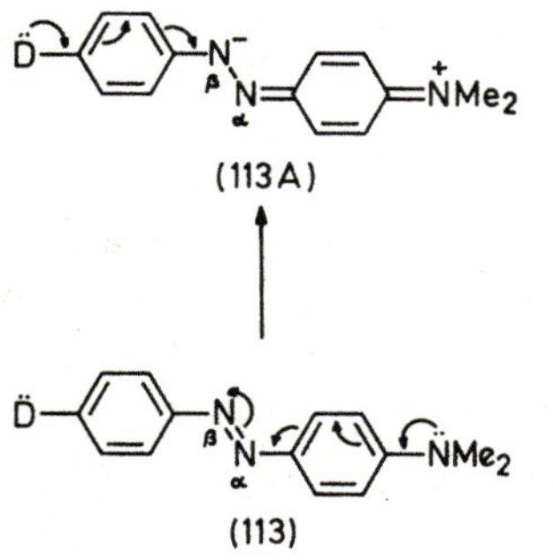

(113A)

(113)

VB theory successfully rationalises the observed bathochromic shift when an electron-accepting group is added to the unsubstituted ring of a donor substituted azobenzene. As mentioned previously, most commercial azo dyes are of this type. As an example we will consider the colours of the *o*-, *m*-, and *p*-cyano derivatives of 4-NN-diethylaminoazobenzene. In the *ortho*- and *para*-cyano dyes the negative charge can reside on the cyano group in addition to the azo group, as depicted in (114B) and (114A) respectively (for the *para*-cyano dye). This sharing of charge markedly stabilises the excited state; therefore ΔE is reduced and a bathochromic shift is predicted. No such stabilisation of the excited state (115A) is possible with a *meta* cyano group since this is not conjugated to the azo group. In this case, only the —I[13] inductive effect of the cyano group is operative in helping to reduce the negative charge density at the azo group. Consequently, the *meta* cyano dye is predicted to be slightly more bathochromic than 4-NN-diethylaminoazobenzene, but much less bathochromic than its *ortho* and *para* cyano isomers. As seen from Table 3.8, this is the case experimentally.

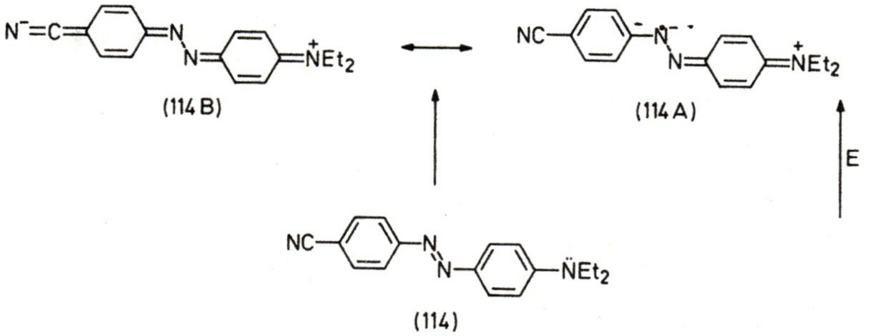

(114 B)

(114 A)

(114)

[13] Throughout the book —I and —M indicate electron-withdrawal whereas +I and +M indicate electron-donation.

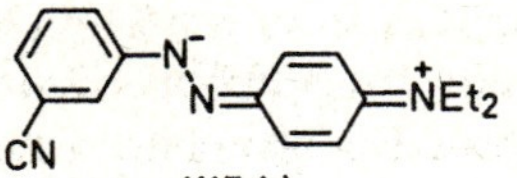

(115A)

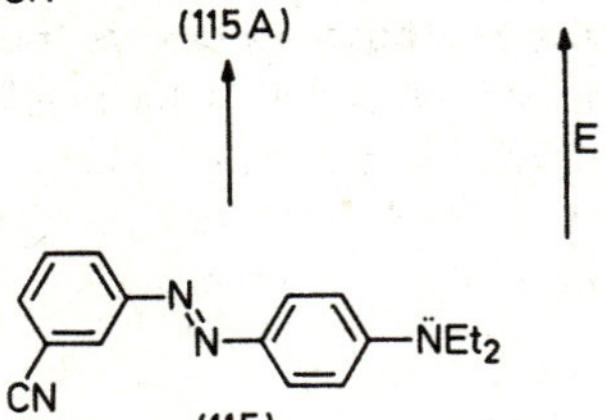

E

(115)

The introduction of electron-withdrawing groups into ring *C* of a donor-acceptor substituted azobenzene should, according to VB theory, produce a hypsochromic shift. This is because the group reduces further the electron density at an already electron deficient nitrogen atom. For substituents *ortho* to the amino group, as in (117), the effect is an inductive one since in the excited state, (117A), the electron-withdrawing group is not conjugated to the positively charged amino nitrogen atom. In contrast, substituents *meta* to the dialkylamino group are conjugated to the positively charged nitrogen atom and so destabilisation could occur both by inductive and mesomeric effects (116A). VB theory is only reasonably successful in accounting for the observed effects since Cl and CF_3 groups produce similar hypsochromic shifts to NO_2 and CN groups (Table 3.9). Thus, the predominant destabilising effect appears to be inductive rather than mesomeric.

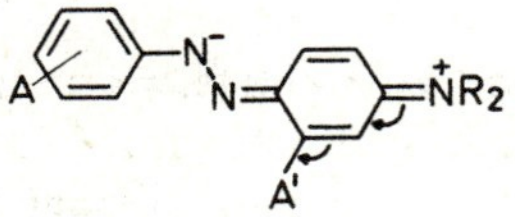

(116A) (117A)

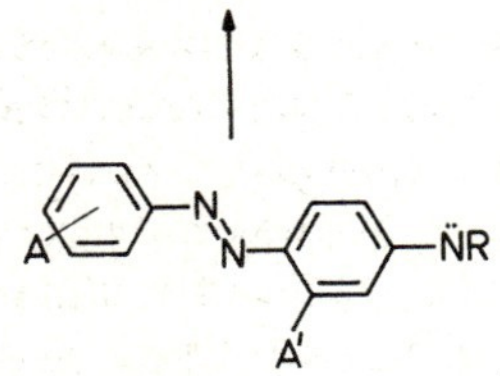

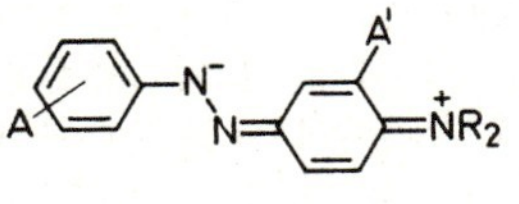

(116) (117)

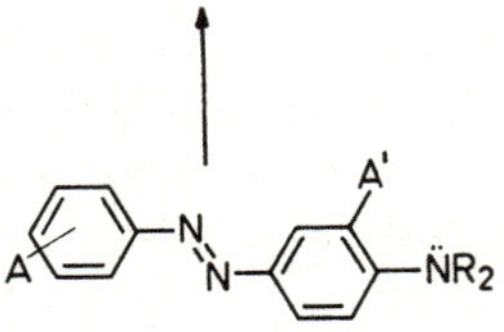

A and A′ are electron-accepting groups

VB theory also explains the bathochromic shift caused by introducing a methoxy group into ring *D* of the dye (97: X=H). The methoxy group stabilises the excited state by sharing the positive charge on the nitro nitrogen atom, as depicted in (97B).

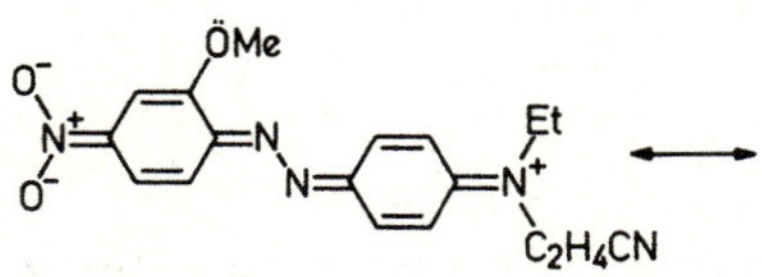

(97A) 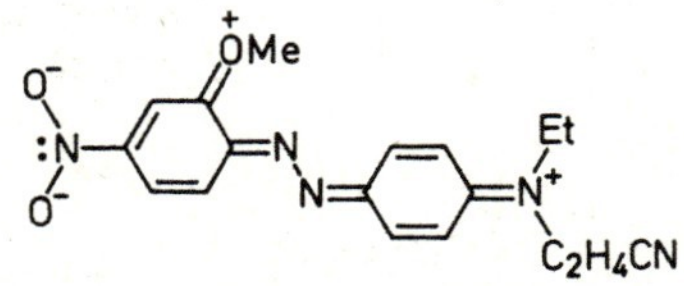 (97B)

VB theory is also reasonably successful in dealing with the commercially important dye structures in which ring *C* contains more than one electron-donating substituent and ring *D* contains more than one electron-withdrawing substituent. Thus, the bathochromic shift caused by the introduction of a methoxy group in the *meta* position to the diethylamino group in ring *C* is explained by its ability to stabilise the excited state by sharing the positive charge on the terminal amino nitrogen atom, as depicted in (118A). However, the bathochromic effect of electron-donating substituents *ortho* to the diethylamino group cannot be explained by VB theory.

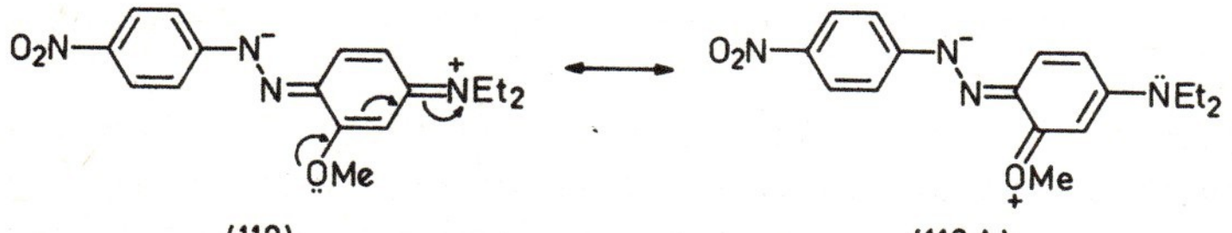

(118) (118A)

The bathochromic effect of additional electron-accepting groups in ring *D* is readily explained. If the groups are conjugated to the azo linkage, *i.e. ortho* or *para* to it, then they stabilise the excited state by sharing the negative charge on the β-azo nitrogen atom both by inductive and, more importantly, mesomeric effects. The mesomeric stabilisation is shown for the 2,4-dicyano dye (119; A and B). This effective sharing of the negative charge results in a large bathochromic shift (Table 3.10).

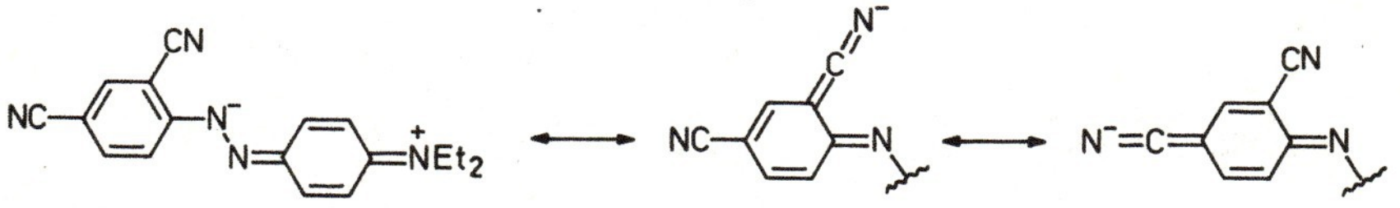

(119) (119A) (119B)

According to VB theory two factors account for the large bathochromic effect of heterocyclic diazo components containing sulphur (the only ones so far studied): one is that the RSE of a heterocyclic ring such as thiophene or thiazole is less than that of benzene, so the loss in RSE in going from the ground state to the excited state is correspondingly lower; in other words, the excited state of a heterocyclic azo dye is destabilised to a lesser extent with respect to the ground state than that of the corresponding benzenoid azo dye. The second factor is that

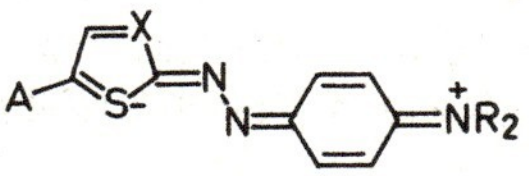

(120A)

(120)

A = acceptor
X = CH, N

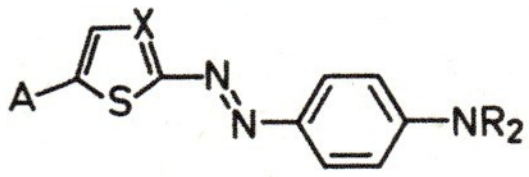

sulphur, a second row element, has available vacant 3*d* orbitals. Therefore, acceptable resonance structures can be written in which the sulphur atom has more than 8 electrons in its valence shell. An important resonance structure which contributes significantly to the stability of the excited state for thiophene and thiazole dyes is that represented by (120A). Here, the sulphur atom is acting as an additional electron-withdrawing group since it is assumed to be capable of accommodating 10 electrons.

In dyes such as (121) the diazo component is locked in a quinonoid configuration in the ground state: in the excited state (121A) it becomes aromatic, *i.e.* it gains in stability, and this contributes towards lowering the energy of the excited state.

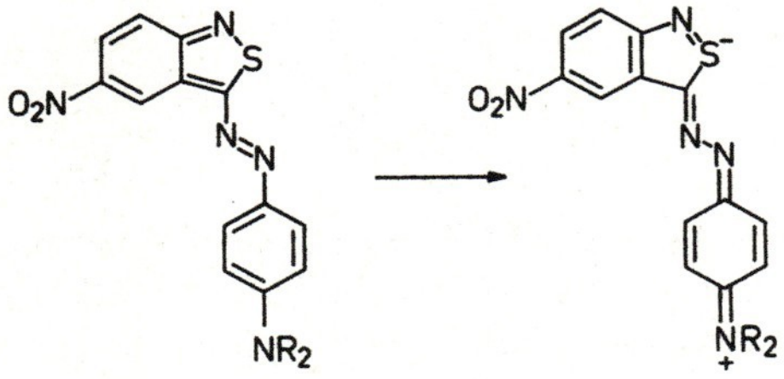

(121) (121A)

MO Theory

We will now consider how MO theory explains the experimental colour-structure relationships in 4-aminoazobenzene dyes. Since the qualitative and quantitative approaches give widely differing results, these are discussed separately.

Qualitative MO Theory — Dewar's Rules

As stated earlier, Dewar formulated a set of rules from PMO theory for predicting qualitatively the effect of substituents on the colour (λ_{max}) of dyes. In the case of azo dyes, these rules were obtained by considering 4-aminoazobenzene dyes as analogues of the odd alternant hydrocarbon anion (122). Alternate atoms are starred so that:

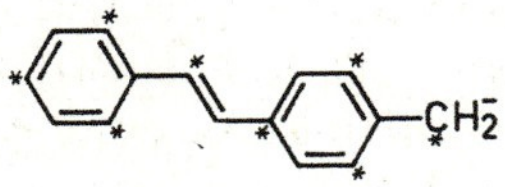

(122)

(i) no two starred atoms are adjacent, and
(ii) that the number of starred atoms exceeds the number of unstarred atoms.

Dewar's Rules then state that:

1. Any electron-withdrawing substituent at a starred position, or any electron-donating substituent at an unstarred position, should cause a large hypsochromic shift.
2. Any electron-withdrawing substituent at an unstarred position, or electron-donating substituent at a starred position, should cause a smaller bathochromic shift.

3. Replacing carbon by nitrogen has the same effect as an electron-withdrawing substituent at that position.
4. Any neutral, unsaturated group such as vinyl or phenyl, anywhere, has a bathochromic effect.

Application of these rules to neutral 4-aminoazobenzene dyes show that they fail as often as they succeed! (Fig. 3.7).[14] The reason is that a basic assumption in PMO theory is that of uniform bond order in alternate hydrocarbon anions: in neutral azo dyes this is not so; the bond order of the azo group tends to 2 and that of the aryl-nitrogen bonds to 1, and it is this lack of uniform bond order which is primarily responsible for the failure of Dewar's Rules in these dyes.

Fig. 3.7. Dewar's Rules predictions for 4-aminoazobenzene dyes

correct prediction ✓
incorrect prediction ×

However, although Dewar's Rules are not very useful for predicting correctly the colours of neutral azo dyes, they are much more applicable to protonated azo dyes (Sect. 3.5.6) and also to cationic dyes (see Chap. 5).

Quantitative MO Theory — PPP Model

The accessibility of PPP MO programs allied with the powerful computers now available has enabled PPP calculations to be performed on an almost routine basis. However, since it is a π-electron only model effects due to σ-electrons are not included. Thus, the weak $n \rightarrow \pi^*$ transition of the azo group ($\lambda_{max}^{C_6H_{12}} \sim 444$ nm, ε_{max} 450), which is primarily responsible for the pale yellow colour of azobenzene, is not calculated by the PPP model. What it does calculate, and very successfully, are the intense $\pi \rightarrow \pi^*$ transitions which characterise all dyestuffs.

The big advantage of the PPP model over the other methods discussed is that it provides *quantitative* values for the important parameters such as the colour (λ_{max}) and the tinctorial strength[15] of dyestuffs. In addition, it generates a lot of other useful data. For example, the dipole moments of the ground state and the various excited states are calculated and this makes it possible not only to predict whether that dye will exhibit solvatochromism, but also, which type. For neutral azo dyes the dipole moment of the first excited state is always greater than that for the ground state so that the dyes exhibit positive solvatochromism. Other useful data are the electron densities of the ground and excited states and also their bond orders. This information is particularly helpful for providing an "electronic picture" of the ground and excited states and also for identifying the principal donor and acceptor groups during excitation.

[14] The reader may verify this for himself by referring to the experimental data in Tables 3.8—3.13.

[15] PPP MO theory calculates a quantity f called the oscillator strength, which is directly proportional to ε_{max} (see Eq. 3.2).

The PPP model has been remarkably successful in calculating the λ_{max} values of neutral azo dyestuffs, particularly the benzenoid types. Most of the results presented in the experimental section have been calculated correctly by the model. For example, a bathochromic shift of 6 nm was calculated[16] for the introduction of a methoxy group into the dye (97; X=H): this compares favourably with the experimental value of 4 nm. In those cases where the PPP model and qualitative VB theory provide conflicting predictions, then it is the PPP model which invariably triumphs. As an example, let us consider the simple case of predicting the order of bathochromicity of azobenzene and its *o*-, *m*- and *p*-amino derivatives. We have already described how the VB method predicted the *ortho* and *para* amino derivatives to be much more bathochromic than the *meta* isomer, which in turn should be slightly more bathochromic than azobenzene itself. The PPP model calculates[17] a different order of bathochromicity. It calculates *ortho*-aminoazobenzene to be the most bathochromic compound and, in contrast to VB theory, *meta*-aminoazobenzene to be more bathochromic than *para*-aminoazobenzene. The PPP calculated and experimental λ_{max} values are shown in Table 3.14.

Table 3.14. Experimental and PPP calculated λ_{max} values for azobenzene and its *o*-, *m*-, and *p*-amino derivatives

Compound	λ_{max} (nm)	
	Exptl. (EtOH)	PPP Calc.[a]
(o-aminoazobenzene; H_2N)	414	424
(m-aminoazobenzene; NH_2)	412	414.5
(p-aminoazobenzene; $-NH_2$)	387	392
(azobenzene)	320	344

[a] Gregory, P., Hutchings, M. G.: Unpublished work.

Comparison of the calculated electron densities in the ground and excited states indicates the origin of the absorption band(s) by highlighting the donor and acceptor groups. For derivatives of 4-aminoazobenzene the PPP model calculates that, on excitation, the terminal amino group is the major electron donor and that the azo group is the principal electron acceptor. This is in agreement with VB theory. However, in contrast to VB theory, the PPP model calculates that in the first

[16] Ref. E4O: p. 35.

[17] Gregory, P., Hutchings, M. G.: Unpublished work: see also Bontschev, D., Ratschin, E.: Monatsh. *101*, 1454 (1970).

excited state (123A) the electron density is greater on the α-azo nitrogen atom: VB theory predicts the electron density to be greatest at the β-azo nitrogen atom since it is the β-, not the α-azo nitrogen atom, which is conjugated to the donor group — the amino nitrogen atom. Indeed, the VB 'picture' of the excited state (124A) resembles closely that calculated for the ground state (123) by the PPP model. These unexpected electronic effects calculated by the PPP model seem to apply to all azo dyes.

(123A) ← (123)

+ = electron deficiency
− = electron excess

(124A) ← (124)

It has already been mentioned that azo dyes derived from 5-membered heterocycles containing sulphur, such as thiophene and thiazole, are much more bathochromic than their benzenoid counterparts. According to VB theory, the sulphur atom plays a decisive role by acting as an efficient electron sink. However, PPP calculations[18] suggest that the bathochromicity arises from the *cis*-diene structure and that the sulphur atom is not important. Thus, dyes of the type (125) should exhibit the same degree of bathochromicity, irrespective of the nature of *X*. A test of this prediction awaits an experimental study of such dyes.

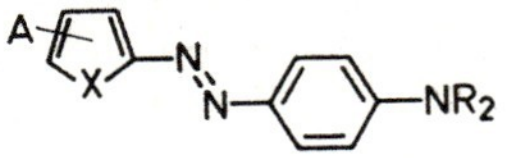

X = O, NH, S, CR_2

(125)

Tinctorial Strength

Being able to predict the tinctorial strength of a dye is arguably of more value than predicting its colour, since the former is directly related to the economic viability of a dye. Thus, suppose it requires 2 g of a yellow dye to colour a given weight of cloth to a desired yellow shade. If a new yellow dye is discovered which has double the tinctorial strength of the former dye and is of a similar molecular weight, then it will only require 1 g of the new dye to obtain the same result. Therefore, so long as the new dye is less than double the cost of the old dye, it is more cost effective.

It is much more difficult to predict correctly the tinctorial strength of a dye than it is to predict its colour. In qualitative VB theory there is no provision for

[18] Ref. E34.: p. 186.

predictions of tinctorial strength. However, an empirical 'rule of thumb' adopted by many practising dyestuffs chemists is that in a given series of dyes, tinctorial strength increases as the dyes become more bathochromic. Thus, blue azo dyes should be stronger than red azo dyes which in turn should be stronger than yellow azo dyes. Though this rule has some validity, there are many exceptions to it. For example, it fails in the basic case of azobenzene and its *o*-, *m*-, and *p*-amino derivatives (Table 3.15).

Table 3.15. Experimental and PPP calculated tinctorial strength of azobenzene and its *o*-, *m*-, and *p*-amino derivatives

	Compound	Exptl.		PPP Calc.[a]
		λ_{max}^{EtOH} (nm)	ε_{max}	'f'
↑	*o*- NH_2	414	6,500	0.67
	m-NH_2	412	1,300	0.17
Batho-	*p*- NH_2	387	24,500	1.31
chromicity	Azobenzene	320	21,000	1.21

[a] Gregory, P., Hutchings, M. G.: Unpublished work.

Of the methods discussed the PPP MO model is the only one which provides a quantitative figure for assessing the tinctorial strength of dyes: it calculates a quantity f called the oscillator strength. Eq. 3.1 shows that f is directly proportional to the square of the transition dipole moment M, but inversely proportional to the wavelength of the absorbed radiation, since ν_M, the mean absorption frequency of the band, is equal to $1/\lambda_M$. In other words, provided the transition dipole moment M remains constant, MO theory predicts that tinctorial strength should decrease as λ_{max} increases: this is in direct contrast to the empirical 'rule of thumb' approach mentioned earlier.

$$f = 4.703 \times 10^{29} \times M^2 \times \nu_M \qquad \text{Eq. 3.1}$$

The oscillator strength f is directly proportional to ε_{max}, as shown by Eq. 3.2.

$$f = 4.32 \times 10^{-9} \times \Delta\nu_{1/2} \times \varepsilon_{max} \qquad \text{Eq. 3.2}$$

where $\Delta\nu_{1/2}$ is the half-band width, *i.e.* the width of the absorption band, in cm^{-1}, at $\varepsilon = {}^1/_2\varepsilon_{max}$

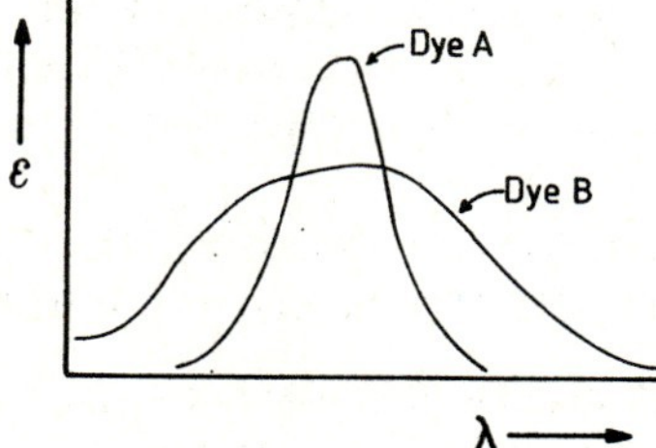

Fig. 3.8. Relationship between tinctorial strength and the area under the absorption curve

Indeed, unlike ε_{max}, f gives a true measure of tinctorial strength since it expresses the area under the absorption curve. Thus, dye *A*, with a high ε_{max} value but narrow absorption curve (*i.e.* low $\Delta\nu_{1/2}$), is tinctorially weaker than dye *B*, which, although it has a lower ε_{max} value, has a broader absorption curve (Fig. 3.8).

In a series of dyes of formula (126), it has been shown that the tinctorial strength, as expressed by the area under the curve, remains essentially constant[19] over a 100 nm range, from yellow (126; X = CF_3; λ_{max}^{EtOH} 467 nm) to violet (126; X = $O^{\ominus}$; λ_{max}^{EtOH} 567 nm). In MO terms, the transition dipole moment *M* must increase sufficiently with increasing wavelength of absorption to offset the decrease in f implicit in Eq. 3.1, *i.e.* $f \propto 1/\lambda$.

O_2N … NEt_2

(126)

In azo dyes, the direction of the transition dipole moment lies along the molecular axis from the donor group *D* to the acceptor group *A*. Hence, the optimum orientation of the molecule for maximum absorption of radiation is that shown in Fig. 3.9.

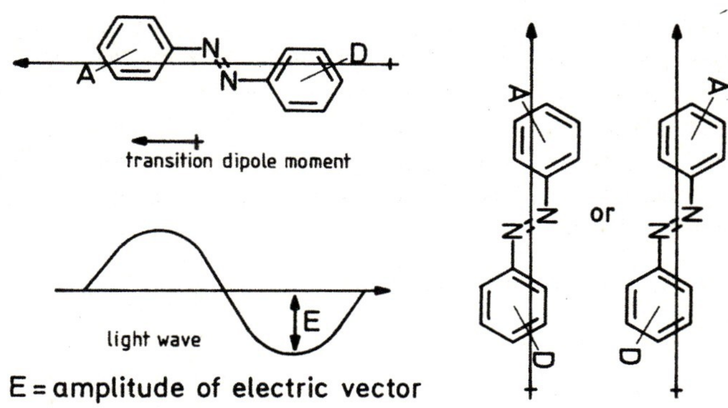

Fig. 3.9. Orientation of azo dyes for maximum absorption of radiation

[19] The dyes (126; *X* = OH and NHAc) had higher tinctorial strength, presumably because of the strong intramolecular hydrogen-bonding between the substituent and the azo group.

3.5.6 Protonated Azo Dyes

As already stated (Sect. 3.3.5), protonation of 4-aminoazobenzene dyes may occur either on the terminal amino group to give the ammonium tautomer (67), or at the β-nitrogen atom of the azo group to give the azonium tautomer (68). The ammonium tautomer, being devoid of the powerful auxochromic amino group, absorbs in the ultra-violet at similar wavelengths to azobenzene (~320 nm). Since the ammonium tautomer is essentially colourless we shall devote our attention to the colour-structure relationships of the azonium tautomer.

The Azonium Tautomer

The azonium tautomer (127) is very similar to the diazahemicyanine dyes (128) — see Chap. 5 — since it contains a delocalised positive charge (127)↔(127A).

(127) (127A)

(128) (128A)

Generally, the azonium tautomer is more bathochromic, tinctorially stronger, and brighter in hue than the neutral azo dye. As seen from Table 3.16, the ε_{max} can be up to double that of the latter. The brightness is a manifestation of the narrower absorption band, as expressed by the half-band width, $\Delta\nu_{1/2}$, — a typical value for an azonium tautomer is 3,500 cm^{-1} compared to ~5000 cm^{-1} for a neutral azo dye.

The effect of substituents on the colour of azonium tautomers is opposite to their effect on the neutral azo dyes. Electron-donating substituents in any position of ring *D* cause a bathochromic shift whereas electron-withdrawing substituents produce

Table 3.16. Spectral data for neutral and protonated azo dyes

X	Neutral Dye			Azonium Tautomer[a]			
	λ_{max}^{EtOH} (nm)	ε_{max}	$\Delta\nu_{1/2}$ (cm^{-1})	λ_{max} (nm)	ε_{max}	$\Delta\nu_{1/2}$ (cm^{-1})	$\Delta\lambda$[b]
H	486	34,000	5,000	515	43,500	3,100	+29
OMe	501	32,800	5,000	486	70,000	3,400	−15
$NHCONH_2$	518	41,000	4,300	500	47,300	3,700	−18
NH_2	514	45,000	4,200	482	63,800	3,500	−32

[a] In EtOH: conc. HCl 2:1 by volume.
[b] Relative to corresponding neutral dye.

a smaller hypsochromic shift. In ring *C*, electron-donating groups generally cause a hypsochromic shift and electron-withdrawing groups a bathochromic shift (Table 3.17). Thus, electron-donating substituents in ring *D* cause the colours of the neutral and protonated dyes to diverge: in contrast, electron-withdrawing groups bring their colours closer together. By choosing the appropriate electron-withdrawing substituent it is possible to obtain a dye in which both the neutral and protonated forms exhibit the same colour. Such a dye, *e.g.* (129), would possess the desirable feature of displaying no colour change with acid.

Table 3.17. Spectral effects in neutral and protonated azo dyes

Dye	Neutral Dye	Azonium tautomer	
	λ_{max}^{EtOH} (nm)	$\lambda_{max}^{EtOH/HCl}$ (nm)	$\Delta\lambda$[a]
C_6H_5–N=N–C_6H_4–NMe_2	408	516	
MeO–C_6H_4–N=N–C_6H_4–NMe_2	405, 440s	548	+32
O_2N–C_6H_4–N=N–C_6H_4–NMe_2	475	508	− 8
O_2N–C_6H_4–N=N–C_6H_3(MeO)–NEt_2	501	486	−30

s = shoulder
[a] Of azonium tautomers

O_2N–C_6H_3(F_3CO_2S)–N=N–C_6H_4–NMe_2

(129)

λ_{max}^{EtOH} 500 nm; ε_{max} 29,800

$\lambda_{max}^{EtOH/HCl}$ 500 nm; ε_{max} 56,300

Table 3.18. Negative halochromism by substituents in ring *D*

X–C_6H_4–N=N–C_6H_4–NR_2

R	X	Neutral Dye	Azonium form	$\Delta\lambda$
		λ_{max}^{EtOH} (nm)	$\lambda_{max}^{EtOH/HCl}$ (nm)	
Me	$-N^{\oplus}\equiv N$	596	506	−90
Et	$(NC)_2C=C(CN)-$	592[a]	586[b]	− 6

[a] Benzene
[b] Dry benzene + gaseous HCl

The phenomenon of negative halochromism, whereby protonation causes a hypsochromic shift, is difficult to achieve by substitution of ring *D*. Very powerful electron-withdrawing substituents are needed and only two have been reported — the diazonium group and the tricyanovinyl group (Table 3.18).

However, most dyes with electron-donating substituents *meta* to the terminal amino group in ring *C* display negative halochromism (Table 3.16). Azo dyes containing heterocyclic rings, such as (130) and (131), also display negative halochromism.

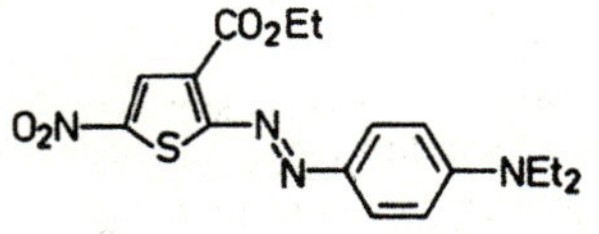

(130)
λ_{max}^{EtOH} 594 nm; 531 nm (EtOH/HCl)

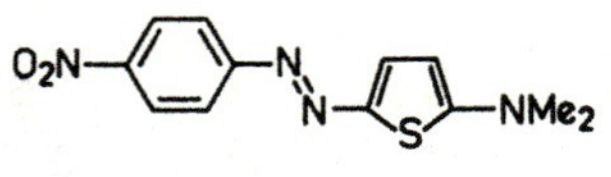

(131)
λ_{max}^{EtOH} 554 nm; 464 nm (EtOH/HCl)

How does the VB and MO theories explain these observed effects? In general VB theory is much less satisfactory in dealing with charged dyes than it is with uncharged (neutral) dyes. However, it can, to some extent, rationalise the observed effects. Thus, the bathochromicity of the azonium tautomer relative to the neutral dye is explained quite simply. Since both the ground state and the first excited state are positively charged, no energetically unfavourable charge separation occurs on excitation from the ground state to the first excited state — only a redistribution of charge. Consequently, the energy separation between the two states is less for the azonium tautomer than it was for the neutral dye. It follows that the ground and first excited state are structurally more similar than in the case of neutral azo dyes, which leads to a narrower absorption band and hence brighter dyes.

The effect of substituents on the colour of azonium tautomers strongly indicates an electron drift, on excitation, from the hydrazo nitrogen atom (β in 132) to ring *C*: this suggests that the quinonoid structure (132) makes the major contribution to the ground state and that the benzenoid structure (132A) makes the major contribution to the excited state! For example, the bathochromic effect of an electron-donating substituent in ring *D* of (132), especially when it is *ortho* or *para* to the azo group, is explained by its ability to stabilise the excited state (132A) by building up the electron density at the ring *D* carbon atom to which the positively charged β-nitrogen atom of the azo group is attached: in contrast, it destabilises the ground state (132), by tending to increase the electron density at the already electron rich β-hydrazo nitrogen atom. Obviously, the converse situation applies when *X* is an electron-withdrawing group. These effects are illustrated in Fig. 3.10.

The empirical 'rule of thumb' for tinctorial strength predictions fails again since the azonium tautomers, whether they are the same colour as the neutral dyes, *e.g.* (129), or more hypsochromic or bathochromic, are tinctorially stronger than the latter dyes.

(132) (132A)

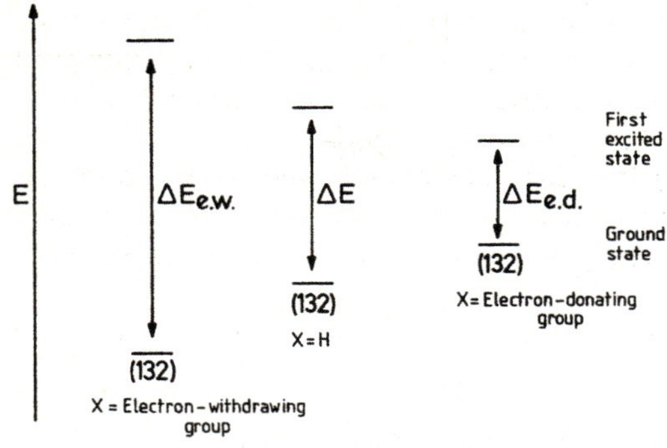

Fig. 3.10. Effect of electron-withdrawing and electron-donating groups on the colour of azonium tautomers

The PPP MO model calculates a higher f value for the azonium tautomer relative to the neutral azo dye: typical values are ~ 1.8 and ~ 1.3 respectively. Thus, it successfully accounts for the higher tinctorial strength of the azonium tautomer. It is also generally successful in calculating the effect of substituents, both in ring *C* and in ring *D*, on colour. In contrast to VB theory, the PPP model indicates an electron drift, on excitation, from the terminal amino nitrogen atom to the α-azo nitrogen atom (Fig. 3.11).

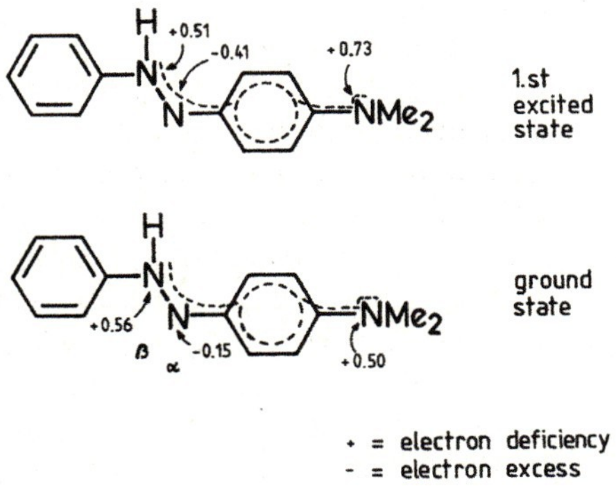

1.st excited state

ground state

+ = electron deficiency
- = electron excess

Fig. 3.11. PPP MO calculations on the azonium tautomer

It is interesting that in the ground state the calculated electron density at the terminal amino nitrogen atom and the β-azo nitrogen atom are approximately the same (+0.5); in VB terms this may be interpreted as a ground state consisting of the resonance hybrid structures (132) and (132A).

The calculations also suggest a more uniform bond order in the ground state of the azonium tautomer than is the case for the neutral azo dyes. In this respect, they are similar to cyanine dyes (see Chap. 5). Hence, Dewar's Rules are more applicable, although they still don't predict with complete accuracy (Fig. 3.12).

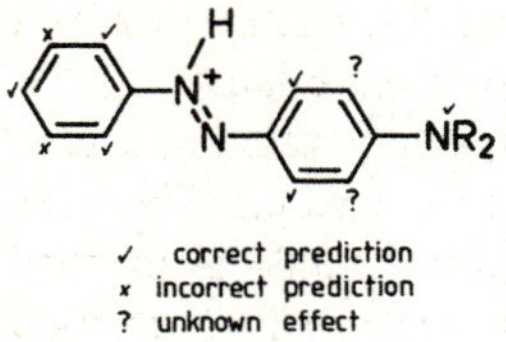

Fig. 3.12. Applicability of Dewar's Rules to azonium tautomers

3.5.7 Azo-Hydrazone Tautomers

As stated in Sect. 3.3.1, the hydrazone tautomer is generally more bathochromic than the azo tautomer. For the 4-phenylazo-1-naphthol system, the hydrazone tautomer (133) is orange (λ_{max} 478 nm) whereas the azo tautomer (134) is yellow (λ_{max} 407 nm).

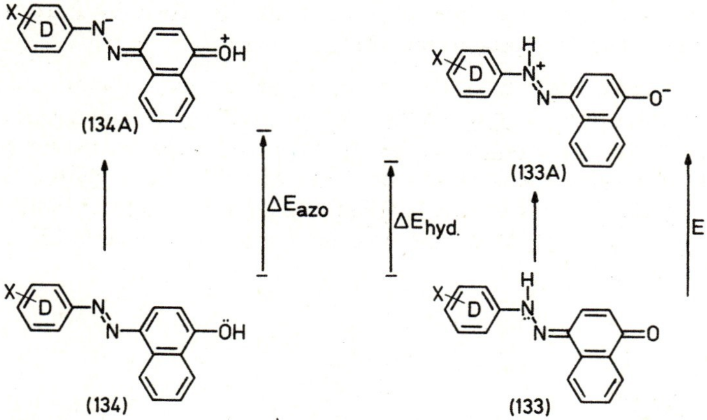

Substituents affect the colour of the azo and hydrazone tautomers in opposite ways. In the azo tautomer (134), electron-withdrawing groups in ring *D* cause a bathochromic shift whereas electron-donating groups cause a smaller hypsochromic shift. These substituent effects are reversed in the hydrazone tautomer (133). Thus, it is electron-donating groups in ring *D* which cause a bathochromic shift and electron-withdrawing groups which produce a hypsochromic shift (Table 3.19). Indeed,

Table 3.19. Visible absorption maxima of the azo (134) and hydrazone (133) tautomers

para X	λ_{max} (nm) Azo (134) in EtOH	$\Delta\lambda$[a]	λ_{max} (nm) Hydrazone (133) in HOAc	$\Delta\lambda$[a]
MeO	404	−3	485	+7
Me	405	−2	489	+11
H	407		478	
CN	424[b]	+17	462	−16
NO_2	432[b]	+25	465	−13

[a] Relative to X=H.

[b] Values quoted are for the methyl ether derivatives.

how substituents influence the colour of a hydroxyazo dye has been proposed as a means of ascertaining whether the azo or hydrazone tautomer predominates.

Electron-donating groups in ring *D* cause the colours of the azo and hydrazone tautomers to diverge; electron-accepting groups cause them to converge. Theoretically, with a powerful electron-accepting group, a 'cross-over' point should be reached where the azo tautomer becomes more bathochromic than the hydrazone tautomer. This has not been observed in the 4-phenylazo-1-naphthol system; even for the *p*-nitro derivative, the hydrazone tautomer is still 33 nm more bathochromic than the azo tautomer (Table 3.19).

Of more interest to dyestuffs chemists are the 2-phenylazo-1-naphthol (135), 1-phenylazo-2-naphthol (136), arylazopyrazolone (137) and arylazohydroxypyridone (138) systems. Since these dyes exist predominantly, if not exclusively, in the hydrazone form, then electron-donating groups *X* in ring *D* cause a bathochromic shift, particularly when they are *ortho* or *para* to the azo linkage, and electron-withdrawing groups cause a hypsochromic shift.

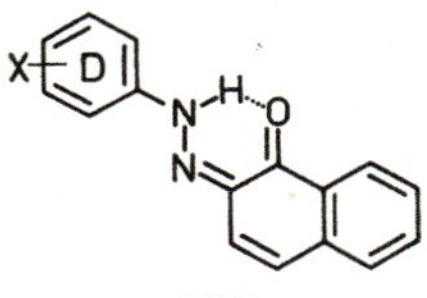

(135)

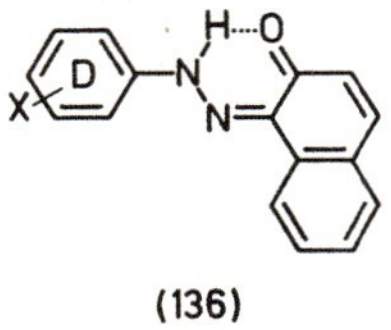

(136)

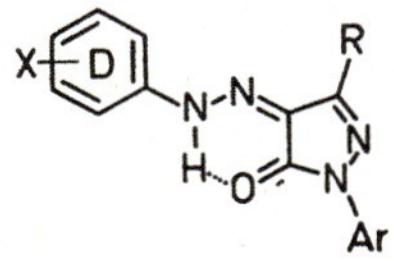

(137)

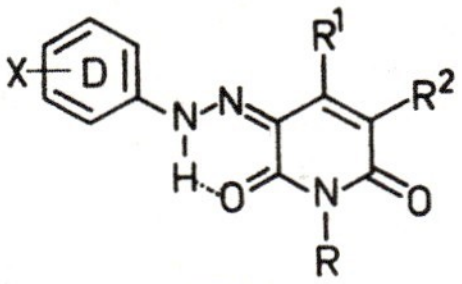

(138)

In VB terms the energy difference between the ground and first excited state of the azo tautomer is relatively large, since the former is a fully aromatic, neutral structure (134) whilst the latter is a charge separated structure (134A) in which aromaticity has been lost. In contrast, the first excited state of the hydrazone tautomer (133A) is more aromatic than the ground state (133) and this partly offsets the loss in stability associated with charge separation. Hence ΔE is smaller and the hydrazone tautomer is predicted to be more bathochromic than the azo tautomer.

In the azo tautomer the hydroxy group is the electron donor and the phenylazo residue is the electron acceptor. Thus, electron-withdrawing groups in ring *D* stabilise the excited state (134A) by delocalising further the negative charge, as depicted by (134B) for the case of a *p*-nitro substituent. In contrast, electron-donating groups

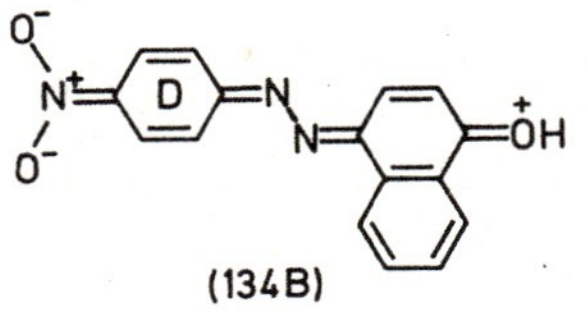

(134B)

destabilise the excited state by tending to increase the electron density at an already electron rich site.

According to VB theory, electron migration in the hydrazone tautomer (133) is from the hydrazo nitrogen atom to the carbonyl group. Consequently, electron-donating groups in ring *D* stabilise the excited state (133A) by tending to reduçe the magnitude of the positive charge on the nitrogen atom. In contrast, electron-withdrawing groups destabilise (133A) by tending to increase the magnitude of the positive charge on the hydrazo nitrogen atom.

The greater tinctorial strength of the hydrazone tautomer over the azo tautomer is again rationalised by its greater bathochromicity.

The electronic effects calculated by the PPP MO model are different to those described by VB theory, especially regarding the acceptor group. For the azo tautomer (139), the PPP model calculates the hydroxy group to be the major electron-donor on excitation, in accord with VB theory, but the major electron-acceptor is now calculated to be the α-, not the β-nitrogen atom, of the azo group.

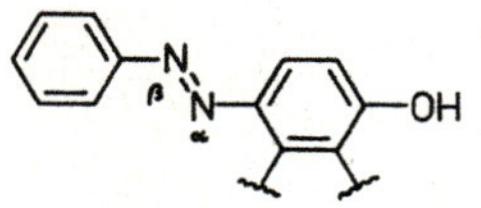

(139)

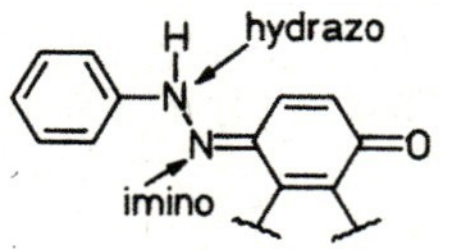

(140)

In the hydrazone tautomer (140) the PPP model calculates that the major donor is the hydrazo nitrogen atom, but that the major acceptor is now the adjacent imino nitrogen atom, not the carbonyl group. Yet again, the ground state calculated by the PPP model for both tautomers corresponds to the VB representation of the excited state. Nonetheless, both theories rationalise the observed effects very well.

The PPP model also calculates the correct order for the tinctorial strength of the azo and hydrazone tautomers. Thus, typical f values for the azo tautomers of both the azophenol and azonaphthol series are 1.2–1.4: these are lower than for the corresponding hydrazone tautomers, $f = 1.4$–1.8.

3.5.8 Polyazo Dyes

Though there are various sub-classes of polyazo dyes, for colour and constitution purposes there are only two types: those in which the azo groups are conjugated to each other and those in which they aren't.

In the latter case the spectra of the polyazo dye approximates to the sum of the spectra of the individual monoazo dyes. Dyes of this type are formed when the delocalised π-electron systems of the individual chromogens are chemically insulated from each other by two or more single bonds. This is normally accomplished by inserting 'insulating' groups between the chromogens (Table 3.20). Obviously, the more effective the insulating group, the closer the spectrum of the polyazo dye to that of the sum of the individual dyes.

Table 3.20. Important insulating groups

—CH_2—	methylene
—HNCONH—	ureido
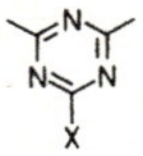	triazinyl
—S—	—
—O—	—
—NH—	—
—CO—	—

The dyes joined together may be the same or different. In the former case, the polyazo dye has a similar colour to the monoazo dye but is n times stronger,[20] where n is the number of azo groups. For example, the disazo dye (141) is double the tinctorial strength of the monoazo dye, (142), but both are yellow (Fig. 3.13).

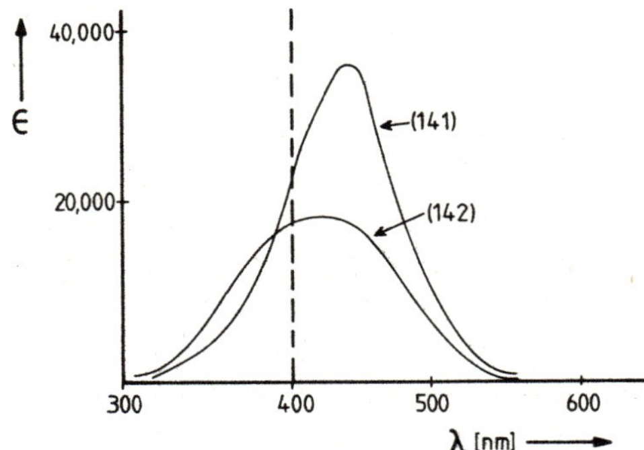

Fig. 3.13. The absorption spectra of an insulated disazo dye and the parent monoazo dye

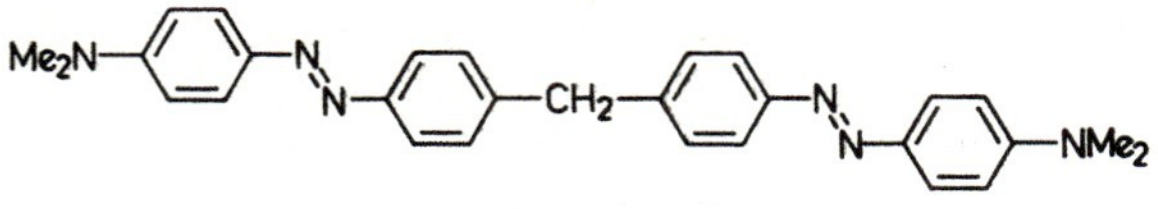

(141)

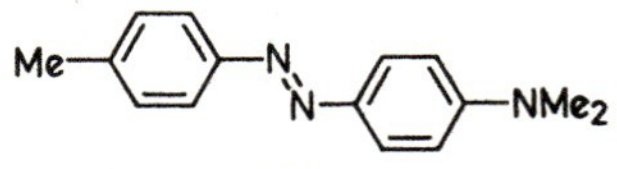

(142)

The technique of linking two different dyes has been used to produce tertiary shades such as brown and green. Thus, brown dyes (143) are obtained by linking a yellow dye with a rubine or purple dye, and green dyes (144) by joining yellow and blue dyes. Such dyes offer no economic advantage over a mixture of the individual dyes.

[20] On a molecular basis: on a weight basis, which is what matters to the dye user, this is not so (see Sect. 3.5.5).

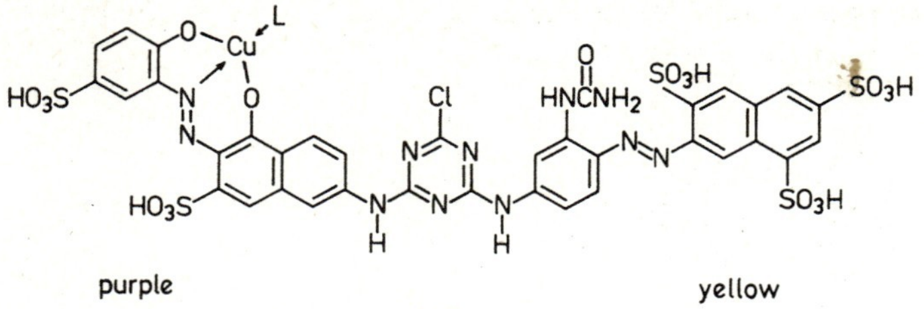

purple yellow

(143)

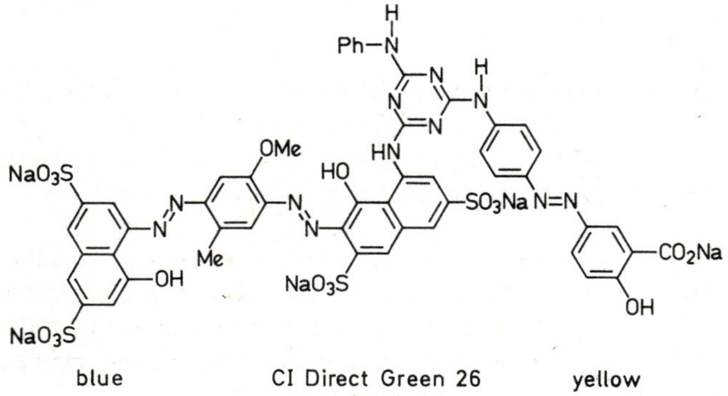

blue CI Direct Green 26 yellow

(144)

Fully conjugated polyazo dyes may be sub-divided into those in which all the azo groups exist as such and those where there is a mixture of azo and hydrazone groupings. In both cases the increased conjugation results in a bathochromic shift although the magnitude of the shift diminishes rapidly with each additional arylazo group — four azo groups is normally the maximum that can be achieved synthetically. The increased conjugation usually causes a broadening of the absorption curve which manifests itself as dullness.

Typical disazo dyes which exist in the azo form are the A→M→E dyes (145). Increasing the electron-donating strength of the terminal group *D*, incorporation of electron-donating groups in *M* or *E*, and changing *M* or *E* from phenyl to naphthyl all cause a bathochromic shift. However, the effect of changes in rings more remote from the one carrying the principal donor group *D*, such as ring *A*, have less effect.

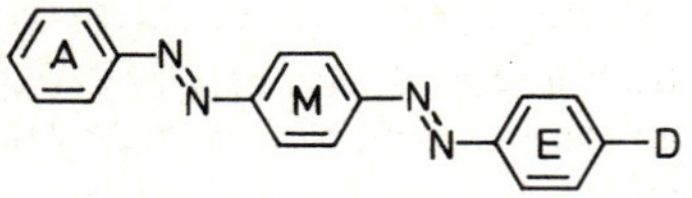

(145)

According to VB theory the bathochromic shift arises because the excited state is stabilised. For example, the naphthyl dyes (146) are more bathochromic than the phenyl dyes because the diiminonaphthoquinone structure (146A) is much more stable than the corresponding benzoquinone structure. This also explains the relative insensitivity of ring *A* on colour since it is not as essential for conjugation to proceed through this ring as it is for the *M* and *E* rings. PPP calculations are also in agreement with this explanation. Thus, electron migration on excitation is calculated to occur from the donor group D to the azo group nearest to the donor group and, to a lesser extent, to the more remote azo group.

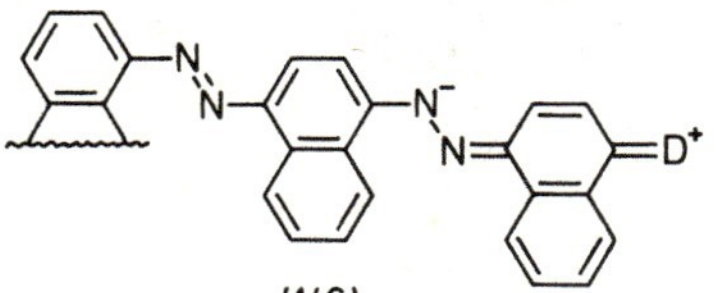

(146)

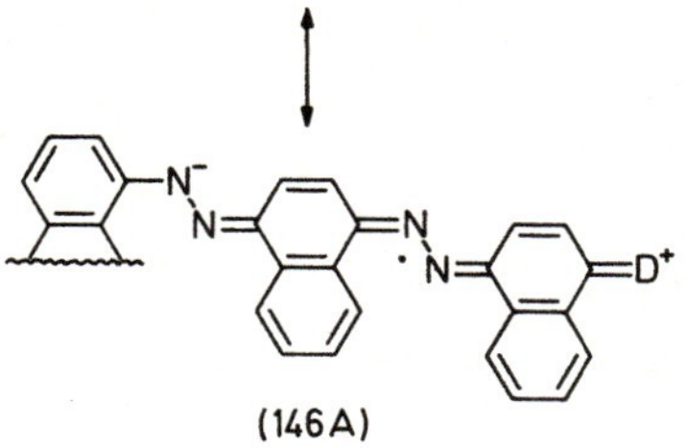

(146A)

Of more interest theoretically are polyazo, particularly disazo, dyes in which one azo group exists in the hydrazone form. An important class is the twice-coupled H-acid dyes (147). Such dyes are dull blues, greens or blacks and are interesting in that their absorption spectra are composed of three peaks in the visible region of the spectrum (Fig. 3.14) — most azo dyes exhibit only one peak.

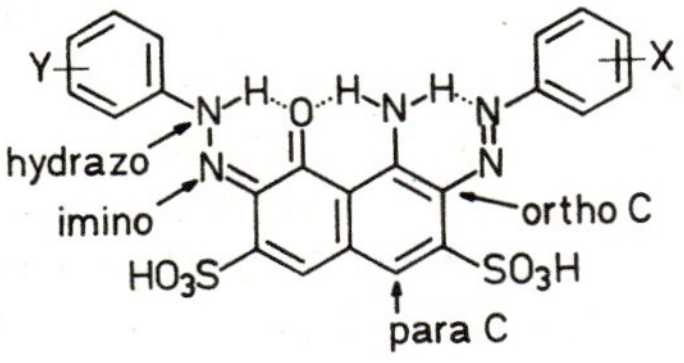

(147)

PPP MO calculations on these dyes[21] calculate three peaks, in excellent agreement with experiment, as shown in Fig. 3.14. From a study of electron migration in the first, second and third excited states, together with substituent effects, it appears that the major blue peak (at λ_{max} *ca.* 550–600 nm) is caused by electron donation from the amino group and the *ortho* and *para* carbon atoms, to the azo and carbonyl groups and also to the imino nitrogen atom. The less intense red (λ_{max} *ca.* 500 nm) and yellow peaks (λ_{max} *ca.* 400 nm) arise from transitions from the hydrazo nitrogen atom and the amino nitrogen atom to the carbonyl group, and from the amino group to the carbonyl group and the imino nitrogen atom respectively.

Further important dyes of this mixed azo-hydrazone type are of formula (148). In general, electron-donating groups X, and particularly Y, in the M-component cause a bathochromic shift. Both electron-donating and electron-withdrawing groups in the A ring cause a bathochromic shift, the former because they enhance the

[21] The calculations were done on the tautomeric form depicted by structure (147).

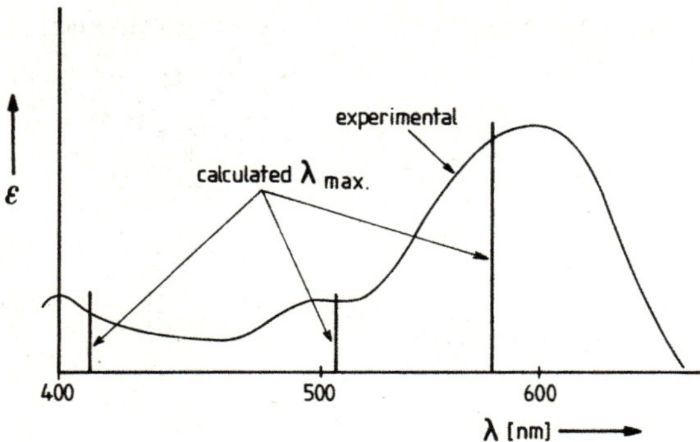

Fig. 3.14. Experimental and calculated ultra-violet/visible spectrum for (147; X = Y = NHAc) in water

electron drift from the hydrazo nitrogen atom to the imino and carbonyl groups, and the latter because they set up a further donor-acceptor chromophoric system (149).

(148) (149)

E = a naphthol, pyrazolone or hydroxypyridone nucleus

3.5.9 Steric Effects

Steric hindrance in dyestuffs molecules can have a profound effect upon the shade and strength of a dye. Thus, it can cause either a hypsochromic or a bathochromic shift of λ_{max}, but it *always lowers* the tinctorial strength. Steric effects have been studied most for monoazo dyes, particularly 4-aminoazobenzene derivatives, and this important class will therefore be used to illustrate the phenomenon, although the principles are applicable to any azo dye.

In azo dyes, three types of steric hindrance are possible: between the azo group and *ortho* substituents; between an auxochrome, *e.g.* a dialkylamino group, and a substituent; and between a chromophore, *e.g.* a nitro group, and a substituent. Steric hindrance about the azo group, especially between that group and ring *D*, is frequently encountered in commercial azo dyes and this type will be discussed first.

Steric Hindrance at the Azo Group

Two planar conformations are possible for an *ortho*-substituted *trans*-azobenzene (Fig. 3.15). Steric hindrance is greatest between the substituent and the lone pair orbital of the β-nitrogen atom, as shown in conformation (a). Such crowding can be relieved by rotation into conformation (b), so that a single substituent will not normally exert much of a steric effect. However, when two *ortho*-substituents are present in the same ring, both conformations are hindered and non-planarity is the likely outcome. Thus, introducing one *ortho* chlorine atom into dye (150; X=Y=H,

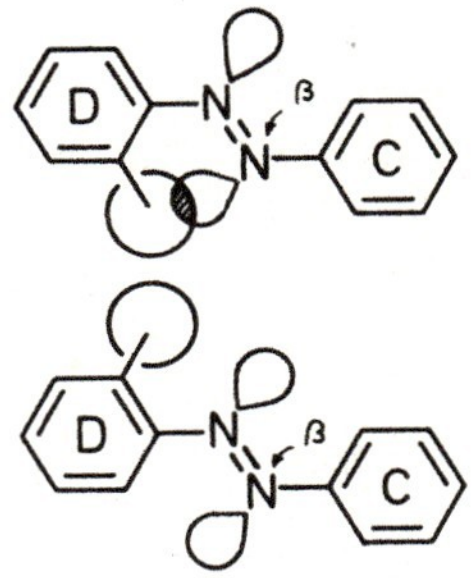

(a)

(b)

Fig. 3.15. Alternative conformations for an *ortho*-substituted azobenzene

$Z=NO_2$) produces a bathochromic shift and only a small reduction in strength because the electronic effect of the substituent predominates. However, a second *ortho* chlorine atom produces a large hypsochromic shift and a notable reduction in strength (Table 3.21). In view of the reduced tinctorial strength, it is surprising that dyes of this type are used widely commercially for providing orange and yellow-brown shades.

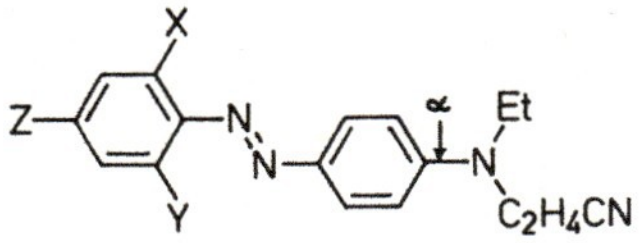

(150)

Table 3.21. Absorption bands of some derivatives of dye (150) in methanol

X	Y	Z	λ_{max} (nm)	ε_{max}	$\Delta\lambda$[a]
H	H	NO_2	453	44,000	
Cl	H	NO_2	475	40,000	+22
Cl	Cl	NO_2	417	31,000	−36
NO_2	H	H	425	36,000	−28
NO_2	H	NO_2	491	38,000	+38
NO_2	NO_2	NO_2	520	48,000	+67
CN	H	H	434	42,000	—
H	H	CN	433	45,000	—
CN	H	NO_2	504	45,000	+51
CN	CN	NO_2	549	38,000	+96

[a] Relative to (150: $X=Y=H$, $Z=NO_2$).

Groups such as nitro exert their optimum electronic effect when they are coplanar with the dye molecule. For a nitro group *ortho* to the azo group a coplanar conformation is not possible, even in the favoured conformation (b), because of steric hindrance between the nitro group and the lone pair orbital on the α-nitrogen atom of the azo group. These steric interactions are minimised by the nitro group rotating out of the molecular plane: as a result, the *o*-nitro dye (150; $X=NO_2$, $Y=Z=H$) is 28 nm less bathochromic than the *p*-nitro isomer. Furthermore, the incorporation of *o*-nitro groups into the parent *p*-nitro dye (150; X=Y=H, Z=NO_2) results in progressively smaller bathochromic shifts of the first band (Table 3.21). In contrast, cyano groups, owing to their rod-like shape, do not have an obvious steric effect and the absorption characteristics of the *ortho* and *para* cyano dyes (150; X=CN or H, Z=H or CN, Y=H) are virtually identical (Table 3.21). Consequently, one of the most bathochromic benzenoid diazo components is 2,6-dicyano-4-nitroaniline.

When 5-membered heterocyclic systems such as thiophenes are used as diazo components, steric hindrance between the *ortho* substituent *X* and the lone pair orbital of the α-azo nitrogen atom is diminished because of the larger bond angles involved (126° v 120°). Thus, groups such as nitro can adopt a more coplanar conformation and their mesomeric effect is increased. Consequently, the dinitrothiophene dye (151; X = NO_2, λ_{max}^{EtOH} 618 nm) absorbs at longer wavelength than the cyanonitro dye (151; X = CN, λ_{max}^{EtOH} 595 nm).

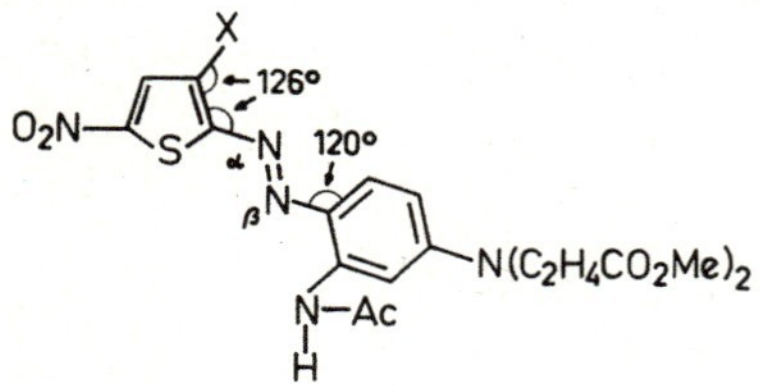

(151)

Steric crowding at the azo group by substituents in ring *C* is encountered less frequently in dyestuffs chemistry. Nonetheless, it has been studied and shown to be significant. Thus, whereas the introduction of one *ortho* methyl group into dye (152; R^1=R^2=H) caused a bathochromic response of 15 nm and a slight increase in intensity, a second methyl group caused a hypsochromic shift of 12 nm and a reduction in intensity (Table 3.22).

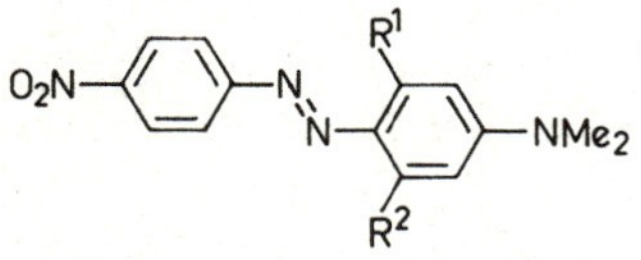

(152)

Table 3.22. Absorption bands of some *ortho*-substituted azo dyes

R^1	R^2	λ_{max}^{EtOH} (nm)	ε_{max}	$\Delta\lambda$ [a]
H	H	480	30,400	
Me	H	495	31,000	+15
Me	Me	468	26,800	−12

[a] Relative to R^1=R^2=H

Steric Hindrance at the Terminal Dialkylamino Group

Methyl groups have been used to hinder terminal dimethylamino groups. For example, a hypsochromic shift of 33 nm together with a fall in intensity is brought about by the introduction of a 3-methyl group into 4-NN-dimethylaminoazobenzene (153; X=Me, Y=Z=H); a second *ortho* methyl group (153; X=Y=Me, Z=H) has little effect on the colour but reduces further the intensity (Table 3.23). The shifts are increased to 57 nm in the corresponding nitro derivatives (153; X=Me, Y=H or Me, Z=NO_2).

Table 3.23 Absorption bands of some dyes with a sterically hindered terminal group

X	Y	Z	λ_{max}^{EtOH} (nm)	ε_{max}	$\Delta\lambda$
H	H	H	408	28,250	
Me	H	H	375	18,200	—33[a]
Me	Me	H	380	10,300	—28[a]
Cl	H	H	377	19,900	—31[a]
Me	H	NO_2	422	19,500	—57[b]
Me	Me	NO_2	423	11,200	—56[b]

[a] Relative to (153; X=Y=Z=H)
[b] Relative to (153; X=Y=H, Z=NO_2)

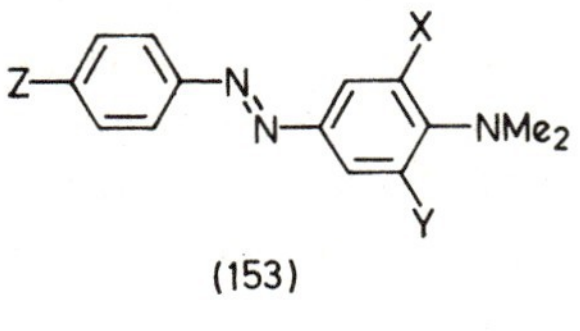

(153)

Similar but increased responses are observed when the size of the crowding group is increased. Thus, the introduction of a 3-methyl group into 4-NN-diethylamino-4′-nitroazobenzene has a greater effect than in the corresponding dimethylamino compound, changing the spectral parameters from λ_{max} 486 nm (ε_{max} 34,000) to λ_{max} 420 nm (ε_{max} 21,300). A *t*-butyl group next to a terminal dimethylamino substituent effectively prevents conjugation. Even a methoxy group *ortho* to a terminal diethylamino group has a pronounced steric effect. In the dye (154; X=OMe), the methoxy group has a negligible effect on λ_{max} suggesting that its electronic and steric effects cancel each other out, but the reduction in ε_{max} from 34,000 for (154; X=H) to 22,600 is almost as great as that caused by a methyl group. As seen earlier (Sect. 3.5.4), several commercial blue dyes contain a methoxy group *ortho* to the terminal amino group since in combination with a *m*-acylamino substituent, a bathochromic shift is obtained which is considerably greater than the sum of those produced by the individual groups. However, this bathochromicity is obtained at the expense of tinctorial strength.

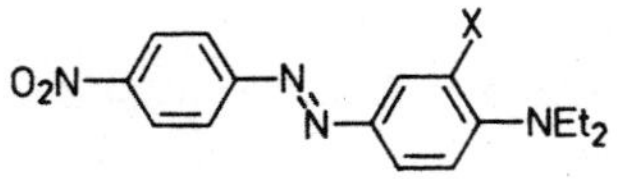

(154)

The bathochromic electronic effect of an electron-donating substituent *ortho* to the terminal dialkylamino group predominates if it forms part of a 5- or 6-membered ring with the amino nitrogen atom. This terminal bridging forces the molecule into a more planar conformation and allows greater overlap of the nitrogen lone pair electrons with the π-system. The optimum effect is observed with derivatives of 9-phenylazojulolidine. For example, the *p*-nitro derivative (155) absorbs at 521 nm (ε_{max} 38,000) in ethanol, compared to λ_{max} 486 nm (ε_{max} 34,000) for the diethylamino dye.

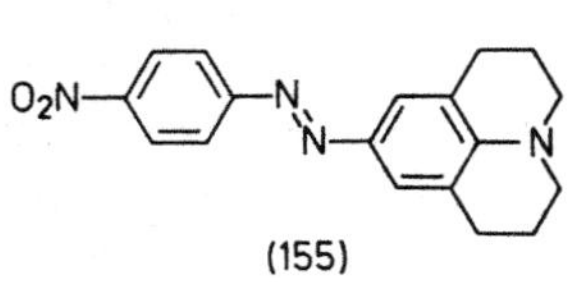

(155)

Similar, but less pronounced behaviour, is shown by various tetrahydroquinoline derivatives. Thus, the dye (156) absorbs at 496 nm (ε_{max} 36,300).

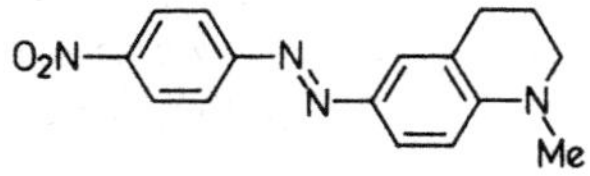

(156)

Steric Hindrance at a Chromophore, e.g. a Nitro Group

There is a paucity of data in this area — the only study appears to have been the effect of an *o*-methyl group on a terminal nitro group. Although the expected hypsochromic response was observed, the drop in intensity was small, even for two *o*-methyl groups (157; X=Y=Me) — Table 3.24. This suggests that the hypsochromic shift may be caused primarily by electronic rather than steric effects; indeed, space-filling molecular models indicate that there is little steric interaction between the methyl group and the nitro group.

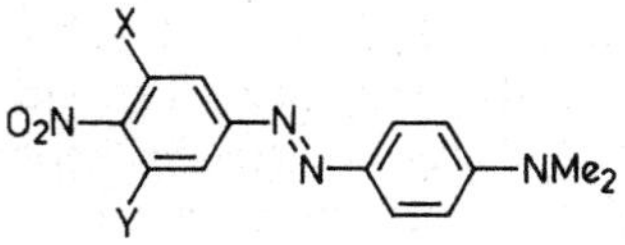

(157)

Table 3.24. Absorption bands of dye (157)

X	Y	λ_{max} (nm)	ε_{max}	$\Delta\lambda$[a]
H	H	479	31,300	
Me	H	460	32,000	−19
Me	Me	443	31,000	−36

[a] Relative to (157; X=Y=H)

Theoretical Explanation of Steric Effects

Can steric effects be explained theoretically and if so, can they be predicted? The answer to both these questions is yes. Since steric crowding is most often relieved by bond rotation to give a non-planar molecule — bond stretching requires far more energy — the orbital overlap between the group and the π-electron system is diminished. This lowers the transition dipole moment and consequently the tinctorial strength is reduced — see Sect. 3.5.5.

To predict the effect of steric hindrance on the colour of a dye requires a knowledge of the relative bond orders for its ground and first excited state: these are obtained from PPP calculations. It is generally true that a bond which is of low π-bond order in the ground state of a molecule has a high order in the first excited state, and *vice-versa*. Twisting about a single bond[22] requires far less energy than does twisting about a double bond.[22] Therefore, if a substituent causes twisting about a single bond in the ground state and about a double bond in the first excited state, then the latter will be raised in energy considerably more than the former: ΔE is therefore increased resulting in a hypsochromic shift (Fig. 3.16a). For the reverse case, the energy of the ground state is raised more than that of the first excited state: ΔE is reduced, resulting in a bathochromic shift (Fig. 3.16b).

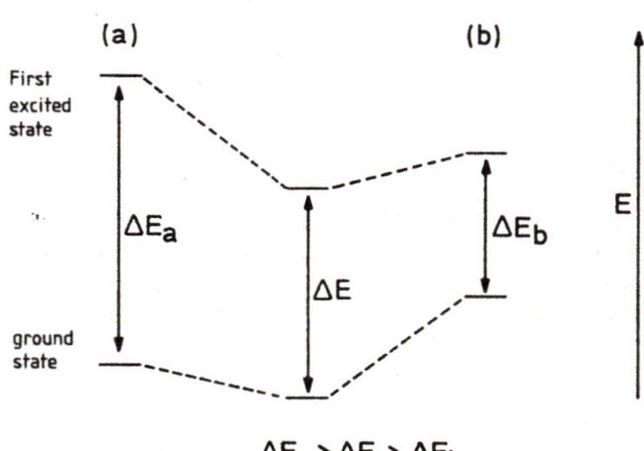

Fig. 3.16. Effect of steric hindrance on the colour of an azo dye

The greater hypsochromic shift of the nitro dye (150; X = Y = H, Z = NO_2) relative to (150; X=Y=Z=H) — Table 3.21 — is also explained. The electron-withdrawing nitro group, by increasing the degree of charge transfer in the excited state, raises the π-bond order of the terminal bond α to a greater extent in the first

[22] These are approximate bond orders.

excited state than in the ground state; this is confirmed by PPP calculations. Consequently, steric crowding will raise the energy of the excited state of the nitro compound more than that of the unsubstituted dye, leading to an enhanced hypsochromic shift.

Steric Hindrance in Azonium Tautomers and Related Compounds

In the azonium tautomers (132) the π-bond orders in both the ground and first excited state are roughly the same (*ca.* 0.6) — see Sect. 3.5.6. Hence, the steric effect of substituents should alter the energy of both states by a similar amount and the λ_{max} should remain essentially the same. However, due to the diminished overlap of the *p*-orbitals, the tinctorial strength should be lower. Unfortunately, such substituents (*e.g.* methyl) *ortho* to the dialkylamino group markedly alter the tautomeric equilibrium in favour of the colourless ammonium tautomer (67) so that it is difficult to test these predictions on protonated azo dyes. However, they have been tested in the analogous diazahemicyanine dyes and found to work! Thus, the introduction of an *o*-methyl group into the blue dye (158; R = H) has a negligible effect on λ_{max} but causes a *ca.* 50% reduction in strength.

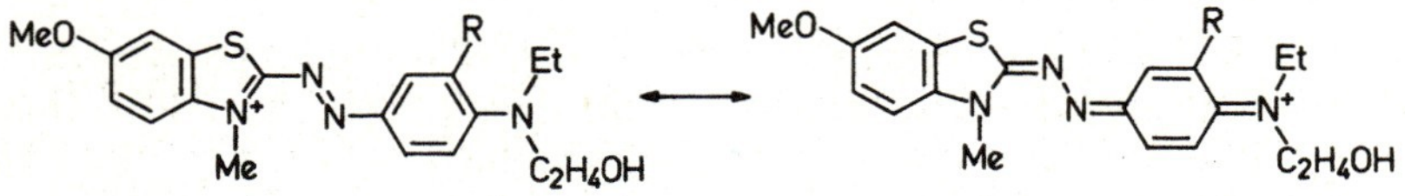

MeO, S, N, N⁺, Me, N, R, Et, N, C_2H_4OH ⟷ MeO, S, N, N, Me, N, R, Et, N⁺, C_2H_4OH

(158) (158A)

3.6 Summary

Azo dyes are the most important class of dye, accounting for over 50% of total world production. Those containing hydroxy groups *ortho* or *para* to the azo linkage are capable of undergoing azo-hydrazone tautomerism. Which form predominates depends on several factors but primarily on the relative thermodynamic stabilities of the two tautomers. Azophenol dyes generally exist in the azo form: in contrast, azoanthrol dyes exist in the hydrazone form. The commercially important 2-phenylazo-1-naphthol and 1-phenylazo-2-naphthol dye systems exist predominantly in the hydrazone form, as do the azopyrazolone and azohydroxypyridone dyes, but for 4-phenylazo-1-naphthol dyes, the tautomeric equilibrium is sensitive to both solvent and substituent effects. In general, the hydrazone form is more bathochromic and tinctorially stronger than the azo form. Aminoazo dyes exist in the azo form.

Addition of acid to azo dyes can result in protonation either at the azo group or at an auxochromic group. In derivatives of 4-aminoazobenzene, protonation at the terminal amino group gives the colourless ammonium tautomer which is more hypsochromic and tinctorially weaker than the neutral dye: in contrast, the azonium tautomer is tinctorially stronger, and also brighter, than the neutral dye and is generally more bathochromic.

Azo dyes having hydroxy, carboxy or amino groups *ortho* to the azo group form complexes with metal ions. The metal alters the properties of the dye, *e.g.* its colour and its resistance to various agents, such as light. Only those metal complex dyes having exceptional stability are suitable as commercial dyes. Generally, they are the copper II, chromium III and cobalt III complexes of *o,o'*-dihydroxy- and *o*-hydroxy-*o'*-carboxyazo dyes.

The colour and strength of a dye in relation to its constitution is a complex subject which depends primarily on the π-electronic make-up of the molecule, although steric effects can also be important. Various theories and rules have been devised for explaining and predicting colour-structure relationships. Of these, the quantitative PPP model currently provides the most comprehensive understanding of colour-structure relationships. It is the only method which provides quantitative values for both λ_{max} and ε_{max} (via f). The latter value is particularly important since it has a direct bearing on the economics of a dye. The limitations of the PPP model are that effects due to σ-electrons, such as hydrogen-bonding and steric hindrance, are excluded, as are solvent effects (the results apply to isolated molecules in the gas phase). In contrast, qualitative MO theory, in the form of Dewar's Rules, is totally unreliable for predicting the effect of substituents on the colour of neutral azo dyes. However, Dewar's Rules are more successful when applied to protonated azo dyes. VB theory, though less successful in the simpler dyes, is useful for explaining and even predicting the colour of more complex neutral dyes.

Thus, where quantitative results, especially of tinctorial strength, and a thorough understanding of colour-structure relationships are required, PPP MO theory should be used. For qualitative colour predictions VB theory is adequate for neutral azo dyes whilst Dewar's Rules work well for protonated azo dyes. However, the experienced dyestuffs chemist still has a role to play, especially for predictions in an established area.

3.7 Bibliography

A. *Basic Structure of Azo Dyes*

1. Allmann, R.: Structural chemistry. In: The chemistry of the hydrazo, azo and azoxy groups — Part 1, Patai, S. (ed.), pp. 44–45. London, New York, Sydney, Toronto: John Wiley and Sons 1975
2. Zollinger, H.: Azo and diazo chemistry, pp. 59 and 293. New York, London: Interscience 1961
3. Fabian, J., Hartmann, H.: Light absorption of organic colorants: Theoretical treatment and empirical rules, pp. 42–50. Berlin, Heidelberg, New York: Springer-Verlag 1980

B. *Tautomerism*

(a) *Comprehensive Reviews*

1. Bershtein, I. Ya., Ginzburg, O. F.: Russian Chem. Rev. *41*, 97 (1972)
2. Ref. A2.: pp. 322–337
3. Venkataraman, K. (ed.): The chemistry of synthetic dyes, Vol. I, pp. 441–447. New York: Academic Press 1952
4. Cox, R. A., Buncel, E.: Azo-hydrazone tautomerism of aromatic azo compounds. In: The chemistry of the hydrazo, azo and azoxy groups — Part 2, Patai, S. (ed.), pp. 838–844. London, New York, Sydney, Toronto: John Wiley and Sons 1975

(b) *Azo-Hydrazone Tautomerism*

5. Zincke, T., Bindewald, H.: Ber. *17*, 3026 (1884)
6. Kuhn, R., Bar, F.: Annalen *516*, 143 (1935)
7. Ospensen, J. N.: Acta Chem. Scand. *5*, 491 (1951)
8. Burawoy, A., Salem, A. G., Thompson, A. R.: J. Chem. Soc. *1952*, 4793
9. Burawoy, A., Thompson, A. R.: J. Chem. Soc. *1953*, 1443
10. Reeves, R. L., Kaiser, R. S.: J. Org. Chem. *35*, 3670 (1970)
11. Juvvik, P., Sundby, B.: Acta Chem. Scand. *27*, 3632 (1973)
12. Jacques, P., et al.: Tetrahedron *35*, 2071 (1979)
13. Kishimoto, S., et al.: J. Org. Chem. *43*, 3882 (1978)
14. Parent, R. A.: J. Soc. Dyers and Colourists *92*, 368 (1976)

(c) *Azonium-ammonium tautomerism*

See protonated azo dyes.

C. *Protonated Azo Dyes*

These are collected together under this heading in the 'Colour and Constitution' Section, E.

D. *Metal Complex Azo Dyes*

(a) *Comprehensive Accounts*

1. Price, R.: The chemistry of metal complex dyestuffs. In: The chemistry of synthetic dyes, Venkataraman, K. (ed.), Vol. III, pp. 303–383. New York, London: Academic Press 1970
2. Ref. A2.: pp. 338–360
3. Drew, H. D. K., Landquist, J. K.: J. Chem. Soc. *1938*, 292
4. Allen, R. L. M.: Colour chemistry, pp. 46–52. London: Nelson 1971
5. Abrahart, E. N.: Dyes and their intermediates, 2nd ed., pp. 108–116. London: Edward Arnold 1977

(b) *Stability/Structure*

6. Snavely, F. A., Fernelius, W. C., Douglas, B. E.: J. Soc. Dyers and Colourists *73*, 491 (1957)
7. Lassettre, E. N.: Chem. Rev. *20*, 267 (1937)
8. Wulf, O. R., Liddel, U.: J. Amer. Chem. Soc. *57*, 1464 (1935)
9. Ooi, S., Fernando, Q., Carter, D.: Chem. Commun. *1967*, 1301
10. Jarvis, J. A. J.: Acta Crysta. *14*, 961 (1961)
11. Snavely, F. A., Fernelius, W. C.: Science *117*, 15 (1952)

E. *Colour and Constitution*

(a) *Early Theories*

(i) *Reviews/Books*

1. Godlove, I. H.: Textile Res. J. *17*, 185 (1947)
2. Curtiss, R. S.: J. Amer. Chem. Soc. *32*, 795 (1910)
3. Watson, E. R.: Colour in relation to chemical constitution. London: Longmans 1918
4. Kauffmann, H., et al.: Z. Farben — Ind. *5*, 417 (1906); Die Auxochrome, Ahrens' Sammlung *12*, 1 (1903)

(ii) *Original Papers*

5. Graebe, C., Liebermann, C.: Ber. *1*, 106 (1868)
6. Witt, O. N.: Ber. *9*, 522 (1876); *21*, 321 (1888)

7. Nietzki, R.: Verhandl, des Vereins zum Beforderung des Gewerbfleisses *58*, 231 (1879)
8. Armstrong, H. E.: Philos. Mag. *23*, 73 (1887), Proc. Chem. Soc. *1888*, 27
9. Gomberg, M.: J. Amer. Chem. Soc. *22*, 757 (1900)
10. Baly, E. C. C.: J. Chem. Soc. *1904*, 1029; *1905*, 766; *1906*, 489; *1907*, 426, 1572
11. Hantzsch, A.: A series of papers published from 1906–16, especially in Chemische Berichte
12. Hewitt, J. T., Mitchell, H. V.: J. Chem. Soc. *1907*, 1251
13. Baeyer, A.: Annalen *354*, 152 (1907)
14. Adams, E. Q., Rosenstein, L.: J. Amer. Chem. Soc. *36*, 1452 (1914)
15. Watson, E. R.: J. Chem. Soc. *1914*, 759
16. Lewis, G. N.: J. Amer. Chem. Soc. *38*, 762 (1916)
17. Dilthey, W., Wizinger, R.: J. Prakt. Chem. *118*, 321 (1928)
18. Bury, C. R.: J. Amer. Chem. Soc. *57*, 2116 (1935)

(b) *Modern Theories (VB and MO)*

19. Hückel, E.: Z. Physik. *70*, 204 (1931), ibid *72*, 310 (1931)
20. Kuhn, H.: J. Chem. Phys. *17*, 1198 (1949)
21. Dewar, M. J. S.: J. Chem. Soc. *1950*, 2329
22. Pariser, R., Parr, R. G.: J. Chem. Phys. *21*, 466 (1953)
23. Pople, J. A.: Trans. Faraday Soc. *49*, 1375 (1953); J. Phys. Chem. *61*, 6 (1957)
24. Roothaan, C. C. J.: Rev. Mod. Phys. *23*, 69 (1951)

(c) *General Accounts (books/reviews)*

25. Lewis, G. N., Calvin, M.: Chem. Rev. *25*, 273 (1939)
26. Shephard, S. E.: Rev. Mod. Phys. *14*, 303 (1942)
27. Maccoll, A.: Quart. Rev. (London) *1*, 16 (1947)
28. Ferguson, L. N.: Chem. Rev. *43*, 385 (1948)
29. Coates, E.: J. Soc. Dyers and Colourists *83*, 95 (1967)
30. Ref. B3.: pp. 323–400
31. Lubs, H. A.: The chemistry of synthetic dyes and pigments, pp. 662–687. New York: Reinhold 1955
32. Ref. A2.: pp. 311–322
33. Brooker, L. G. S., Van Lare, E. J.: Color and constitution of organic dyes. In: Encyclopedia of chemical technology, Kirk-Othmer (eds.), 2nd ed., Vol. 5, pp. 763–788. New York, London, Sydney: Interscience 1964
34. Griffiths, J.: Colour and constitution of organic molecules. London, New York, San Francisco: Academic Press 1976
35. Klessinger, M.: Chemie in Unserer Zeit. *12*, 1 (1978)
36. Dahne, S.: Science *199*, 1163 (1978)
37. Fabian, J.: Chimia *32*, 323 (1978)
38 Ref. D4.: pp. 14–20
39. Hallas, G.: J. Soc. Dyers and Colourists *95*, 285 (1979)
40. Griffiths, J.: Rev. Prog. in Colouration. In press

(d) *4-Aminoazobenzene Derivatives*

41. Bridgeman, I., Peters, A. T.: J. Soc. Dyers and Colourists *86*, 519 (1970)
42. Sawicki, E.: J. Org. Chem. *22*, 915 (1957)
43. Ross, W. C. J., Warwick, G. P.: J. Chem. Soc. *1956*, 1719
44. Griffiths, J., Roozpeikar, B.: J. Chem. Soc. Perkin I *1976*, 42
45. Castelino, R. W., Hallas, G.: J. Chem. Soc. (B) *1971*, 793
46. Hallas, G., Ho, W. L., Todd, R.: J. Soc. Dyers and Colourists *90*, 121 (1974)
47. Martynoff, M.: Compt. rend. *235*, 54 (1952)
48. Gregory, P., Thorp, D.: J. Chem. Soc. Perkin I *1979*, 1990

(e) *Protonated Azo Dyes*

49. Lewis, G. E.: Tetrahedron *10*, 129 (1960)
50. Liler, M.: Adv. in Phys. Org. Chem. *11*, 308 (1975)
51. Sawicki, E.: J. Org. Chem. *22*, 1084 (1957)
52. Yagupol'skii, L. M., Gandels'man, L. Z.: J. Gen. Chem. USSR *37*, 1992 (1967); *35*, 1259 (1964)
53. Zhmurova, N.: J. Gen. Chem. USSR *27*, 2745 (1957)
54. Yamamoto, S., Tenno, Y., Nishimura, N.: Austral. J. Chem. *32*, 41 (1979)
55. Chu, K. Y., Griffiths, J.: Tetrahedron Letters *1976*, 405

(f) *Azo-hydrazone Tautomers*

56. Griffiths, J.: J. Soc. Dyers and Colourists *88*, 106 (1972)

(g) *Heterocyclic Azo Dyes*

57. Dickey, J. B., et al.: J. Org. Chem. *24*, 187 (1959)
58. Dickey, J. B., et al.: J. Soc. Dyers and Colourists *74*, 123 (1958)
59. Mikhailenko, F. A., Shevchuk, L. I., Kiprianov, A. I.: Khim. Getero Soed. *1973*, 923
60. Mikhailenko, F. A., Shevchuk, L. I.: ibid *1974*, 1325

(h) *Azo Naphthylamine Dyes*

61. Brode, W. R., Eberhart, D. R.: J. Org. Chem. *5*, 157 (1940)

(i) *Steric Effects*

62. Hoyer, E., Schickfluss, R., Steckelberg, W.: Angew. Chem. Internat. Edit. *12*, 926 (1973)
63. Kiprianov, A. I., Zhmurova, N.: J. Gen. Chem. USSR *23*, 915, 653 (1953)
64. Yamamoto, S.: Bull. Chem. Soc. Japan *46*, 3139 (1973)

Chapter 4

Anthraquinone Dyes

4.1 Introduction

Anthraquinone dyes are the second most important class of dyes. In contrast to the azo dyes (Chap. 3), which have no natural counterparts, the anthraquinone chromogen provided all the important natural red dyes. Indeed, these were renowned for their outstanding fastness properties, especially to light (see Sect. 1.2.3).

The principal reason why anthraquinone dyes are less widely used commercially than azo dyes is simply that they are less cost effective. Two factors are responsible for this: the more important one is the inherently lower tinctorial strength of anthraquinone dyes relative to azo dyes (anthraquinone dyes are usually less than half the strength of azo dyes); the second factor is the reduced versatility in the synthesis of anthraquinone dyes, since, in contrast to azo dye synthesis, the substituents generally have to be introduced into the preformed anthraquinone nucleus (see Sect. 2.4.2).

Azo dyes dominate the yellow/orange/red shade areas of manufacturers' dye ranges because they can match the excellent properties of the anthraquinone dyes whilst possessing considerable advantages in cost effectiveness. However, anthraquinone dyes predominate in the bright blue/turquoise region since azo dyes having comparable properties are not currently available. Thus, the two types tend to complement each other in providing a range of dyes which cover the whole shade gamut, *viz*. yellow → green.

4.2 Structure of Anthraquinone Dyes

Anthraquinones may be considered as ring annelated derivatives of *p*-benzoquinones. X-ray and electron diffraction studies show that *p*-benzoquinone has a planar structure (1) in which the bond length *a* approximates to that of a carbon-carbon single bond, and those of *b* and *c* to olefinic and ketonic double bonds respectively. The bond angles are approximately 120° which suggests that the carbon and oxygen

$a \sim 1.50$	aliphatic	C—C 1.54
$b \sim 1.32$	olefinic	C=C 1.34
$c \sim 1.22$	carbonyl	C=O 1.22

(1) (all bond lengths in Å)

atoms are sp^2 hybridised. The lack of π-electron delocalisation indicated by these results is supported by SCF-MO calculations which indicate a bond order of less than 1.15 for *a* and of greater than 1.98 for *b*. *p*-Benzoquinone thus resembles an α,β-unsaturated carbonyl compound in reactivity. In contrast, anthraquinone (2) is much less reactive. Crystallographic data show that the bond lengths in the two fused rings *A* are similar to those in benzene, including the *b* bonds: the *a* bonds again approximate to carbon-carbon single bonds (2). Consequently, anthraquinone may be considered as two benzene rings linked by two carbonyl groups which interact only weakly with the rings. The molecule is planar with bond angles of approximately 120°, indicating sp^2 hybridisation of the carbon and oxygen atoms.

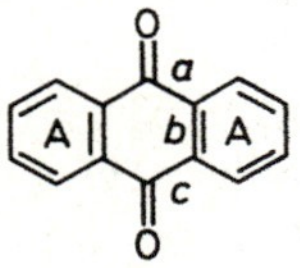

a 1.50
b ~1.39 (benzene 1.39)
c 1.21

(2) (all bond lengths in Å)

The introduction of amino and hydroxy groups into anthraquinone results in a lengthening of the carbonyl carbon-oxygen bond and a shortening of the aryl-carbon-heteroatom bond length compared to the expected values. For example, in 1,5- and 1,8-diphenylaminoanthraquinone the carbonyl bond length *c* is 1.23 to 1.25 Å and the carbon-nitrogen bond length *d* is 1.36 Å versus the expected values of 1.22 and 1.43 Å respectively. The effect is most pronounced in the 1,8-diphenylamino derivative (3; D=NHPh) where bond *c* is appreciably longer than *c'* (1.25 v 1.22 Å). These results suggest some contribution from resonance forms such as (3A) and (3B).

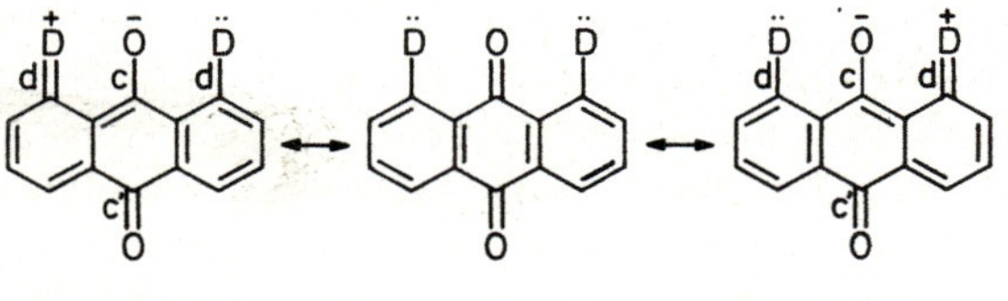

D=OH/NHR

(3B) (3) (3A)

4.3 Tautomerism

4.3.1 Tautomerism of Hydroxyanthraquinone Dyes

In contrast to the hydroxyazo dyes (see Sect. 3.3.1), there has been little published work on the tautomerism of hydroxyanthraquinone dyes. Indeed, it is only in the last few years that the topic has been studied.

The technique that has been used most for investigating azo-hydrazone tautomerism is ultra-violet/visible absorption spectroscopy. This is because the tautomers absorb at sufficiently different wavelengths (*ca.* 70 to 80 nm) so that the characteristic spectrum of each tautomer is clearly observed (see Sect. 3.3.2). However, studies on hydroxyanthraquinones have aimed at elucidating the structure of the actual tautomer since, in most cases, only one tautomeric form is observed. One of the most valuable tools for this purpose is ^{13}C nmr spectroscopy. This technique provides values for the chemical shifts of carbon atoms in a molecule relative to some internal standard: for hydroxyanthraquinone compounds, the crucial chemical shifts are those for the carbonyl carbon atom and the carbon atom to which the hydroxy group is attached. These values can then be compared to those obtained from authentic quinones and phenols and inferences drawn therefrom.

Comparison of the ^{13}C nmr spectra of 1-hydroxyanthraquinone and anthraquinone shows that the signal from the C9 carbonyl carbon atom in 1-hydroxyanthraquinone appears 5.6 ppm downfield relative to that in anthraquinone, whereas the signal from the C10 carbonyl carbon atom is displaced slightly upfield (—0.7 ppm). The signal from the bridgehead carbon atom, 9a, is shifted significantly upfield (—17.4 ppm), although the upfield shift disappears when the hydroxy group is methylated.

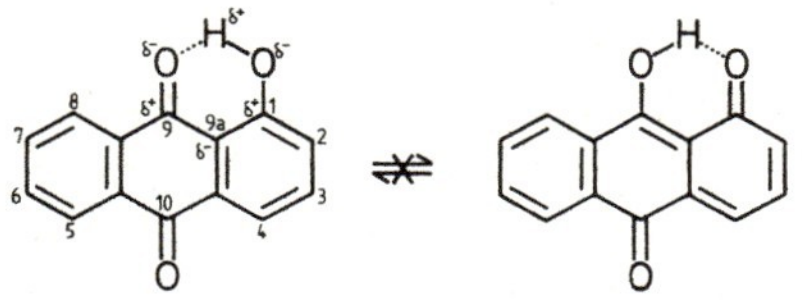

(4) (5)

These results are entirely consistent with the intramolecularly hydrogen-bonded structure (4). Thus, chelation increases the polarisation of the carbonyl group causing a reduction in the electron density at the C9 carbonyl carbon atom, *i.e.* it becomes deshielded. Chelation also increases the electron-donating ability of the hydroxy group which is reflected by an increased electron density at the bridgehead carbon atom 9a, and a reduced electron density at C1. Thus, in the 6-membered chelate ring there is an alternation of electron density; see structure (4). Deshielding of the C9 carbonyl carbon atom and shielding of the bridgehead 9a carbon atom are not observed for 1-methoxyanthraquinone, which hasn't any facility for hydrogen-bonding.

There is no evidence for tautomer (5) since the chemical shift values of C9 and C1 are characteristic for quinonoid carbonyl and a carbon to which a hydroxy group is attached.

For 2-hydroxyanthraquinone the ^{13}C nmr data is consistent with structure (6): there is no evidence for tautomer (7).

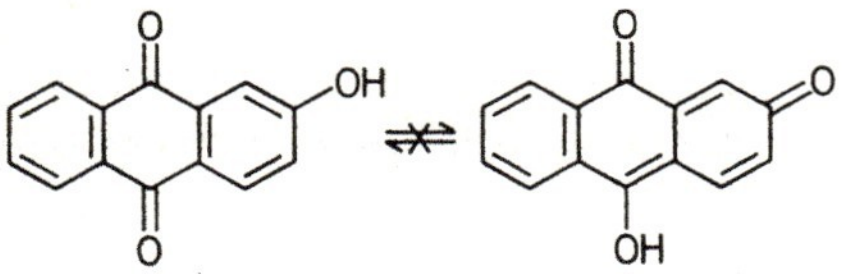

(6) (7)

Four of the dihydroxyanthraquinone compounds are used widely either as dyes themselves or as dye intermediates: they are 1,2-(alizarin), 1,4-(quinizarin), 1,5-(anthrarufin) and 1,8-dihydroxyanthraquinone (chrysazin) — see Sect. 2.4.2.

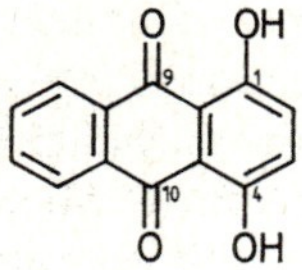

meso (8) vraie (9) amphi (10)

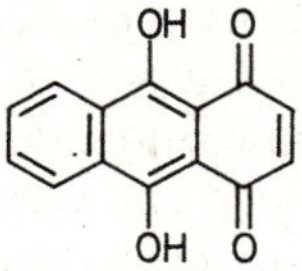

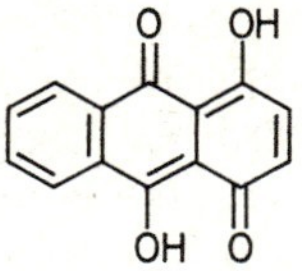

For the most important dihydroxyanthraquinone, quinizarin, three tautomers are possible (8, 9 or 10). In quinizarin, both carbonyl groups have bond lengths of *ca.* 1.29 Å. This value lies between those for an aromatic C—O single bond (1.36 Å) and a C=O double bond (1.22 Å). Consequently, they are of little value for differentiating between the possible tautomers. However, a comparison of the calculated dipole moments of each tautomer with the experimentally determined value for quinizarin suggests that it exists as the *meso* form (8). The strongest evidence comes from ^{13}C nmr studies. These clearly show that the chemical shifts of the carbon atoms at the 9/10 and 1/4 positions are characteristic of a quinonoid carbon atom and a carbon atom to which a hydroxy group is attached respectively. Therefore, it can be concluded that quinizarin exists solely in the *meso* form (8).

Similarly, ^{13}C nmr spectroscopy and bond length data offer no evidence for tautomerism in the other three important dihydroxyanthraquinone derivatives, alizarin (11), anthrarufin (12), and chrysazin (13). Additional evidence for the structure of anthrarufin was obtained from the ^{1}H nmr spectra which did not indicate any proton delocalisation.

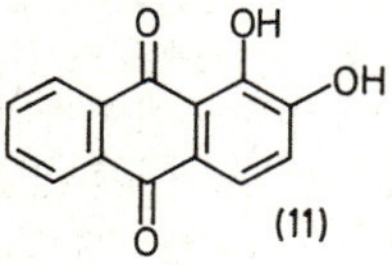

(11) 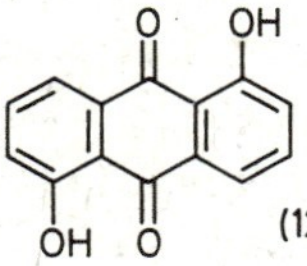 (12) 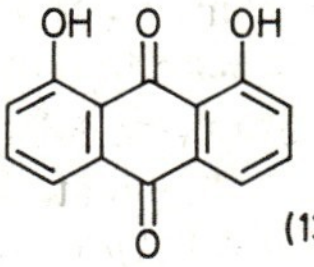 (13)

4.3.2 Reduced Hydroxyanthraquinone Dyes — *Leuco*-Quinizarin

As seen from Sect. 2.4.2, the commercially important 1,4-diaminoanthraquinone dyes are obtained by nucleophilic substitution reactions on *leuco*-quinizarin. Three tautomeric structures can be written for *leuco*-quinizarin, (14, 15 and 16). *Leuco*-quinizarin was assigned structure (14) by Zahn and Ochwat[1] in 1928 who obtained it by condensation of 1,4-dihydroxynaphthalene with succinic anhydride, Eq. 4.1. Using infra-red spectroscopy Flett, in 1948, could not differentiate between tautomers (14) and (15) although a band at 2950 cm^{-1} indicated the presence of a methylene group, thus suggesting the presence of tautomer (14) and/or (15) rather than (16).

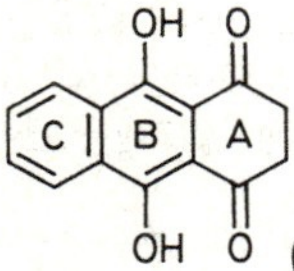

(14) 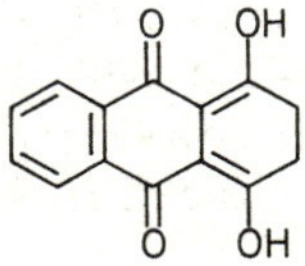 (15) 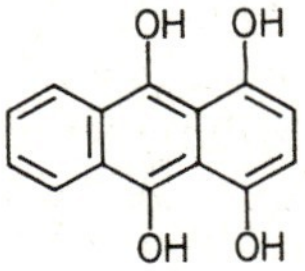 (16)

[1] Annalen *462*, 72 (1928).

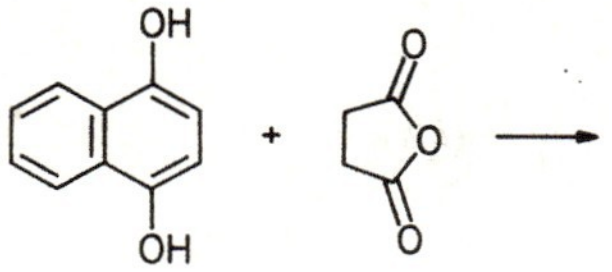

OH OH + O O O → (14) Eq. 4.1

The problem was resolved by Bloom and Hutton in 1963 who thoroughly investigated the structure of *leuco*-quinizarin by ^{1}H nmr spectroscopy. They noted that *leuco*-quinizarin lacked the singlet aromatic signal (7.27 ppm) of the *A* ring of quinizarin. However, they observed a new sharp singlet at 3.03 ppm (methylene), which was absent in quinizarin and which appeared at the same frequency as that of the methylene groups in 2,3-dihydro-1,4-naphthoquinone (17; R=H) and its dihydroxy derivative (17; R=OH). Thus, *leuco*-quinizarin was assigned the structure (14).

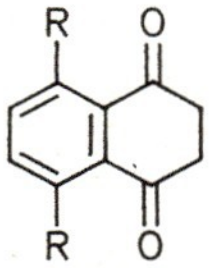

R O R O R=H or OH

(17)

As in the case of azo-hydrazone tautomerism it is the relative thermodynamic stabilities that determine which tautomer predominates. For 1- and 2-hydroxyanthraquinone, tautomers (4) and (6) are clearly more stable than (5) and (7) because of their greater aromaticity — they contain two benzenoid rings rather than one benzenoid and one quinonoid ring. As seen from Table 3.1 the RSE[2] of a benzene ring is *ca.* 134 kJ mol^{-1} greater than that for a quinonoid ring, an amount sufficient to ensure that 1- and 2-hydroxyanthraquinone exist solely as tautomers (4) and (6) respectively.

The thermodynamic stabilities of the three possible tautomers of quinizarin increase in the order (10) < (9) < (8) since two benzenoid rings confer more stability than one naphthalene ring which in turn confers more stability than a benzenoid and an *o*-quinonoid ring combined (see Table 3.1). Similar reasoning explains the sole existence of tautomer (11) for alizarin, (12) for anthrarufin, and (13) for chrysazin: the reader may verify this for himself by drawing the possible tautomers for these compounds and comparing their thermodynamic stabilities.

For the three possible tautomers of *leuco*-quinizarin (14–16), (14) with an intact naphthalene nucleus, and the fully aromatic tautomer (16), are obviously more stable than (15). Using the RSE data in Table 3.1 and standard bond energies, tautomer (14) is *ca.* 40 kJ mol^{-1} more stable than (16). Since reduction of the dimethyl ether (18) of quinizarin leads to (19) and not (20), it would seem that the two strong intramolecular hydrogen-bonds between the hydroxy hydrogen atom and the carbonyl oxygen atom must also be a significant factor in favour of (14).

[2] Resonance Stabilisation Energy.

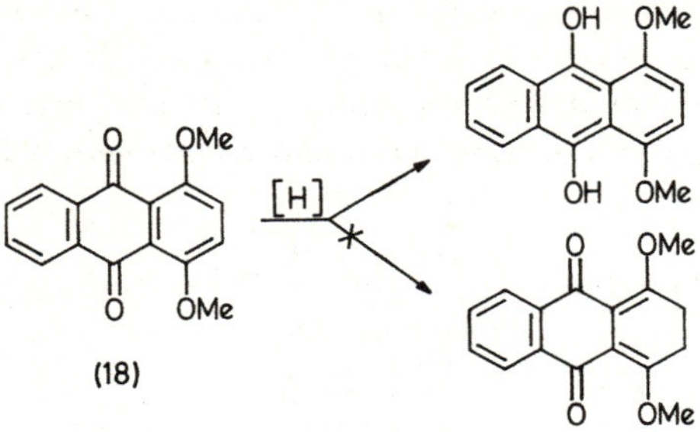

(18) (19) (20)

4.3.3 Aminoanthraquinone Dyes

These are the most important of the anthraquinone derivatives since most commercial anthraquinone dyes contain at least one amino substituent. For example, 1,4-diaminoanthraquinone (21) is CI Disperse Violet 1.

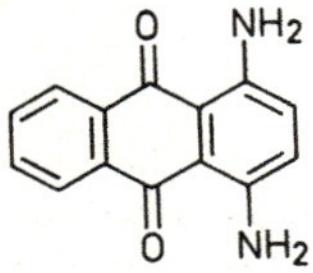

(21)

As mentioned in Sect. 4.2 the structure of two of the aminoanthraquinone derivatives, 1,5-diphenylaminoanthraquinone (22A) and 1,8-diphenylaminoanthraquinone (22B) have been determined by X-ray crystallography. In both cases the C—N and C—O bond lengths are characteristic of amino and carbonyl bonds in which there is some delocalisation of π-electrons between the amino and carbonyl groups (see Sect. 4.2). Although the amino hydrogen atom was not located experimentally, the geometry of the molecules is consistent with intramolecular hydrogen-bond formation. Consequently, it may be concluded that aminoanthraquinone compounds, like their aminoazo counterparts (see Sect. 3.3.4), exist totally in the amino form (22); the lack of the imino form (23) is again probably due to the relative instability of the imino (C=N) bond.[3]

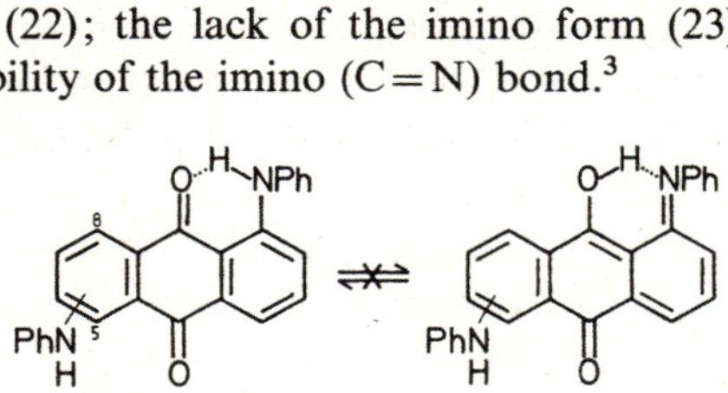

(22) A = 5-PhNH B = 8-PhNH (23)

4.3.4 Reduced Aminoanthraquinone Dyes

These are intermediates for the manufacture of dyes of industrial importance. Taking *leuco*-1,4-dibenzylaminoanthraquinone as an example, three tautomers (24, 25 and 26) are possible. A thorough investigation using ^{1}H nmr spectroscopy showed that

[3] See Dudek, G. O., Holm, R. H.: J. Amer. Chem. Soc. *84*, 2691 (1962).

only one tautomer (24) was present. The preference for this tautomer over the fully aromatic tautomer (26) has been ascribed to the strong intramolecular hydrogen-bonding between the amino proton and the oxygen atom of the carbonyl group. Tautomer (25) is most unlikely because it contains two imino groups and, indeed, it has not been detected.

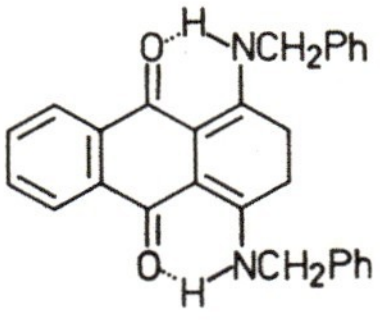

(24) 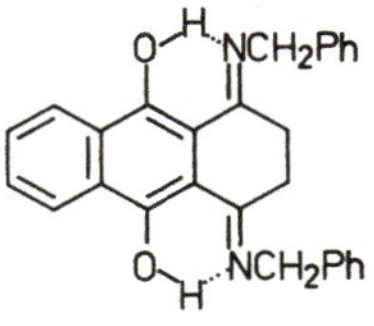 (25) 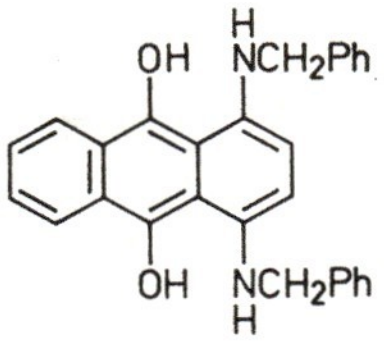 (26)

4.4 Protonated and Ionised Anthraquinone Dyes

4.4.1 Introduction

It is most important that commercial anthraquinone dyes are insensitive to pH variations — a housewife would not appreciate a dye which changed colour if vinegar (acetic acid) was spilled on it or one which changed colour and washed out in the presence of washing soda (sodium carbonate)! Consequently, we shall consider the effect of both acids and alkalis on the properties of anthraquinone dyes. Acids have most effect on anthraquinones containing amino groups whereas hydroxyanthraquinone dyes are affected more by alkalis. However, we shall consider first the parent molecule, anthraquinone.

4.4.2 Anthraquinone

Anthraquinone itself is very resistant to both acids and alkalis. Indeed, it is completely unaffected by alkalis. However, under very forcing conditions (conc. sulphuric acid) anthraquinone protonates at a carbonyl oxygen atom to give the cationic species (27). The latter is red compared to the pale yellow colour of anthraquinone due to the enhanced electron-accepting ability of the protonated carbonyl group (see Sect. 4.6.3).

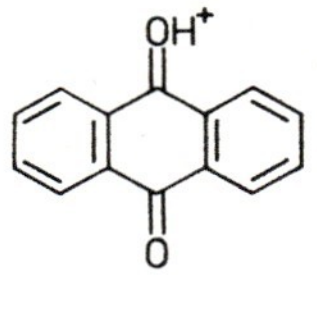

(27)

4.4.3 Aminoanthraquinone Dyes

The acid sensitivity of aminoanthraquinone dyes depends both on the position and the type of amino group.

Dyes containing a primary or secondary amino group in the 1-position are less sensitive to acid than their 2-isomers. As mentioned in Sect. 4.6.3 substituents at the 1-position capable of forming an intramolecular hydrogen-bond with the carbonyl group do so. This reduces the basicity of the substituent, as seen in Table 4.1.

Table 4.1. Basicities of aminoanthraquinones (in MeOH)

Substituent	pK_b	Substituent	pK_b	Substituent	pK_b[a]
1-NH_2	16.1	2-NH_2	15.1	1,5-diNH_2	~15.7
1-NHMe	16.5	2-NHMe	16.0	1,4-diNH_2	14.3
1-NMe_2	~12.3	2-NMe_2	16.2		

[a] 1st dissociation
NB. Smaller pK_b values indicate a higher basicity

For anthraquinones substituted in the 1-position with a tertiary amino group, intramolecular hydrogen-bonding is not possible. Space filling molecular models indicate that such groups, *e.g.* NMe_2, cannot lie coplanar with the anthraquinone molecule because of a *peri* interaction with the anthraquinone carbonyl oxygen atom. Consequently the overlap of the lone pair $2p_z$ orbital of the amino group and the π-electron system of anthraquinone is diminished and this results in a very significant increase in the basicity of the dimethylamino group, *e.g.* 1-NN-dimethylaminoanthraquinone is *ca.* 10,000 times more basic than 1-aminoanthraquinone (Table 4.1) and thus protonates much more readily.

In theory, protonation of aminoanthraquinone dyes (28) can occur either at nitrogen to give (29) or at oxygen to give the resonance stabilised cation (30). Surprisingly, only the N-protonated cation (29) is observed experimentally. This species is analogous to the ammonium tautomer of an aminoazo dye (see Sect. 3.5.6) — both are practically colourless. Indeed, one of the reasons why both 2-(substituted)-amino- and 1-NN-dialkylaminoanthraquinone compounds are rarely used commercially is because of the relative ease with which they protonate.

(28)

R = H, Me, Ph, etc.

H_3O^+

(29)

(30A)

(30B)

Many important commercial blue dyes are derivatives of 1,4-diaminoanthraquinone. The gradual addition of acid to the parent compound results eventually in a species whose visible absorption spectrum resembles that of 1-aminoanthraquinone (Fig. 4.1). Thus, protonation of one of the amino groups has taken place to give (31). Furthermore, since the curves pass through a well defined isosbestic point, only two species are present — 1,4-diaminoanthraquinone and (31).

(31)

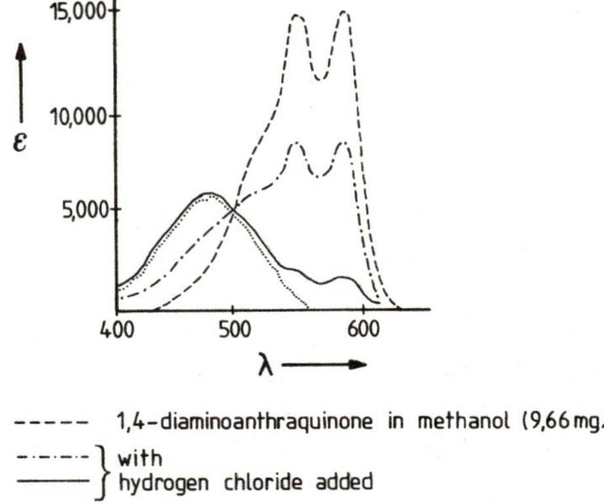

Fig. 4.1. Effect of acid on 1,4-diaminoanthraquinone

Thus, unlike the azonium-ammonium tautomerism observed in protonated azo dyes (see Sect. 3.3.5), no comparable equilibrium exists for protonated aminoanthraquinone dyes.

4.4.4 Hydroxyanthraquinone Dyes

Because the hydroxy group in 2-hydroxyanthraquinone is not intramolecularly hydrogen-bonded it is more easily ionised than that in 1-hydroxyanthraquinone. This is reflected by their respective pK_a values of 7.6 and 11.5.

More important commercially are dihydroxyanthraquinone dyes, especially those in which both hydroxy groups are adjacent to the carbonyl groups, such as 1,4-, 1,5-, and 1,8-dihydroxyanthraquinones. The pK_a values of such dyes are higher than those of corresponding dyes such as 2,6-dihydroxyanthraquinone in which neither hydroxy group is intramolecularly hydrogen-bonded (Table 4.2). In dyes such as

Table 4.2. pK_a Values of some dihydroxyanthraquinone dyes

Compound	1,4-diOH	1,5-diOH	1,8-diOH	2,6-diOH	1,2-diOH
pK_a1	9.9	9.5	8.3	6.2	7.5
pK_a2	11.2	11.1	12.5	8.3	11.8

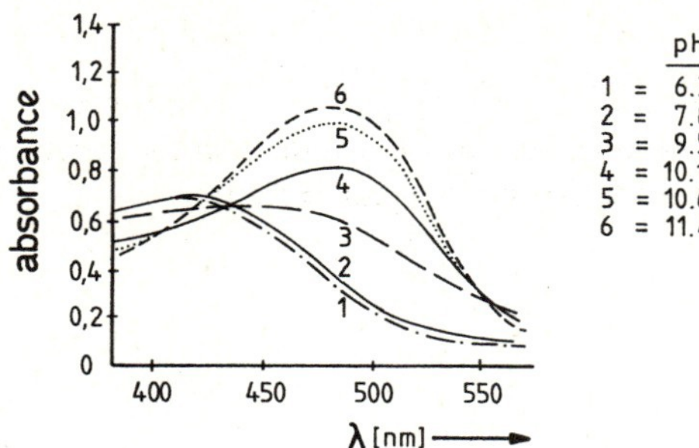

Fig. 4.2a. The visible absorption spectra of anthrarufin in solutions of varying pH's

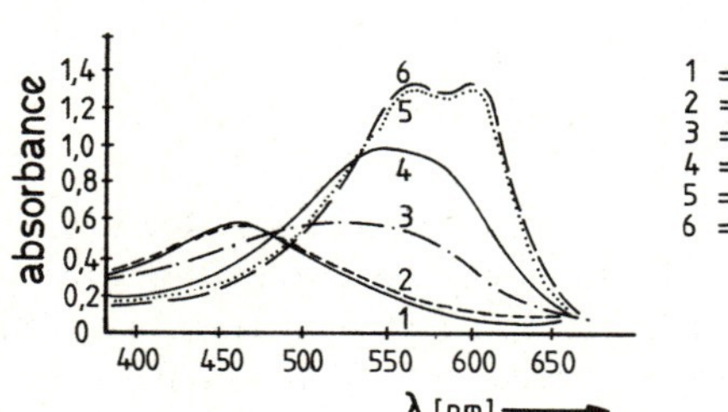

Fig. 4.2b. The absorption spectrum of quinizarin in solutions of varying pH's

1,2-dihydroxyanthraquinone, as expected, the 2-hydroxy group ionises at a much lower pH than the intramolecularly hydrogen-bonded 1-hydroxy group.

In all the dihydroxyanthraquinone dyes ionisation of the first hydroxy group reduces the acidity of the second hydroxy group because of electron donation by the negatively charged oxygen atom. Since the oxido group is a more powerful electron donor than the hydroxy group, ionisation of a hydroxyanthraquinone dye results in a bathochromic shift. This is illustrated in Fig. 4.2a and b. Using the pK_a values in Table 4.2, the absorption curves corresponding to the dihydroxy dye and its monobasic and dibasic ions can be deduced, *e.g.* Figs. 4.2a 1, 4.2a 3 and 4.2a 6 respectively for 1,5-dihydroxyanthraquinone. An interesting feature of the spectrum of 1,4-dihydroxyanthraquinone at high pH values is the presence of a double peak for the dibasic anion, see (32) and Fig. 4.2b 6; the unionised dye exhibits only a single peak. This phenomenon is discussed in Sect. 4.6.3.

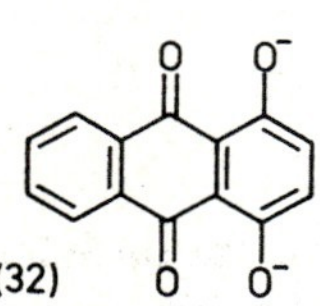

(32)

1-Hydroxyanthraquinone derivatives are important for providing metal complex dyes whereby a metal ion replaces hydrogen in the 6-membered chelate ring (see Sect. 4.5).

4.4.5 Aminohydroxyanthraquinone Dyes

Anthraquinone dyes containing both amino and hydroxy groups are amphoteric: they undergo a hypsochromic shift on the addition of acid and a bathochromic shift on the addition of alkali. CI Acid Blue 69 (33) is a typical example.

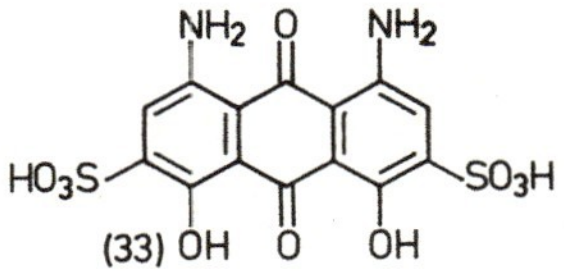

(33)

4.5 Metal Complexed Anthraquinone Dyes

4.5.1 Introduction

In azo dyes metallisation is employed to improve certain properties of the dye, in particular its fastness to light (see Sect. 3.4.6). Since anthraquinone dyes already possess excellent fastness to light metallisation offers no advantages in this respect. Indeed, metallisation would result in dullness and hence nullify one of the major advantages of anthraquinone dyes compared to azo dyes — their superior brightness. Consequently, there are very few commercially important metallised anthraquinone dyes.

4.5.2 Commercial Dyes

The majority of metallised azo dyes are synthesised as discrete compounds *before* application to the fibre, (*i.e.* they are pre-metallised dyes). In contrast, the few commercial metallised anthraquinone dyes are mordant dyes, *i.e.* they are metallised as part of the dyeing process *after* the metal free dye has been applied to the fibre (see Sects. 1.2.2 and 3.4.1).

The most famous metal complex anthraquinone dye is Turkey Red, an aluminium-calcium complex of alizarin (see Sect. 1.2.3). Other typical examples are CI Mordant Red 3 (34) and CI Mordant Brown 42 (35).

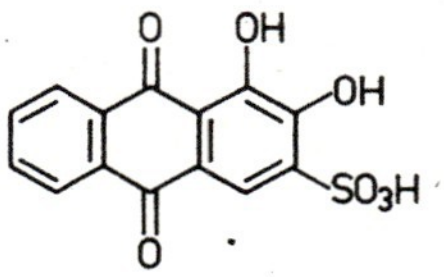

Cr[III] complex (34) 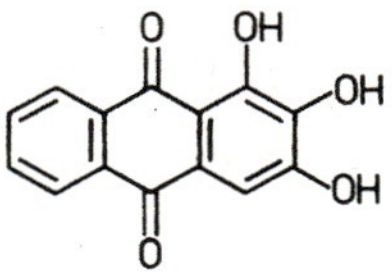   Cr[III] complex (35)

By varying the metal ion different colours can be obtained from the same hydroxy-anthraquinone. This property is known as polygenesis. Thus, the following colours have been obtained from alizarin:

aluminium III — red
iron II — deep violet
iron III — brownish-black

tin II — reddish-violet
tin IV — violet
chromium III — brownish-violet

The few anthraquinone pigments that are used commercially are mainly red to violet in colour and, unlike the mordant dyes, they are metallised before use. Typical examples are CI Pigment Red 83, a metal complex of alizarin, and CI Pigment Violet 5.1, an aluminium complex of (36).

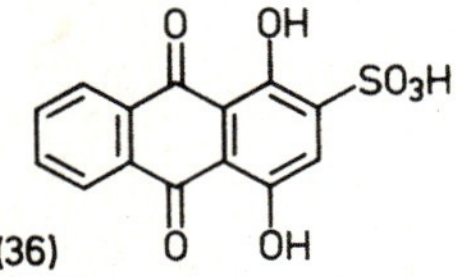

(36)

4.5.3 Structure and Properties

All the dyes are metal complexes of 1-hydroxyanthraquinone derivatives. By analogy with Turkey Red, whose structure has been elucidated (see Sect. 1.2.3), they probably contain, as a basic unit, a 6-membered chelate ring in which the metal ion is coordinated to the oxygen atom of the 1-hydroxy group and the quinone oxygen atom (37). Thus, they resemble the 2:1 azo dye: metal complexes (see Sect. 3.4.2) in which the carbonyl group has replaced the azo group.

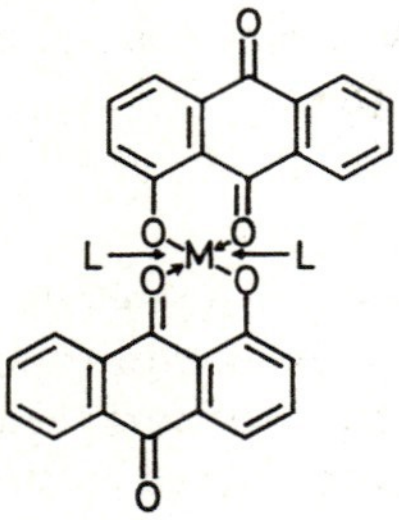

M = metal
L = monodentate ligand

(37)

As in the azo dyes, metallisation causes the colour of a dye to undergo an appreciable bathochromic shift since the donor properties of the hydroxy group and the acceptor properties of the carbonyl group are both enhanced. The aqueous solubility of the metallised dye is generally lower than that of the metal free dye.

4.6 Colour and Constitution

4.6.1 Introduction

A knowledge of colour-structure relationships is important because not only does it enable dyes of a desired hue to be synthesised, it also allows tinctorial strength, a parameter of the utmost impotance in determining the economic viability of a dye, to be correlated with structure.

The first major attempts to rationalise the colour and constitution of anthraquinone dyes were made by Peters and Sumner in 1953 and by Labhart in 1957: the latter worker examined the spectra of some 68 dyes. However, general studies of colour-structure effects in anthraquinone dyes have been relatively few in number, especially by comparison with similar studies on azo dyes (see Sects. 3.5.4 to 8).

In this section the experimental results are presented first; these are then discussed in terms of the VB/resonance theory and PPP MO theory. Steric effects are considered separately (see Sect. 4.6.4).

4.6.2 Experimental Results

Spectrum of Anthraquinone

The electronic absorption spectrum of anthraquinone consists of three broad bands (Fig. 4.3): the long wavelength band at *ca.* 405 nm is responsible for the pale yellow colour of anthraquinone. It is so weak ($\varepsilon_{max} \sim 60$) that it has often gone undetected by many workers who have mistakenly ascribed the second band (at ~325 nm) as the long wavelength absorption band. The finely structured short wavelength band is the most intense of the three and because of its fine structure — it has peaks at 246, 252, 263 and 272 nm — it is more correct to speak of six absorption bands for anthraquinone.

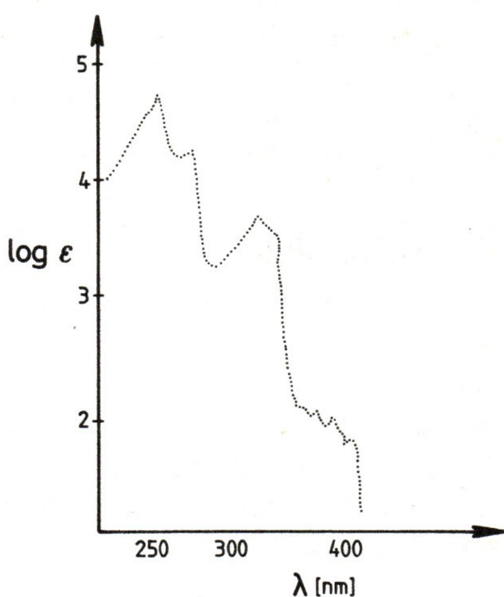

Fig. 4.3. Electronic absorption spectrum of anthraquinone

Anthraquinone Derivatives — General Patterns

A study of the effects of substituents on the electronic spectrum of anthraquinone shows that the nature and number of the substituents, as well as their position in the anthraquinone nucleus, play a major role.

There is a marked difference between the effects of electron-donating and electron-accepting substituents. The introduction of electron-accepting substituents has little effect, as was the case in azo dyes (see Sect. 3.5.4), and thus the spectra of the derived dyes resemble that of anthraquinone. In contrast, electron-donating substituents can have a significant effect, producing dyes which vary in colour from yellow, through red and blue, to green.

The position of the substituent in the molecule is also important, *e.g.* 1-substituted anthraquinone compounds are more bathochromic than their 2-substituted isomers. Intramolecular hydrogen-bonding in 1-substituted anthraquinone compounds, and steric effects generally, can also affect the colour significantly. In the ensuing presentation of the experimental data it will become apparent how these factors influence the spectrum of anthraquinone. Only the visible absorption bands are considered since these are the ones primarily responsible for the colour of a dye.

Monosubstituted Anthraquinones

Because of the symmetry of the molecule only two isomers are possible — 1-substituted or 2-substituted anthraquinone. As already mentioned, electron-accepting groups have little effect on the absorption spectrum. However, they do weaken the $n \rightarrow \pi^*$ transition of the carbonyl group so that the band at *ca.* 405 nm is usually missing: if it is observed, it occurs at longer wavelengths (Table 4.3).

The nature of the electron-donating group determines whether it exerts a major or a minor effect on the spectrum. Those groups which possess a $2p_z$ lone pair of electrons, such as OR or NR_2 groups, are the ones which produce a profound effect. In anthraquinone derivatives containing such groups, a relatively intense *new* band appears in the visible region of the spectrum, termed a charge-transfer band, and it is this band which is responsible for the colour of anthraquinone dyes. Thus, all commercial anthraquinone dyes contain at least one donor group of this type. Other types of electron-donating groups such as alkyls or phenyls exert only a small bathochromic effect on the long wavelength $\pi \rightarrow \pi^*$ band of anthraquinone at *ca.* 325 nm (see Fig. 4.3); since this band is still in the ultra-violet, we shall concentrate on anthraquinone compounds substituted by donor groups of the former type.

For 2-substituted anthraquinones, bathochromicity increases in line with the electron-donating ability of the substituent. This is also generally true for 1-substituted

Table 4.3. Spectral data for 1- and 2-substituted anthraquinones (in MeOH)

Substituent	2-Substituted		1-Substituted	
	λ_{max} (nm)	ε_{max}	λ_{max} (nm)	ε_{max}
H	~410[a]	~60[a]	~410[a]	~ 60[a]
NO_2	~420[a]	~50[a]	—	—
Cl	—	—	~415[a]	~100[a]
OMe	363	3,950	378	5,200
$NHCOCH_3$	367	4,200	400	5,600
OH	368	3,900	402	5,500
NH_2	440	4,500	475	6,300
NHMe	462	5,700	503	7,100
NHPh	467	7,100	500	7,250
NMe_2	472	5,900	503	4,900

[a] $n \rightarrow \pi^*$ transitions

anthraquinones but in this case there are exceptions due either to intramolecular hydrogen-bonding or steric hindrance (see Table 4.3). The former causes a larger than expected bathochromic shift and the latter a smaller than expected bathochromic shift. Thus, whereas 1-hydroxyanthraquinone (38) absorbs at considerably longer wavelengths than 1-methoxyanthraquinone, the corresponding pair of 2-isomers absorb at similar wavelengths (Table 4.3). Steric effects are discussed later (Sect. 4.6.4).

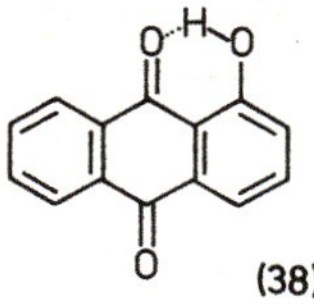

(38)

The 1-substituted compounds are generally tinctorially stronger than the 2-isomers (Table 4.3), but since both are relatively weak ($\varepsilon_{max} < 10{,}000$), monosubstituted anthraquinone compounds are rarely used commercially. 1-Methylaminoanthraquinone (39) — CI Disperse Red 9 — one of the strongest dyes of this type, is an exception.

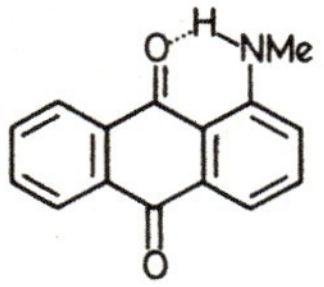

(39)

Disubstituted Anthraquinones

Disubstituted anthraquinones may be classified into two distinct types; those in which the substituents are in different benzene rings and those in which they are in the same ring. Each type has quite different spectral features.

Substituents in Different Rings

As mentioned in Sect. 4.2 anthraquinone may be considered as consisting of two isolated benzoyl chromogens, as depicted in (40). Consequently, little or no interaction would be expected between substituents located in different rings, *e.g.* X and Y in (40). In support of this postulate the ε_{max} values for both 1,5- and 1,8-diaminoanthraquinone are approximately twice that of 1-aminoanthraquinone but their absorption maxima are similar to the latter compound (Table 4.4).

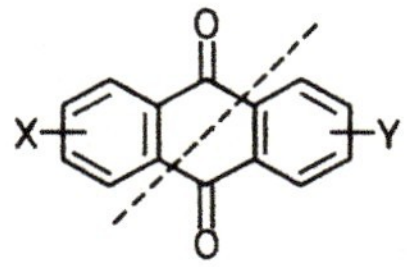

(40)

Table 4.4. Spectral data for diaminoanthraquinones with substituents in different rings

Compound	λ_{max}^{MeOH}	ε_{max}
1-NH_2 AQ	475	6,300
1,5-diNH_2 AQ	487	12,600
1,8-diNH_2 AQ	507	10,000

Dyes of this type have found only limited commercial application. A typical dye is CI Acid Violet 34 (41).

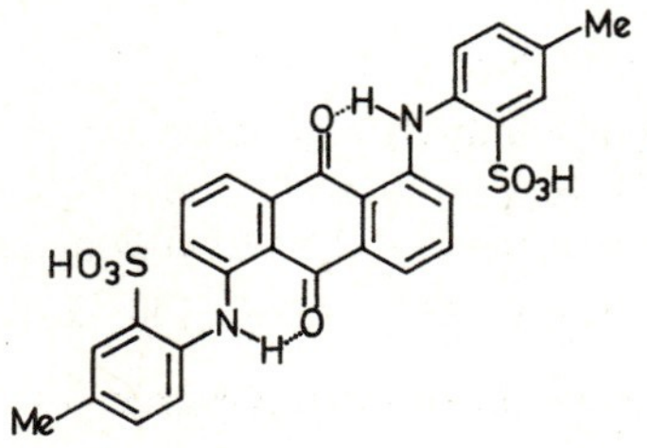

(41)

Substituents in the Same Ring

The 1,4- and 1,2-disubstituted anthraquinones are the most important dyes of this type: indeed, the majority of commercial anthraquinone dyes are based on 1,4-bis-(donor-substituted)-anthraquinones and we shall discuss this type first.

1,4-Disubstituted Anthraquinones

By varying the donor groups shades from yellow to turquoise can be obtained. Thus, when the two donor groups are of moderate electron-donating ability, such as hydroxy, orange dyes result. For example, 1,4-dihydroxyanthraquinone (quinizarin) (8) has been used as an orange coloured smoke dye.

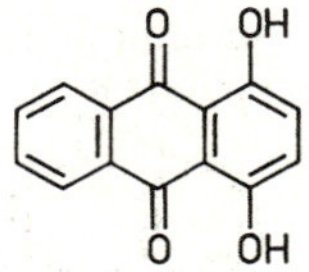

λ_{max}^{MeOH} 470 nm, ε_{max} 17,000

(8)

With two strong donor groups such as amino or substituted amino groups blue to turquoise colours are obtained. This type of system is employed extensively for providing commercial dyes, *e.g.* CI Disperse Blue 3 (42) and CI Acid Blue 25 (43).

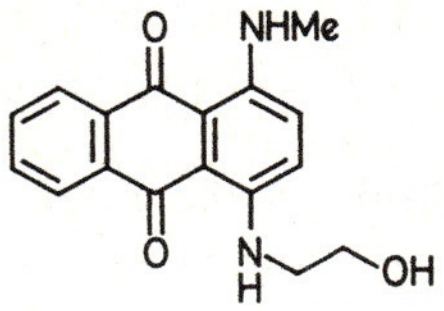

(42) 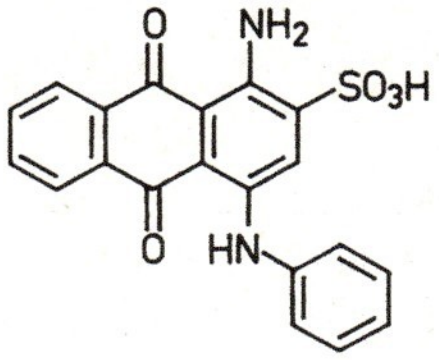 (43)

Almost any intermediate shade can then be produced by incorporating two donor groups of different strengths in the 1- and 4-positions. Thus, 1-amino-4-hydroxy-anthraquinone (44) is CI Disperse Red 15 and 1-amino-4-methylaminoanthraquinone (45) is CI Disperse Violet 4.

(44) (45)

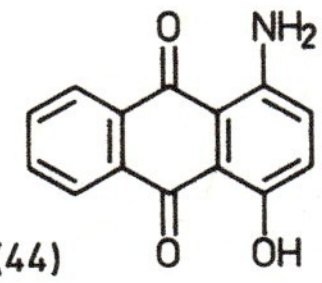

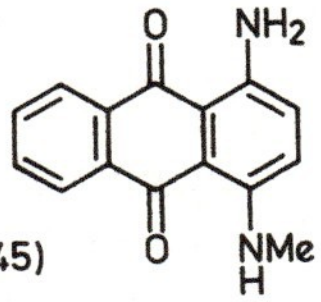

The 1,4-disubstituted anthraquinones are much more bathochromic than their 1,5- or 1,8-isomers. The difference is striking, as shown in Table 4.5.

Table 4.5. Colours of 1,4- versus 1,5- and 1,8-isomers (in MeOH)

Compound	Colour	λ_{max}	ε_{max}
1,4-diNH_2 AQ	Blue	550, 590	15,850, 15,850
1,5-diNH_2 AQ	Red	487	12,600
1,8-diNH_2 AQ	Red	507	10,000
1,4-diOH AQ	Orange	470	17,000
1,5-diOH AQ	Yellow	425	10,000
1,8-diOH AQ	Yellow	430	10,960

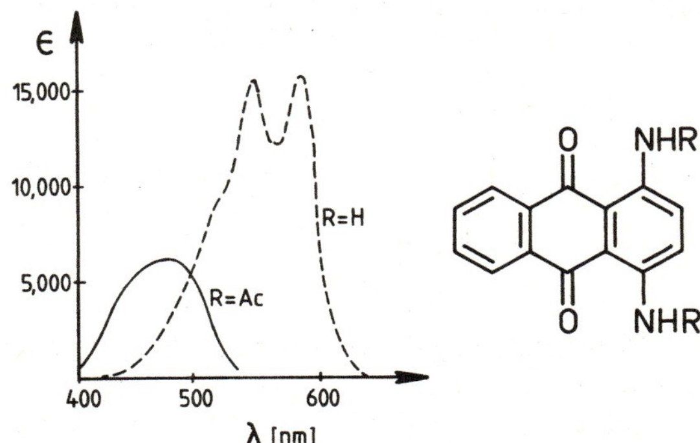

Fig. 4.4. Twin absorption peaks of 1,4-diaminoanthraquinones

Anthraquinone dyes substituted in the 1- and 4-positions with primary or secondary amino groups are unusual in that they exhibit two absorption peaks, usually of equal or similar intensity (see Table 4.5 and Fig. 4.4).

The peaks do not change in position or intensity on dilution, so the phenomenon is not caused by aggregation. Since they are also observed in the vapour state, it is not a peculiarity of the solvent either. However, acylation destroys the fine structure and also produces a large hypsochromic shift, *e.g.* the dye (46) is orange (Fig. 4.4). As discussed later (Sect. 4.6.3) the effect is thought to be due to vibrational fine structure.

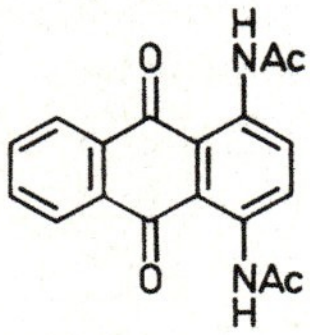

$\lambda_{max}^{o\text{-chlorophenol}}$ 474 nm, ε_{max} 6,700

(46)

In the benzoylamino dyes (47) the colour can be modified slightly by varying the *X* substituent in the benzene ring: electron-withdrawing substituents cause a hypsochromic shift and electron-donating substituents a bathochromic shift as shown in Table 4.6. Similar trends are observed for 1,4-dibenzoylaminoanthraquinone dyes. Principally because of low tinctorial strength, dyes of this type have found only limited commercial applications, usually as pigments, *e.g.* CI Pigment Red 89 (48).

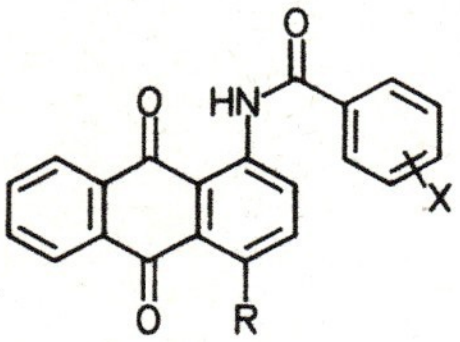

(47)

Table 4.6. Effect of substituent X on the spectra of (47; R=H)

X	λ_{max} (nm)	ε_{max}
p-NO_2	407	6,500
H	415	6,350
p-MeO	424	6,200

Solvent: pyridine

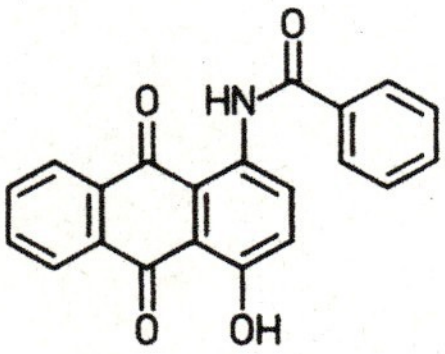

(48)

Much more important commercially are the arylaminoanthraquinones of type (49). Again the colour can be 'fine-tuned' by varying the substituent. In general, when *X* is an electron-donating group a bathochromic shift ensues whereas a hypsochromic shift is observed when *X* is electron-withdrawing. These effects are shown in Table 4.7.

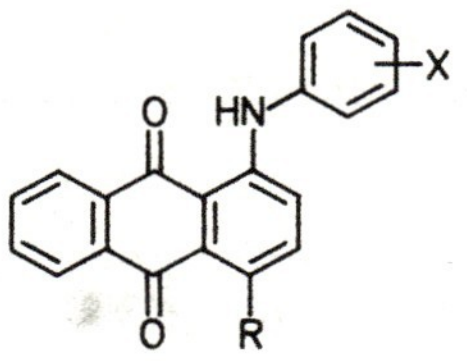

(49)

Table 4.7. Effect of substituents on 1-anilinoanthraquinone (49; *R*=H)

X	λ_{max} (nm)	ε_{max}
p-NO_2[a]	487	10,700
H[b]	500	7,600
p-NH_2[a]	525	6,000

[a] In trichlorobenzene
[b] In MeOH

A typical commercial dye of this type is CI Acid Violet 43 (50).

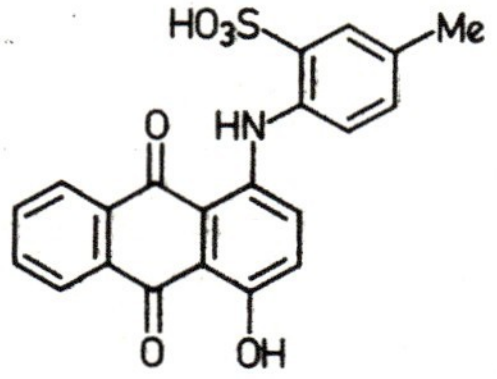

(50)

Most 1,4-diarylaminoanthraquinone dyes are green. This is because of an additional weak absorption band at *ca.* 410 nm which imparts a yellow component to the usual blue colour of a 1,4-diaminoanthraquinone. Thus, in water, CI Acid Green 25 (51) absorbs at 410 nm as well as 608 nm and 646 nm (see Fig. 4.5). Because of the

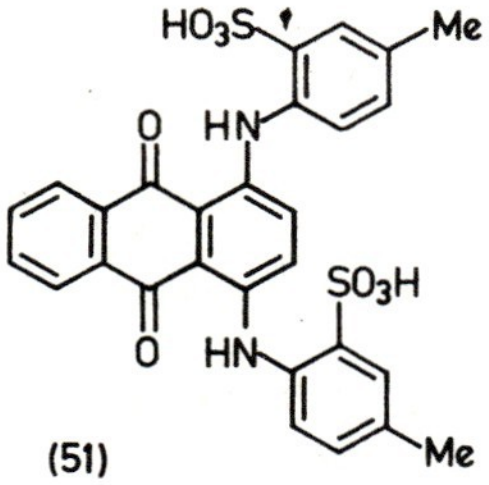

(51)

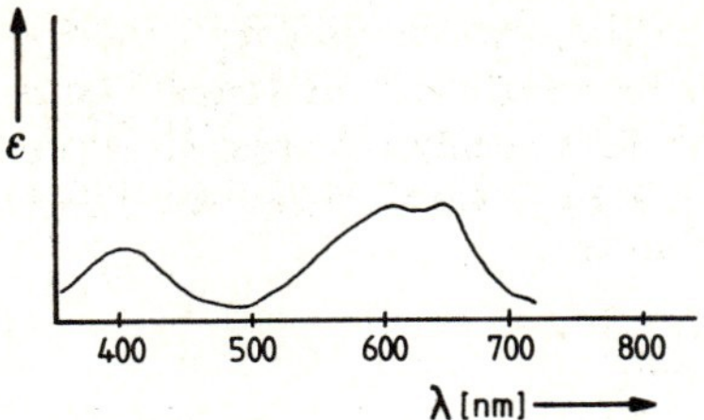

Fig. 4.5. Absorption spectrum of CI Acid Green 25

lower intensity of the shorter wavelength band, the greens generally have a bluish tone. However, the incorporation of nitro groups, especially in the *para* position, or extending the conjugation of the aryl ring, increase the intensity of the band at *ca.* 410 nm thus producing more neutral green dyes.

Annelation of benzene rings to the 1,4-bis-(donor)-anthraquinone system (52) can lead to either a bathochromic or a hypsochromic shift depending on the positions at which annelation occurs. Thus, annelation across the 5,6-positions with a benzene ring causes a bathochromic shift of some 20–30 nm. In contrast, annelation across the 2,3 or 6,7 positions causes a small hypsochromic shift: these are exceptions to the rule that increased conjugation results in a bathochromic shift (see Sect. 3.5.2). Annelation of two rings across the 2,3 and 5,6 positions is bathochromic whereas similar annelation across the 2,3 and 6,7 positions is hypsochromic.

(52)

1,2-Disubstituted Anthraquinones

These compounds have been studied much less than the corresponding 1,4-disubstituted anthraquinones primarily because they have not achieved any commercial importance. The main reason is that 1,2-disubstituted anthraquinones are generally tinctorially weaker and also more hypsochromic than their 1,4-isomers. Compare, for example, (11) with (8). These effects are caused partly by the increased sensitivity of the 1,2-disubstituted compounds to steric hindrance, an aspect which is discussed more fully in Sect. 4.6.4.

(11) λ_{max}^{MeOH} 430 nm, ε_{max} 5,500

(8) λ_{max}^{MeOH} 470 nm, ε_{max} 17,000

However, it should not be forgotten that the famous natural red dye, Madder, was 1,2-dihydroxyanthraquinone (alizarin) — see Sect. 1.2.3. Indeed, the old mordant dye, Turkey Red, was a metal complex derivative of alizarin (see Sect. 1.2.3). Most commercial 1,2-disubstituted anthraquinone dyes are used in the form of their metal complex derivatives (Sect. 4.3.2), especially those of alizarin.

Other Disubstituted and Polysubstituted Anthraquinones

None of the other possible substitution patterns for disubstituted anthraquinone compounds is of any importance commercially. Indeed, there is a dearth of spectral data for these compounds. However, spectral data are available for five of the ten possible positional isomers of the diaminoanthraquinone compounds. As seen from Table 4.8, the 2,3 derivative is the most hypsochromic and the 1,4 derivative the most bathochromic member of the series.

Table 4.8. Spectral data for isomeric diaminoanthraquinones

Substitution pattern	$\lambda_{max}^{CH_2Cl_2}$ (nm)
2,3	442
1,2 1,5	480
1,8	492
1,4	550, 592

Anthraquinones containing three or more substituents are used extensively as commercial dyes. Particularly important are the 1,2,4-trisubstituted and especially the 1,4,5,8-tetrasubstituted compounds because they are tinctorially the strongest and most bathochromic of the anthraquinone dyes.

Many commercial 1,2,4-trisubstituted dyes are derived from 1-amino-4-bromo-anthraquinone-2-sulphonic acid (Bromamine Acid) by condensing the latter with an amine (see Sect. 2.4.2). Thus, condensation with aniline gives the blue dye (53). The long wavelength band of this dye exhibits three peaks (there are also two shoulders) and it has been suggested that the additional peak is caused by intramolecular hydrogen-bonding between the sulphonic acid group and the amino

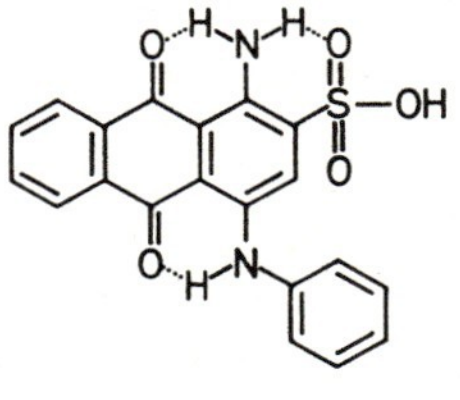

(53)

$\lambda_{max}^{H_2O}$ 604 623 640

ε_{max} 11,500 11,200 11,200

Shoulders at 570 (8,300) and 660 (10,000)

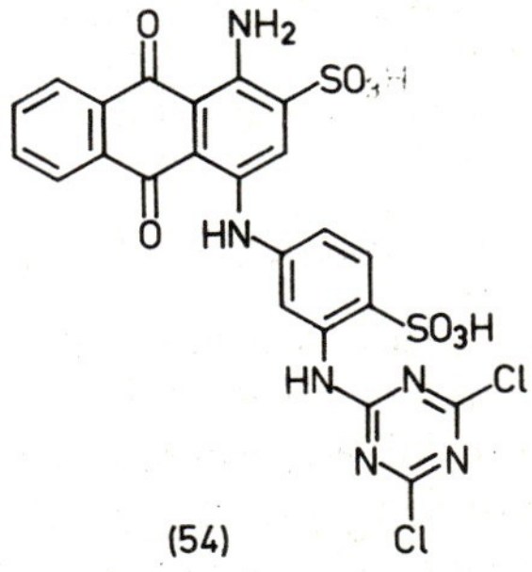

(54)

group. CI Reactive Blue 4 (54) is just one example of the many commercial dyes based on this type of chromogen.

An electron-donating group in the 2-position of a 1,4-bis-(donor-substituted)-anthraquinone causes a hypsochromic shift. Indeed, the majority of red commercial anthraquinone dyes are represented by structure (55).

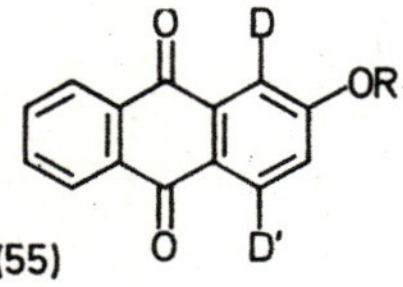

(55)

R = alkyl or aryl
D,D' = donor groups

The most important tetrasubstituted anthraquinone dyes contain four donor groups, usually amino or a combination of amino and hydroxy groups, in the 1,4,5 and 8 positions. They provide blue dyes such as CI Disperse Blue 1 (56) and CI Acid Blue 45 (57).

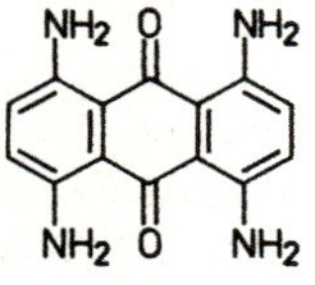

(56)

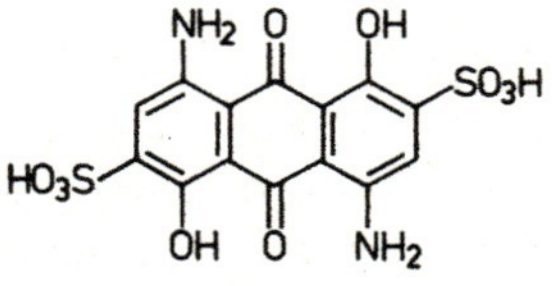

(57)

Many important commercial blue dyes are mixtures, *e.g.* partially brominated or methylated derivatives of (58) and partially methylated derivatives of (59).

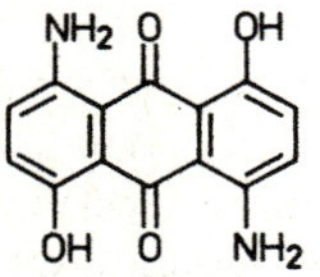

(58)

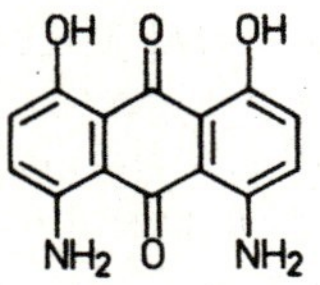

(59)

As discussed in Sect. 6.4, it is the outstanding thermal and (photo)chemical stability of the anthraquinone dyes, especially the blues, which account for their commercial importance.

Effect of Electron-withdrawing groups on Donor Substituted Anthraquinones

An electron-withdrawing group in a donor substituted anthraquinone can cause either a hypsochromic or a bathochromic shift depending on the number and positions of the donor groups and its location relative to them. Generally, however, a bathochromic shift is observed (Table 4.9). For example, in 1,4-bis-(donor-substituted)-anthraquinones, a 2-cyano group causes a bathochromic shift and blue dyes are obtained. These effects are discussed later (Sect. 4.6.3).

Table 4.9. Visible absorption maxima of some donor-acceptor substituted anthraquinones

Substituents	$\lambda_{max}^{CH_2Cl_2}$ (nm)	$\Delta\lambda$ (nm)
1-NH_2	465	0
1-NH_2-6-Cl	470	+ 5
1-NH_2-6,7-diCl	477	+12
1-NH_2-4-NO_2	460	− 5
2-NH_2	410	0
2-NH_2-1-Cl	405	− 5
2-NH_2-1-NO_2	410	0
1,8-diNH_2-4,5-diOH	590	0
1,8-diNH_2-4,5-diOH-2,3,6,7-tetraBr	654[a]	+64

[a] In chlorobenzene

Heterocyclic Anthraquinone Dyes

Few heterocyclic analogues of anthraquinone have been investigated as dyes. Heterocyclic dyes of structure (60) have similar tinctorial strengths to the corresponding anthraquinone dyes but they are generally more bathochromic, the

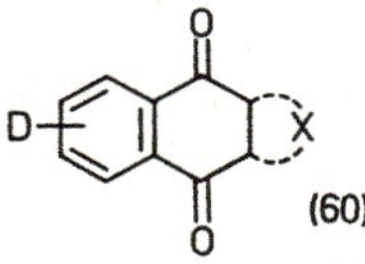

(60)

D = donor group(s)
X = heteroatom

bathochromicity increasing as the heterocyclic ring becomes more electron-withdrawing. Dyes of this type are not used commercially. In contrast, the 2,3-dicarboxyimide derivatives (61) of 1,4-diaminoanthraquinone provide commercially important bright turquoise dyes.

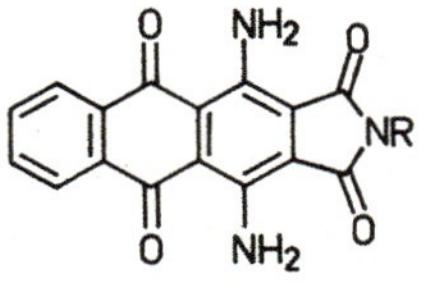

(61)

R = H $\lambda_{max}^{trichlorobenzene}$ 685 nm

Polyanthraquinone Dyes

In contrast to the polyazo dyes (see Sect. 3.5.8) polyanthraquinone dyes (excluding vat dyes — see Sect. 5.2) are of little importance. If two identical anthraquinone dyes are linked together by an insulating group, *e.g.* methylene or *m*-phenylene, the resultant dye (63) has the same λ_{max} as the parent dye (62) but double the intensity (Fig. 4.6). Of course, the molecular weight has roughly doubled, so such dyes offer no economic advantages.

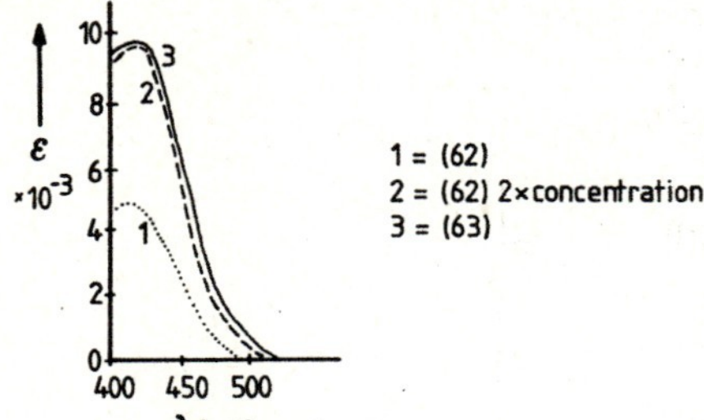

Fig. 4.6. Effect of linking two anthraquinone dyes

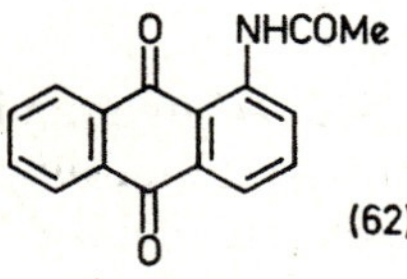

NHCOMe

(62)

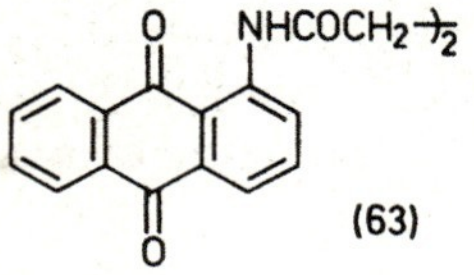

$NHCOCH_2 \frac{}{}\!)_2$

(63)

As in azo dyes, the combination of a yellow dye and a blue dye produces a green dye (64).

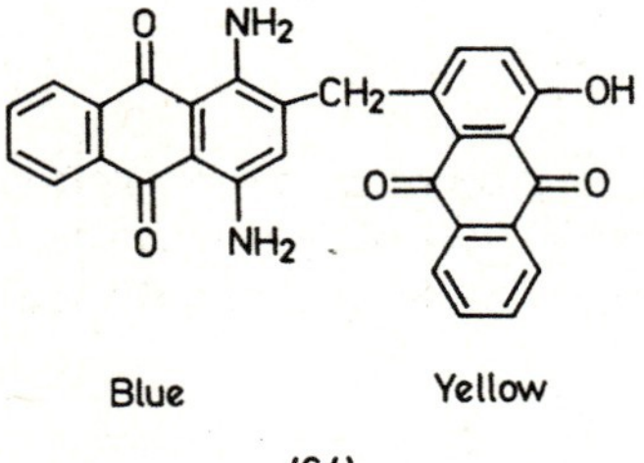

$\lambda_{max}^{C_6H_6}$ 412; 585 nm

ε_{max} 7,250; 1,800

Blue Yellow

(64)

4.6.3 VB/MO Explanation of Colour and Constitution

The VB and MO theories of colour and constitution of anthraquinone dyes will now be considered separately, beginning with VB theory.

VB Theory

As in azobenzene (see Sect. 3.5.4) the intense long wavelength transition of anthraquinone (65) — neglecting the weak $n \rightarrow \pi^*$ transition at *ca.* 405 nm — occurs in the ultra-violet (Fig. 4.3). This is because the charge separated structures (65A and 65B), which represent crudely the excited state, are relatively unstable — they have lost the RSE of a benzene ring and contain a positively charged carbon atom having only six valence electrons. Therefore the energy gap between the ground state and the first excited state is large.

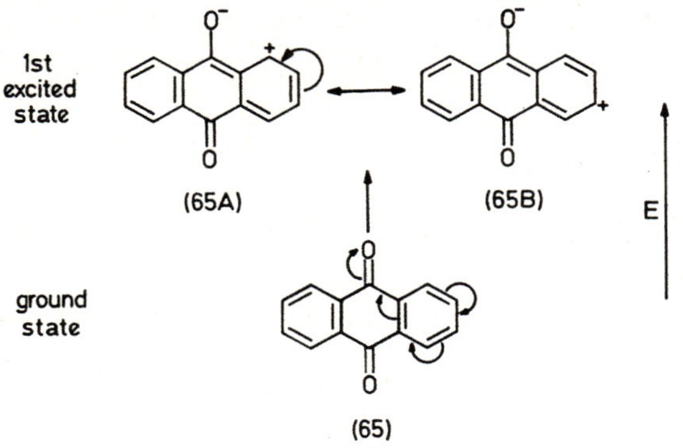

1st excited state

(65A) (65B)

E

ground state

(65)

Monosubstituted Anthraquinones

An electron-withdrawing substituent has a negligible effect since an anthraquinonoid carbon atom is still the principal electron donor group. However, a donor substituent[4] markedly stabilises the excited state since all the atoms now possess 8 electrons in their valency shell. The magnitude of the bathochromic shift depends on the electron-donating power of the group, *i.e.* NHMe > NH_2 > NHAc > OMe (Table 4.3).

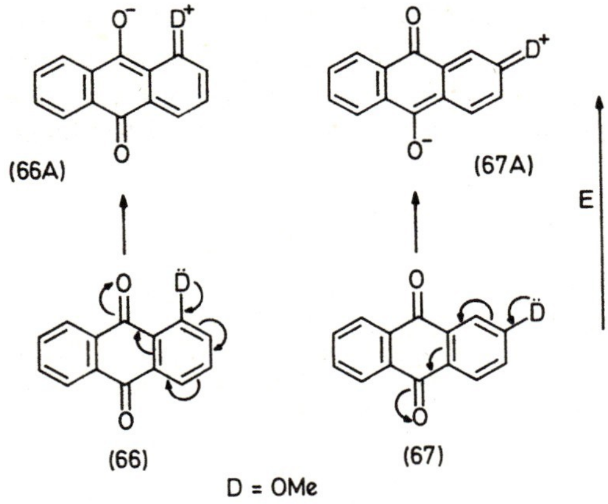

(66A) (67A)

E

(66) (67)

D = OMe

λ_{max}^{MeOH} 378 nm λ_{max}^{MeOH} 363 nm

The greater bathochromicity of 1-substituted anthraquinones (66) compared to their 2-substituted isomers (67) arises from two effects. First, the charged atoms are in close proximity in the 1-substituted compounds and this results in a greater electrostatic (+, − attraction) stabilisation. The second factor, which only applies to groups containing a labile hydrogen atom such as OH, NH_2 and NHR, is intramolecular hydrogen-bonding. This has a profound effect since it not only enhances both the electron-donating ability of the donor group and the electron-withdrawing ability of the carbonyl group, it also holds that group in a planar conformation thereby maximising the orbital overlap. These effects are shown in Fig. 4.7. Protonation of the carbonyl group and ionisation of the hydroxy group

[4] Only those with $2p_z$ lone pair electrons are considered.

take these effects to their extreme limits and explains the bathochromic shift observed for anthraquinone in an acid medium and hydroxyanthraquinones in an alkaline medium.

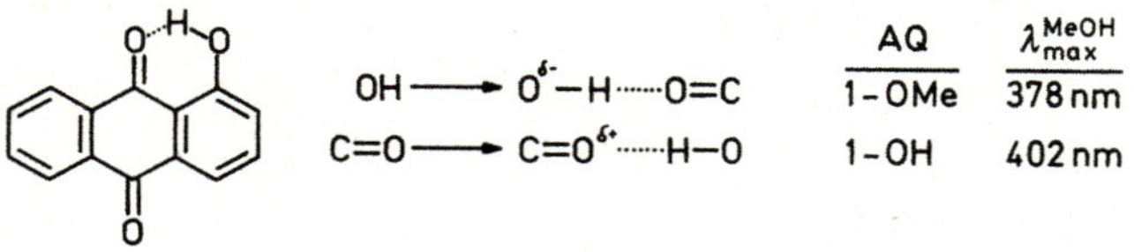

OH ⟶ $O^{\delta-}$—H······O=C

C=O ⟶ C=$O^{\delta+}$······H—O

AQ	λ_{max}^{MeOH}
1-OMe	378 nm
1-OH	402 nm

Fig. 4.7. Intramolecular hydrogen-bonding in 1-hydroxyanthraquinone

Disubstituted Anthraquinones

The most striking feature of the commercially important disubstituted anthraquinones is the marked bathochromicity of the 1,4-compounds compared to their 1,5-, 1,8- or 1,2-isomers.

In 1,5-disubstituted anthraquinones the excited state may be crudely represented by the charge separated structures (68A, of which there are two; and 68B). There is no interaction between the two halves of the chromogen (separated by the dashed line): thus, 1,5-disubstituted anthraquinones resemble 'doubled-up' 1-substituted anthraquinones — they have a similar λ_{max} but double the intensity of the latter compounds (see Table 4.4).

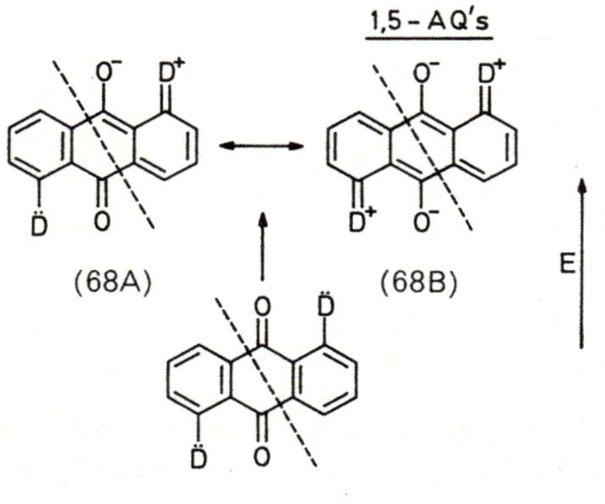

1,5 - AQ's

(68A) (68B)

E

D = OR, NHR, etc.

(68)

In contrast, the corresponding charge separated structures for 1,4-disubstituted anthraquinones (69B) are predicted to be exceptionally stable since they contain a fully aromatic naphthalene nucleus (*cf.* 68B). Hence, the energy gap between the ground and first excited state of 1,4-disubstituted anthraquinones is much smaller than that for the 1,5-isomers and the former compounds would therefore be expected to absorb at longer wavelengths.

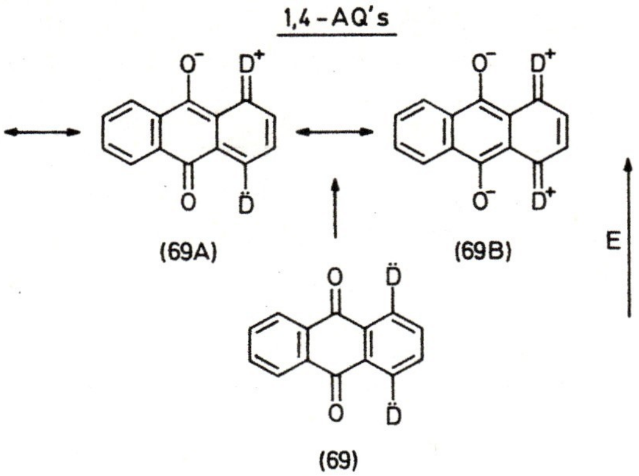

1,4-AQ's

(69A) (69B) E

(69)

There is more conjugation in 1,8-disubstituted anthraquinones than in the 1,5-isomers but less than in the 1,4-isomers. Thus, although the donor groups can interact with each other, they are only conjugated to one carbonyl group (70A and 70B). None of the structures contributing to the excited state contains a naphthalene nucleus.

1,8-AQ's

(70A) (70B) E

(70)

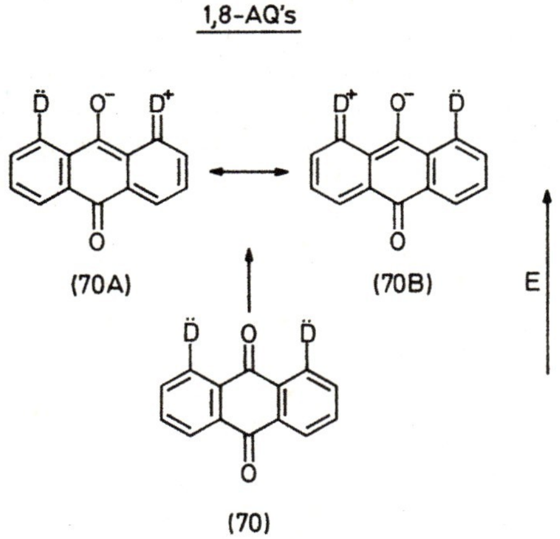

In 1,2-disubstituted anthraquinones two of the charge separated structures (71A and 71B) contributing to the excited state are analogous to those for 1- and 2-substituted anthraquinones (66A and 67A respectively). Like the 1,4-disubstituted anthraquinones, structure (71C) also contains a naphthalene nucleus and hence makes the major contribution to the excited state. The excited state is of higher energy (*i.e.* less stable) than that of the 1,4-isomer since it contains an *ortho* rather than a *para*-quinonoid ring and because of the lower electrostatic attraction involved (*e.g.* one attractive and one repulsive interaction in 71C).

1,4,5,8-Tetrasubstituted anthraquinones are the most bathochromic of the anthraquinone dyes because the major contributing structures (72A and 72B) to the excited state each contain a naphthalene ring: therefore, the dyes may be considered as being composed of two 1,4-bis-(donor-substituted)-anthraquinone units within one molecule.

1,2-AQ's

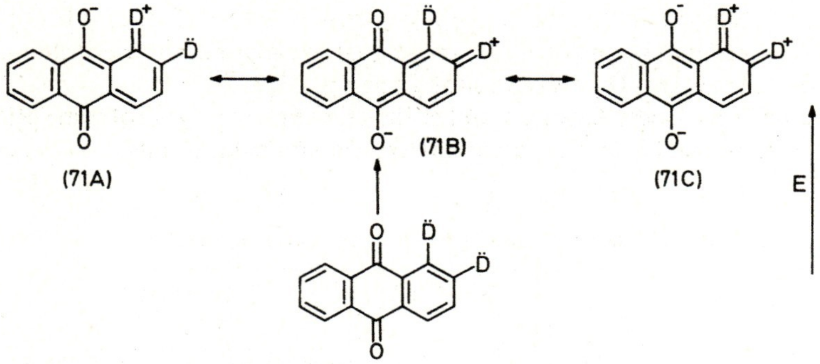

(71A) (71B) (71C) (71)

E

1,4,5,8-AQ's

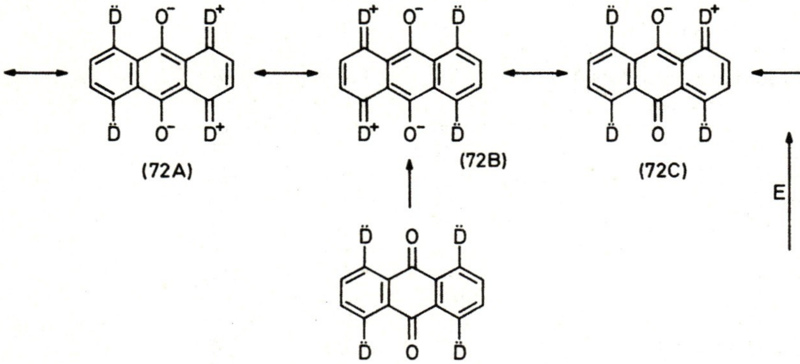

(72A) (72B) (72C) (72)

E

VB theory therefore predicts the following order of bathochromicity for the important donor substituted anthraquinone dyes: — 2 < 1 < 1,5 < 1,8 < 1,2 < 1,4 < < 1,4,5,8. As seen from Table 4.10, this is in excellent agreement with experiment.

Table 4.10. Relative bathochromicities of donor-substituted anthraquinones

AQ	λ_{max}^{MeOH} (nm)
2-NH_2	440
1-NH_2	475
1,5-di-NH_2	487
1,8-di-NH_2	507
1,2-di-NH_2	509
1,4-di-NH_2	550, 590
1,4,5,8-tetra-NH_2	600, 629

MO Theory

Although Dewar's Rules are, in theory, applicable to anthraquinone dyes, they do not appear to have been used for predicting qualitatively the effect of substituents on the colour of anthraquinone dyes. Thus, we shall concentrate on the application of quantitative PPP MO theory to anthraquinone dyes.

Such studies have been fewer in number than those on azo dyes but some calculated results are shown in Table 4.11, alongside the experimental results.

Table 4.11. PPP MO calculated results for some anthraquinone dyes[a]

AQ	Calculated		Experimental (in CH_2Cl_2)
	λ_{max}	f	λ_{max}
1-OMe	381	0.23	380
1-OH (intramolecularly) (H-bonded)	409	0.25	405
1-NH_2 (intramolecularly) (H-bonded)	465	0.25	465
1-NHMe (intramolecularly) (H-bonded)	509	0.25	508
2-OH	360	0.11	365
2-NH_2	409	0.17	410
1,2-di-OH	414	0.26	416
1,4-di-OH	472	0.33	476
1,5-di-OH	418	0.40	428
1,8-di-OH	424	0.37	430
1,4-di-NH_2	562	0.35	550
1,4-di-NHMe	619	0.40	620
1-OH-4-NH_2	522	0.33	520
1,5-di-OH-4,8-diNH_2	552	0.53	590
1,4,5,8-tetra-NH_2	609	0.56	610

[a] Kogo, Y., et al.: J. Soc. Dyers and Colourists *96*, 475 (1980).

The calculated electron density changes[5] for key positions in the first excited state of 1-aminoanthraquinone (73) and 2-aminoanthraquinone (74) are shown in Fig. 4.8. Though the calculations indicate that the amino group is the principal donor, the remainder of the changes are complex. In particular, the carbonyl groups are not the sole electron acceptors: the unsubstituted ring shows a significant overall increase in electron density. It is noteworthy that for 1-aminoanthraquinone the increase of π-electron density accompanying oxygen atom 10, which is not conjugated to the amino group, is greater than at oxygen atom 9, which is. Therefore, as was the case in the azo dyes, it is concluded that the VB representation of the first excited state corresponds to that of the ground state as calculated by PPP MO theory.

Electron-withdrawing groups have either a bathochromic or hypsochromic effect depending on their position within the anthraquinone nucleus and the predicted electron density change at that position. Thus, electron-withdrawing groups in any

[5] See ref. E4.: p. 177.

	(73)	(74)
1	−0.43	−0.08
2	−0.06	−0.41
3	+0.03	−0.10
4	+0.11	+0.10
5	+0.02	+0.02
6	+0.03	+0.04
7	+0.04	+0.04
8	+0.02	+0.03
9	+0.08	+0.15
10	+0.13	+0.07

Fig. 4.8. Changes in π-electron density accompanying electronic excitation to the first excited state for (73) and (74). (A positive sign represents a build up of π-electron density and a negative sign a decrease in π-electron density).

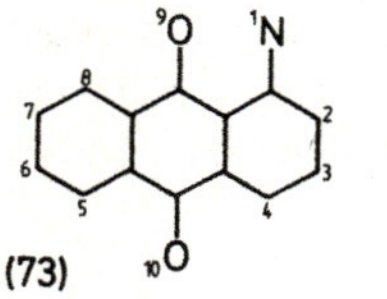

(73)

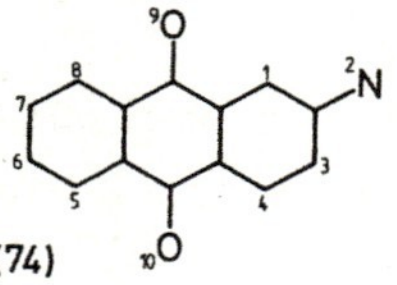

(74)

position of the unsubstituted ring in 1- and 2-aminoanthraquinone are predicted to give a bathochromic shift since all these carbon atoms are electron rich in the excited state. A bathochromic shift is observed experimentally (see Tables 4.9 and 4.12). In contrast, for an electron-withdrawing group at an electron deficient position, *e.g.* the 4-position in 1-aminoanthraquinone and the 1-position in 2-aminoanthraquinone, a hypsochromic shift is predicted: this is the case experimentally (Table 4.9). However, the predicted bathochromic shift of the 1-nitro group in 2-aminoanthraquinone is incorrect.

Table 4.12. Donor/acceptor substituted anthraquinones

AQ	$\lambda_{max}^{CH_2Cl_2}$ (nm)	Δλ (nm)	Prediction
1-NH_2	465		
1-NH_2-6-Cl	470	+ 5	✓
1-NH_2-6,7-diCl	477	+11	✓
1-NH_2-4-NO_2	460	− 5	✓
2-NH_2	410		
2-NH_2-1-Cl	405	− 5	✓
2-NH_2-1-NO_2	410	0	×

Several explanations have been proposed for the twin peaks observed in anthraquinones substituted in the 1- and 4-positions by powerful electron donor groups, *e.g.* NH_2, NHMe, NMe_2 and $O^{\ominus}$. Since twin peaks are not observed for the weaker electron-donating groups such as OH and NHAc, then intramolecular hydrogen-bonding is not the cause. Neither is it a solvent or aggregation effect since the twin peaks persist in the vapour phase as already stated. The suggestion that the peaks correspond to two distinct electronic transitions is not supported by PPP MO calculations which clearly indicate only *one* electronic transition in the visible region of the spectrum for such compounds. On this basis, it is thought that the phenomenon is caused by vibrational fine structure.

PPP calculations also account for the green colour of the 1,4-diarylaminoanthraquinone dyes such as (51). Two bands are calculated in the visible region of the the spectrum, one at λ_{max} 625 nm ($\lambda_{max}^{CCl_4}$ 638 nm) which produces a blue colour, and one at λ_{max} 407 nm ($\lambda_{max}^{CCl_4}$ 402 nm) responsible for a yellow colour — hence the observed green colour (see Fig. 4.5).

Both the VB and MO theories are supported by the effect of solvents on the colour of anthraquinone dyes. Increasing the polarity of the solvent causes a bathochromic shift (positive solvatochromism) which is typical of compounds in which the excited state is more polar (*i.e.* more charge separated) than the ground state. These effects are illustrated in Table 4.13.

Table 4.13. Positive solvatochromism of anthraquinone dyes

AQ Dye	λ_{max} (nm)	
	CH_2Cl_2	MeOH
2-NH_2	410	440
1-NH_2	465	475

Tinctorial Strength

As discussed in Sect. 3.5.5 VB theory cannot predict tinctorial strength. The empirical rule of thumb approach that tinctorial strength increases in line with increasing bathochromicity is generally true, although there are notable exceptions. For example, the red dye, 1,8-diaminoanthraquinone (λ_{max}^{MeOH} 507 nm, ε_{max} 10,000) is significantly weaker than the orange dye, 1,4-dihydroxyanthraquinone (λ_{max}^{MeOH} 470 nm, ε_{max} 17,000).

In contrast to VB theory, PPP MO theory provides quantitative values for tinctorial strength via the oscillator strength f (see Sect. 3.5.5). This enables comparisons to be made not only with different anthraquinone dyes but also with other dye classes, *e.g.* azo dyes. The f values for anthraquinone dyes are in reasonable agreement with experimental tinctorial strength values, ranging from $f = 0.25$ for mono-(donor-substituted)-anthraquinones such as 1-aminoanthraquinone, to $f = \sim 0.6$ for 1,4,5,8-tetra-(donor-substituted)-anthraquinones (see Table 4.11). They therefore rationalise the observed lower tinctorial strength of anthraquinone dyes relative to azo dyes — the latter have f values ranging from ~ 1 to ~ 1.5 (see Sect. 3.5.5).

As mentioned in Sect. 4.1, anthraquinone dyes, especially reds and blues, are generally brighter than azo dyes. Fig. 4.9 shows the absorption spectrum of a typical anthraquinone dye, CI Basic Blue 22 (75). The absorption curve is as broad ($\Delta v_{1/2}$ = 3300 cm^{-1}) as that of a typical blue azo dye (*e.g.* structure 120 in Chap. 3) and these are dull! Obviously some other factor is responsible for the 'brightness' of anthraquinone dyes: this other factor is fluorescence. In nearly all dyes fluorescence markedly increases the brightness of the dye. Though the fluorescence of most anthraquinone dyes is small, it is nonetheless sufficient to be detected by the human eye. Fluorescence is dealt with in more detail in Chap. 6.

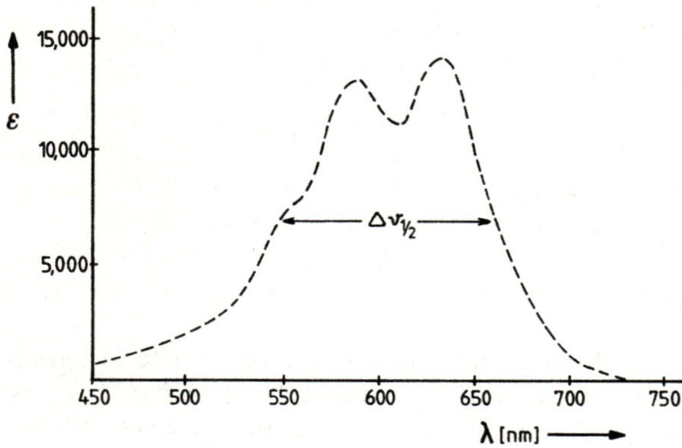

Fig. 4.9. Absorption spectrum of CI Basic Blue 22 (75) (in water)

(75) O NHMe, O HN–(CH$_2$)$_3$–$\overset{+}{N}$Me$_3$ · A$^-$

Polarisation of Absorption Bands

Experiments using dyes in stretched films indicate that for 1-(donor-substituted)-anthraquinone dyes the long wavelength absorption band in the visible region of the

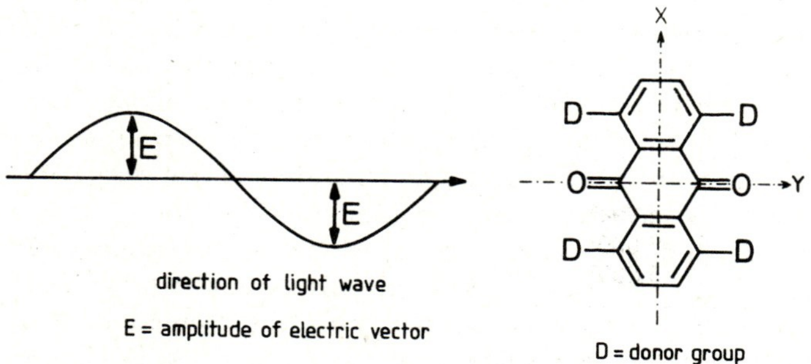

Fig. 4.10. Orientation of anthraquinone dyes for maximum absorption of visible light

spectrum is polarised along the molecular axis, *i.e.* in the x-direction. This agrees with the electron drift on excitation predicted by both the VB and MO theories. The optimum orientation of the molecule for maximum absorption of light is therefore that depicted in Fig. 4.10. In 2-(donor-substituted)-anthraquinones the long wavelength band is polarised along the y-axis. In both 1- and 2-(donor-substituted)-anthraquinones the bands in the ultra-violet are polarised alternately in the x and y directions, as shown in Table 4.14.

Table 4.14. Polarisation of the ultra-violet bands of anthraquinone dyes

λ_{max} of $\pi \rightarrow \pi^*$ band	Direction of polarisation
~324	x
~272	y
~252	x
~249	y

4.6.4 Steric Effects

As discussed in Sect. 3.5.9 steric hindrance *always* causes a reduction in tinctorial strength but may cause either a hypsochromic or a bathochromic shift of λ_{max}. A hypsochromic shift is observed when the bond about which twisting occurs to relieve the steric hindrance has a higher bond order in the excited state than in the ground state. A bathochromic shift results for the reverse case (see Sect. 3.5.9). Both PPP MO calculations and qualitative VB theory indicate that the former situation applies to neutral anthraquinone dyes and steric hindrance should therefore cause a hypsochromic shift.

In anthraquinone dyes steric hindrance can be between a bulky 1-substituent and a carbonyl group or between two *ortho* substituents. 1-NN-Dimethylaminoanthraquinone provides an example of the first type. This dye is slightly more hypsochromic than 1-N-methylaminoanthraquinone (in CH_2Cl_2) and significantly weaker (Table 4.15). Intramolecular hydrogen-bonding holds the NHMe group coplanar with the anthraquinone nucleus (76). Hence the overlap between the $2p_z$ lone pair and the

Table 4.15. Examples of steric hindrance in anthraquinone dyes

AQ	λ_{max} (nm)	ε_{max}
1-NHMe	508[a]	7,100[b]
1-NMe_2	504[a]	4,900[b]
1-NHAc	414[c]	5,000[c]
1-NMeAc	337[c]	5,000[c]

[a] in CH_2Cl_2
[b] in MeOH
[c] in *o*-chlorophenol

π-system is at a maximum. Replacing the hydrogen atom by the more bulky methyl group destroys both the intramolecular hydrogen-bonding and the coplanarity because of steric hindrance between the NMe_2 group and the adjacent C=O group. The steric interactions are minimised by the NMe_2 group rotating out of the plane of the molecule, as depicted in (77). Consequently, the overlap of the nitrogen $2p_z$ orbital with the π-system is considerably diminished — hence the lower tinctorial strength. This latter effect is the prime reason why 1-NN-dialkylaminoanthraquinones have not found use as commercial dyes.

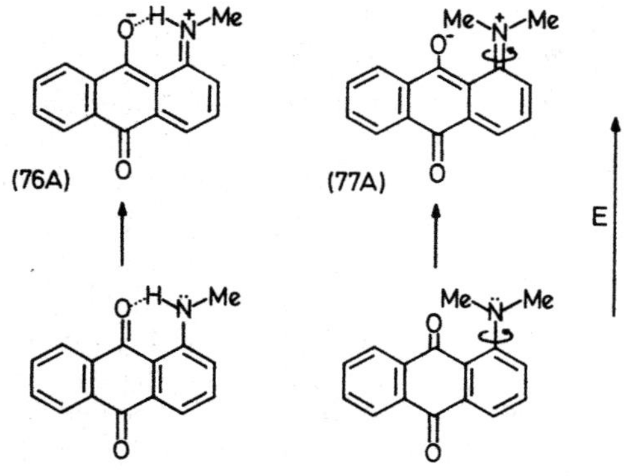

(76A) (77A)

E

(76) (77)

The more bulky N-methylacetylamino group apparently causes a large hypsochromic shift (Table 4.15). However, since the ε_{max} remains unaltered it is probable that the band at *ca.* 337 nm corresponds to the long wavelength $\pi \rightarrow \pi^*$ transition of anthraquinone (see Fig. 4.3). This infers that the steric hindrance between the substituent and the adjacent carbonyl group is so great that the charge transfer band is not observed at all!

The second type of steric hindrance is exemplified by the dye (78; R=Me). The 2-methyl group twists the cyclohexylamino group out of planarity, as shown by the drop in ε_{max} and the hypsochromic shift. Steric interactions of this kind are a contributing factor to the hypsochromicity and lower tinctorial strength of 1,2-bis-(donor-substituted)-anthraquinone dyes relative to their 1,4-isomers (compare dye 11 with 8).

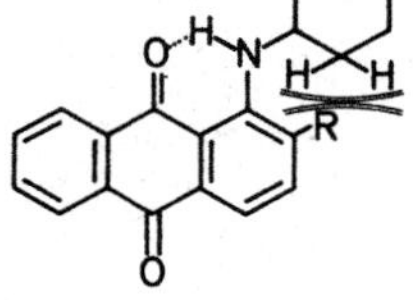

R	$\lambda_{max}^{toluene}$	ε_{max}
H	510 nm	3,020
Me	476 nm	2,188

(78)

4.7 Summary

Anthraquinone dyes comprise the next most important class after the azo dyes. They provide mainly blue and turquoise (and, to a lesser extent, red) colours since azo dyes generally cannot equal the excellent properties of such dyes, especially their brightness and light fastness. However, the inherently lower tinctorial strength of anthraquinone dyes makes them uneconomic against azo dyes and this is the prime reason why azo dyes are gradually replacing anthraquinone dyes.

In contrast to hydroxyazo dyes, tautomerism has not been observed in hydroxyanthraquinone dyes. Investigation of such compounds by ^{13}C nmr spectroscopy has shown that they exist as hydroxy derivatives of 9,10-anthraquinone. Similarly, aminoanthraquinone dyes exist as amino derivatives of 9,10-anthraquinone.

Protonation of aminoanthraquinone dyes occurs exclusively on the nitrogen atom of the amino group: surprisingly, there is no protonation on the oxygen atom of the carbonyl group to give a resonance stabilised cation. Hence, protonation results in a hypsochromic shift.

Addition of alkali to hydroxyanthraquinone dyes causes ionisation of the hydroxy groups and a bathochromic shift ensues. Hydroxy groups in the 1-position are intramolecularly hydrogen-bonded with the carbonyl group and are therefore more difficult to ionise than 2-hydroxy groups.

In contrast to metallised azo dyes, metal complex anthraquinone dyes are not important commercially. Azo dyes are metallised primarily to improve certain properties, particularly light fastness. Since anthraquinone dyes already exhibit excellent light fastness, metallisation offers no advantages — indeed, it nullifies the brightness of such dyes relative to azo dyes.

Colour-structure relationships are complex but depend primarily on the π-electronic make-up of the molecule, although steric effects can also be important. Only anthraquinones containing donor substituents having a filled $2p_z$ orbital such as NH_2 or OH have intense absorption bands in the visible region of the spectrum — hence all commercial anthraquinone dyes are of this type. In general, 1-(donor-substituted)-anthraquinones are stronger and more bathochromic than their 2-substituted isomers. The most favourable substitution patterns for high tinctorial strength and bathochromicity are 1, 4 and 1, 4, 5, 8. Consequently the majority of commercial anthraquinone dyes are based on these systems.

As in the azo dyes, the VB and MO theories are used for studying colour-structure effects. Qualitative VB theory rationalises and even predicts the colour of anthraquinone dyes very well indeed. However, it cannot deal with tinctorial strength. PPP MO theory provides quantitative values for both the colour and the tinctorial strength of the dyes and this theory should be used for a fundamental understanding of the phenomenon.

Steric hindrance in anthraquinone dyes *always* lowers the tinctorial strength and generally causes a hypsochromic shift.

In conclusion, although azo dyes pose a serious threat to existing anthraquinone dyes, this will be countered to some extent by the continuing search for quinone dyes with increased tinctorial strength, and by research into cheaper routes to existing anthraquinone dyes, a pattern which is already apparent in quinizarin chemistry (see Chap. 2).

4.8 Bibliography

A. *Structure of Anthraquinone Dyes*

1. Gleicher, G. J.: Theoretical and general aspects. In: The chemistry of the quinonoid compounds, Patai, S. (ed.), Part 1, pp. 8–13. London, New York, Sydney, Toronto: John Wiley and Sons 1974
2. Bernstein, J., Cohen, M. D., Leiserowitz, L.: The structural chemistry of quinones. In: The chemistry of the quinonoid compounds, Patai, S. (ed.), Part I, pp. 70–77. London, New York, Sydney, Toronto: John Wiley and Sons 1974
3. Bowen, H. J. M. (comp.) et al.: Tables of interatomic distances and configuration in molecules and ions. Chemical Society Special Publication No. 11, pp. M238. London: The Chemical Society 1958

B. *Tautomerism*

(a) *Hydroxyanthraquinone Dyes*

1. Arnone, A., et al.: J. Mag. Reson. *28*, 69 (1977)
2. Berger, Y., Castonguay, A.: Org. Mag. Reson. *11*, 375 (1978)
3. Ref. A2.: p. 76
4. Graf, H., et al.: Chem. Phys. Letters *59*, 217 (1978)
5. Dumas, J., Cohen, A., Gomel, M.: Bull. Chem. Soc. Fr. *1972*, 1340

(b) *Reduced Hydroxyanthraquinone Dyes*

6. Bloom, S. M., Hutton, R. F.: Tetrahedron Letters *1963*, 1993

(c) *Aminoanthraquinone Dyes*

7. Ref. A2.: pp. 76–77

(d) *Reduced Aminoanthraquinone Dyes*

8. Ref. B6.: p. 1996

C. *Protonated and Ionised Anthraquinone Dyes*

1. Peters, R. H., Sumner, H. H.: J. Soc. Dyers and Colourists *69*, 2101 (1953); *72*, 77 (1956)
2. Issa, R. M., El-Ezaby, M. S., Zewail, A. H.: J. Phys. Chem. *244*, 155 (1970)
3. Gillet, H., Pariaud, J.-C.: Bull. Chem. Soc. Fr. *1966*, 2624
4. Fabian, J., Hartmann, H.: Light absorption of organic colorants. Theoretical treatment and empirical rules. pp. 112–114. Berlin, Heidelberg, New York: Springer-Verlag 1980

D. *Metal Complexed Anthraquinone Dyes*

1. Allen, R. L. M.: Colour chemistry, pp. 143–148. London: Nelson 1971
2. Colour Index, 3rd ed., Vol. 4, pp. 4511–4590. The Society of Dyers and Colourists, Bradford; The American Association of Textile Chemists and Colorists, Lowell, Mass., U.S.A.
3. Chung, R. H., Farris, R. E.: Dyes, anthraquinone. In: Encyclopedia of chemical technology, Kirk-Othmer (eds.), 3rd ed., Vol. 8, pp. 212–279. New York, Chichester, Brisbane, Toronto: John Wiley and Sons 1979.

E. *Colour and Constitution*

(a) *Neutral Anthraquinone Dyes*

1. Fain, V. Ya.: The electronic spectra of anthraquinones, 1970. Translated by R. J. Ramsden and P. J. Rhodes, edited by R. W. A. Oliver, 1974. (R. W. A. Oliver, 71 Harboro Road, Sale, Cheshire).
2. Ref. C1
3. Labhart, H.: Helv. Chim. Acta *40*, 1410 (1957)
4. Griffiths, J.: Colour and constitution of organic molecules. pp. 176–180. London, New York, San Francisco: Academic Press 1976
5. Ref. C4.: pp. 100–110
6. Hida, M.: Dyestuffs and Chemicals *8*, 493 (1963)
7. Griffiths, J.: Rev. Prog. in Colouration: In Press
8. Tanizaki, Y., et al.: Bull. Chem. Soc. Japan *45*, 1018 (1972)
9. Egerton, G. S., Roach, A. G.: J. Soc. Dyers and Colourists *74*, 401 (1958)
10. Matsuoka, M., Kishimoto, M., Kitao, T.: J. Soc. Dyers and Colourists *94*, 435 (1978)

(b) *Protonated and Ionised Anthraquinone Dyes*

11. Ref. C2
12. Ref. C1
13. Ref. C3

(c) *Fluorescence of Anthraquinone Dyes*

14. Allen, N. S., McKellar, J. F.: J. of Photochemistry *5*, 317 (1976); *7*, 107 (1977)
15. Allen, N. S., Bentley, P., McKellar, J. F.: J. of Photochemistry *5*, 225 (1976)
16. Allen, N. S., Harwood, B., McKellar, J. F.: J. of Photochemistry *9*, 559, 565 (1978)
17. Nakashima, M., Roach, J. F.: Spectroscopy Letters *12*, 139 (1979)

Chapter 5

Miscellaneous Dyes

5.1 Introduction

Having discussed the two most important dye classes, namely azos and anthraquinones, we shall now consider six other important dye types. These are: —
1. Vat
2. Indigoid
3. Phthalocyanines
4. Polymethines
5. Aryl carboniums, and
6. Nitro and Nitroso dyes.

Together, these eight dye classes cover the vast majority of synthetic organic dyestuffs used today.

To keep the chapter manageable only the more important and interesting aspects of each class will be discussed. The order used in Chapter 2 is repeated here. Hence, vat dyes, which are related structurally to the anthraquinone dyes discussed in Chapter 4, are considered first.

5.2 Vat Dyes

5.2.1 Introduction

Vat dyes are dyes of any chemical class that are applied to the fibre by a vatting process. They are insoluble in water and cannot be used for dyeing the fibre directly because of their lack of affinity. However, upon reduction a vat dye is converted into its water soluble *leuco* form which does have affinity for cellulosic fibres. On exposure of the dyed fabric to the atmosphere the *leuco* form is oxidised back to the original water insoluble form which is thus trapped within the fibre pores (see Sect. 6.3.4 for further details). Hence, vat dyes generally have excellent fastness to washing treatments.

Structurally, most vat dyes contain one or more carbonyl groups linked by a quinonoid system. Many vat dyes contain no further substituents and it is because of the lack of powerful donor and acceptor groups that vat dyes generally have excellent chemical and photochemical stability. (This aspect is discussed more fully in Sect. 6.4.4). Indeed, the fastness properties of vat dyes are surpassed by no other class of dyestuffs.

The two most important chemical classes of vat dyes are the polycyclic aromatic carbocycles containing one or more carbonyl groups (often called the anthraquinonoid vat dyes), and the indigoid dyes. For such a small molecule (1), the indigoid dyes have surprisingly bathochromic shades, (*i.e.* blue). Hence, they are discussed separately (Sect. 5.3). This section is therefore restricted to the most important type of vat dyes — the anthraquinonoid vat dyes.

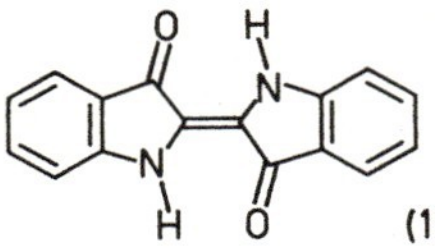

(1)

5.2.2 The Anthraquinonoid Vat Dyes

Over 200 anthraquinonoid vat dyes are at present in commercial use and their hues cover the whole of the visible spectrum, *i.e.* from yellow through blue to black. The most important commercial dyes are the blues, greens, browns and blacks, since these dyes, unlike many of the yellow and orange dyes, do not cause phototendering[1] of the fibre (see Sect. 6.4.7). Because the *leuco* form has most affinity for cellulosic fibres, the latter fibres constitute the major outlet for the vat dyes.

The constitutions of the anthraquinonoid vat dyes vary from simple structures, *e.g.* CI Vat Yellow 3 (2), resembling the anthraquinone disperse dyes discussed in Chapter 5, to very complex structures containing up to 19 condensed rings, *e.g.* CI Vat Green 8 (3).

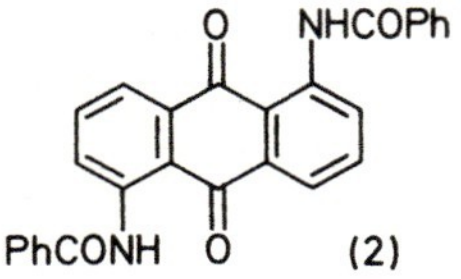

O NHCOPh

PhCONH O (2)

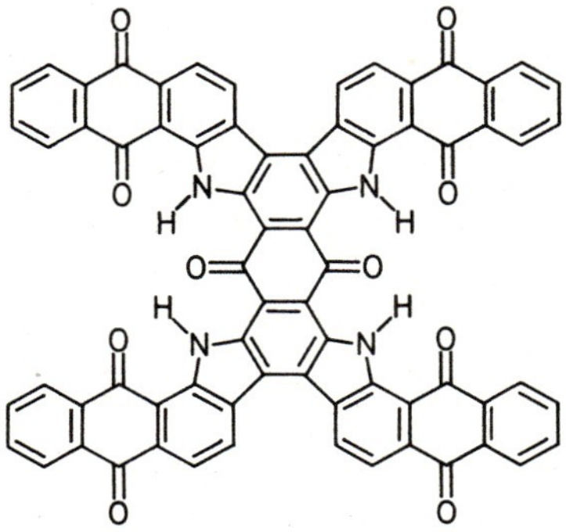

(3)

[1] Phototendering is the name given to the phenomenon whereby a dyed fabric undergoes adverse structural changes in the presence of light.

Some of the more important types of anthraquinonoid vat dyes are now discussed. Obviously, only the more significant and interesting aspects can be considered in such a short treatise. Hopefully, the flavour of the chemistry involved will become apparent.

Acylaminoanthraquinones

The members of this class are structurally the simplest vat dyes. They provide mostly yellow to red shades. However, their importance has declined due to competition from more cost effective products from other dye types, *e.g.* azo dyes. A typical example of an acylaminoanthraquinone vat dye is CI Vat Yellow 3 (2).

Indanthrones

The first vat dye of practical importance was discovered by René Bohn in 1901. He endeavoured to make diphthaloylindigo (5) by the alkaline fusion of the anthraquinone (4) — Scheme 5.1. However, instead of the anticipated compound, he obtained a brilliant blue vat dyestuff with exceptionally high fastness properties. Its structure was later elucidated as (6). Bohn named his product Indanthren (derived from *ind*igo and *anthracene*): it was later marketed by BASF[2] as Indanthrene Blue R (CI Vat Blue 4). Subsequently, it was given the chemical name indanthrone, thereby avoiding the trade name and simultaneously indicating the quinonoid structure.

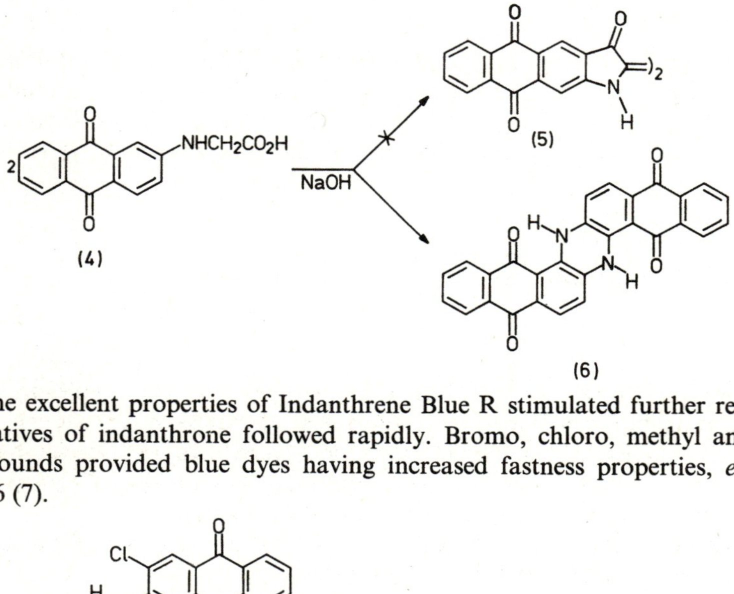

2 $NHCH_2CO_2H$ (4) NaOH (5) (6)

Scheme 5.1

The excellent properties of Indanthrene Blue R stimulated further research and derivatives of indanthrone followed rapidly. Bromo, chloro, methyl and hydroxy compounds provided blue dyes having increased fastness properties, *e.g.* CI Vat Blue 6 (7).

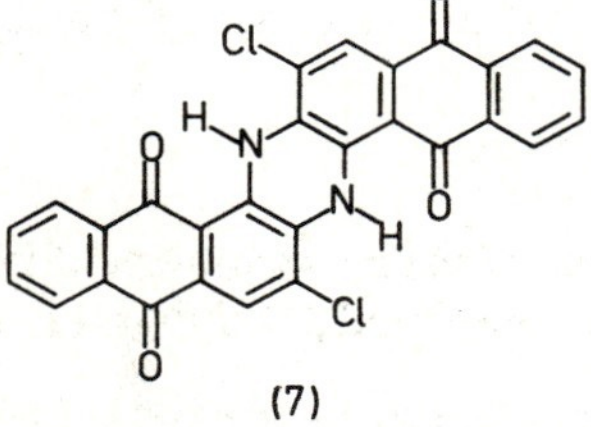

Cl Cl (7)

[2] Even today BASF are the leading manufacturers of vat dyes.

Benzanthrones

The benzanthrones are undoubtedly the single most important group of vat dyestuffs since a large number of very important violet, blue and green dyes are benzanthrone derivatives.

The first of these is dibenzanthrone (also called violanthrone), CI Vat Blue 20 (9), prepared by the alkaline fusion of benzanthrone (8). Dibenzanthrone was first marketed by BASF in 1904. It gives dark blue shades on cellulosic fibres and silk, and displays good fastness to light. Dibenzanthrone is still used extensively today.

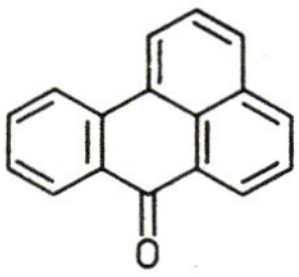

(8)

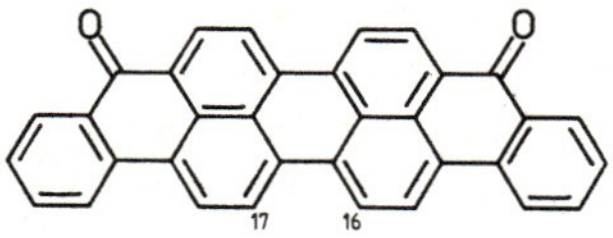

17 16

(9)

The introduction of methoxy groups at the 16 and 17 positions of dibenzanthrone (9) causes a significant bathochromic shift and gives rise to *the* outstanding vat dyestuff — Caledon[3] Jade Green XBN (10) — CI Vat Green 1. As discussed earlier (Sect. 2.4.3), (10) is actually prepared by methylating dihydroxydibenzanthrone.

OMe OMe

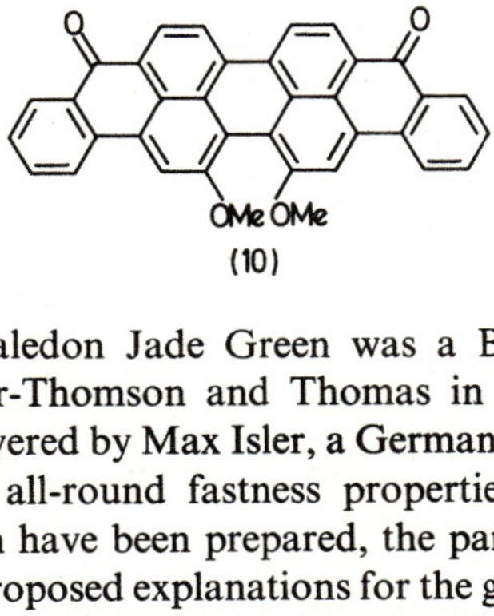

(10)

Caledon Jade Green was a British discovery and was first made by Davies, Fraser-Thomson and Thomas in 1920 from dihydroxydibenzanthrone which was discovered by Max Isler, a German. Caledon Jade Green has a bright green shade and good all-round fastness properties. Although many derivatives of Caledon Jade Green have been prepared, the parent dye still reigns supreme.

Proposed explanations for the green colour of this dye compared to the blue colour of dibenzanthrone are discussed later (Sect. 5.2.4).

Acridones

This group provides a number of useful vat dyes. One of the more important is CI Vat Violet 14, whose main constituent is the trichloro compound (11). It gives reddish-

[3] Caledon is the trademark of ICI.

violet shades which have exceptionally good fastness to light. The colour of the parent molecule is extremely sensitive to the number and position of the substituents. Thus, the hues of the mono-, di- and trichloro derivatives range from orange, through scarlet and red, to violet (Table 5.1).

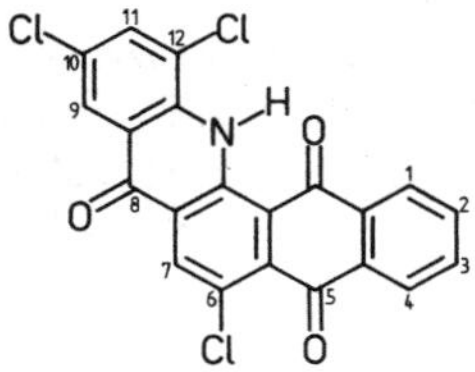

(11)

Table 5.1. Effect of substituents on acridone vat dyes

Position of Cl	Hue
12	Bluish-red
9, 12	Scarlet
9, 11	Orange
10, 11	Violet
9, 10, 12	Scarlet
9, 11, 12	Orange

Another important member of this group is CI Vat Green 3 (12). It has excellent all-round fastness properties but is noted particularly for its outstanding light fastness, rated at 8 on a 1–8 scale!

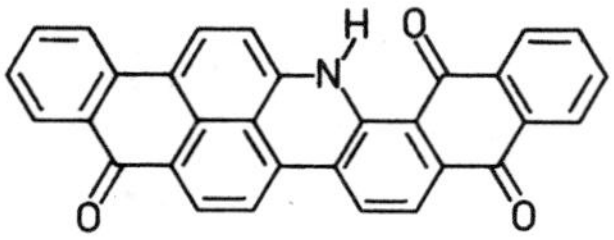

(12)

Carbazoles

Vat dyes containing a carbazole group are very important; they provide shades of good all-round fastness properties ranging from yellow to black. An unusual feature of this class of vat dyes is that many of the commercial products are isomers. This is illustrated by the three dyes: — CI Vat Orange 15 (13), CI Vat Brown 3 (14), and CI Vat Black 27 (15). It should be noted that (13) is a bis-1,5-diaminoanthraquinone derivative (these are orange), (14) is a mixed 1,5- and 1,4-diaminoanthraquinone derivative (orange and violet respectively — orange + violet = brown) and that (15) is a bis-1,4-diaminoanthraquinone derivative (these are blue) — see Sect. 4.6.2.

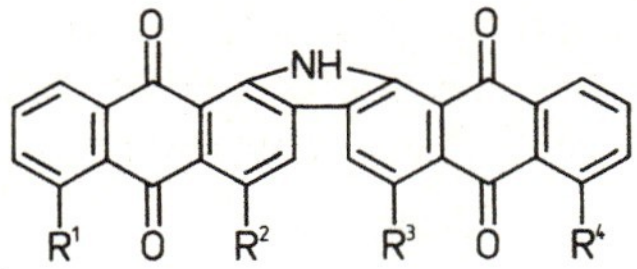

(13) $R^2 = R^3 = H$; $R^1 = R^4 = PhCONH$

(14) $R^2 = R^4 = H$; $R^1 = R^3 = PhCONH$

(15) $R^1 = R^4 = H$; $R^2 = R^3 = PhCONH$

5.2.3 Sulphur-containing Vat Dyes

When attempting to distinguish vat dyes from sulphur dyes[4] it is found that there are many borderline cases. Thus, there are anthraquinonoid vat dyes which contain sulphur: on the other hand, there are compounds which may be dyed by either a vat process or by the sulphide process used for dyeing the sulphur dyes. Dyes such as these have been classified as the 'sulphurised vat dyes', a class which is subdivided further into anthraquinone derivatives and indophenol derivatives.

Sulphurised Vat Dyes

(a) Anthraquinone Derivatives

CI Vat Yellow 2 (16), a bisthiazolo compound, is a member of this class. Its fastness to light is only moderate and it causes marked tendering of the fibre. In spite of these defects, this greenish-yellow dye is still used extensively, especially for producing bright green shades in combination with Caledon Jade Green (10).

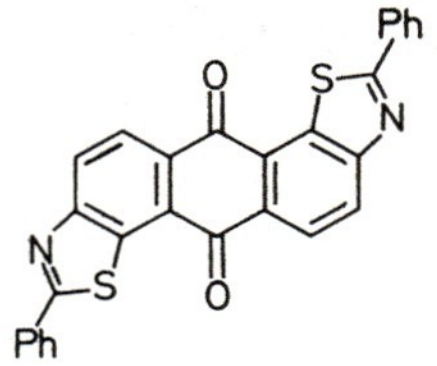

(16)

(b) Indophenol Derivatives

The most important vat dye of this type is undoubtedly CI Vat Blue 43, first introduced by Cassella in 1909 as Hydron Blue R. It is obtained by condensing

[4] Although sulphur dyes, especially sulphur blacks, represent a commercially important class of dyes (they are used primarily for providing black shades on cellulosic fibres), they are not discussed. The main reason is that even now, more than 100 years after their discovery, very little is known about their constitution. However, useful reviews on sulphur dyes are: —
1. Orton D. G.: Sulphur dyes. In: The chemistry of synthetic dyes, Venkataraman, K. (ed.), Vol. VII, pp. 1–34. New York, San Francisco, London: Academic Press 1974, and
2. Venkataraman, K.: The chemistry of synthetic dyes, Vol. II, pp. 1059–1100. New York: Academic Press 1952.

p-nitrosophenol with carbazole in sulphuric acid to form the indophenol (17): refluxing this (or its *leuco* form) with sodium polysulphide in butanol gives the dye proper, of probable structure (18).

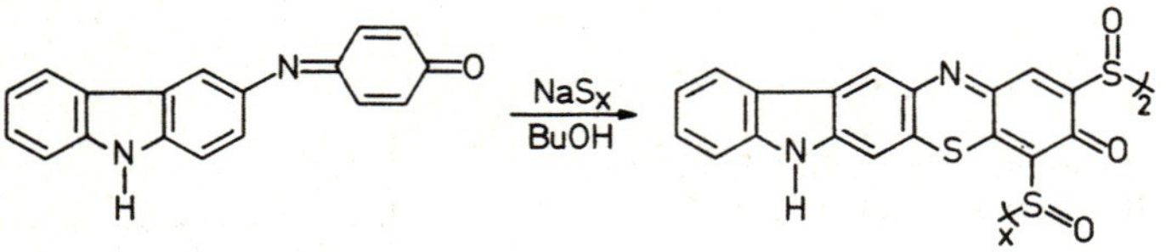

(17) $\xrightarrow[\text{BuOH}]{\text{NaS}_x}$ (18)

CI Vat Blue 43 gives reddish-blue shades. It is valuable in that it has better fastness properties than most blue sulphur dyes and is used as an inexpensive substitute for indigo (see Sect. 5.3).

5.2.4 Colour and Constitution of Anthraquinonoid Vat Dyes

The colour and constitution of those vat dyes containing amino (NH) groups, *e.g.* the arylaminoanthraquinones (2) and the carbazoles (13, 14 and 15), can be explained by analogy with the amino substituted anthraquinone dyes described in Chapter 4. Consequently, this section is devoted to discussing briefly the colour-structure effects of vat dyes composed only of condensed carbocyclic rings and carbonyl groups, although the intriguing case of Caledon Jade Green (10) (which contains two methoxy groups) is also included.

Many vat dyes can exist in both a *cis* and a *trans* form. Invariably, the *cis* isomer absorbs at longer wavelengths than the *trans* isomer. For example, *cis*-dibenzopyrenequinone (19) absorbs at longer wavelengths than its commercially important counterpart, the reddish-yellow *trans* isomer (20).

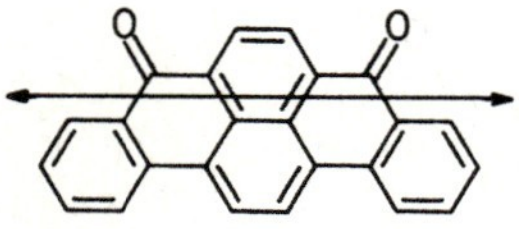

(19) 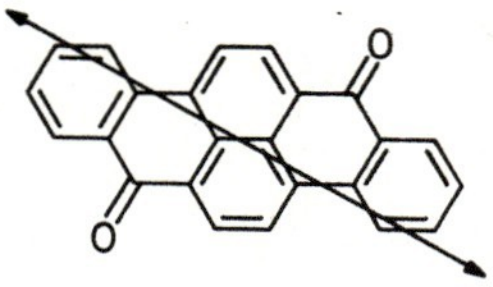 (20)

The PPP MO model calculates these effects. It also indicates that the long wavelength absorption band of both isomers arises from a single $\pi \rightarrow \pi^*$ transition polarised along the long molecular axis, as indicated in (19) and (20). As was the case for typical donor-acceptor dyes belonging to the azo and anthraquinone classes (see Chaps. 3 and 4 respectively), the transition has some charge transfer character. The naphthalene ring acts as the donor and the carbonyl groups as the acceptor part of the molecule.

The usefulness of the PPP model is also evident from its application to anthanthrone vat dyes. As seen from Fig. 5.1, the absorption spectrum of anthanthrone (21) is excellently reproduced by PPP MO calculations.

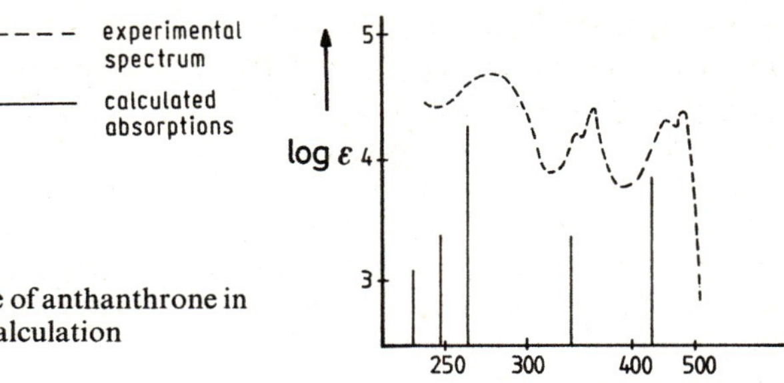

Fig. 5.1. Spectral absorption curve of anthanthrone in dioxane and results of the PPP calculation

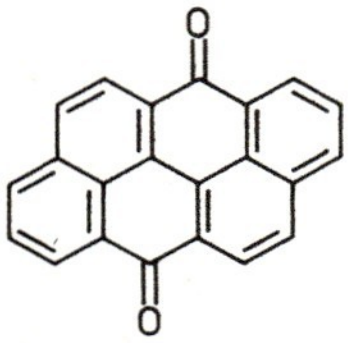

(21)

The most studied class of vat dyes is the dibenzanthrones since that most important vat dye, Caledon Jade Green (10), belongs to this class. As mentioned earlier, violanthrone (9), which is the *cis* isomer of dibenzanthrone, is still used widely as a blue vat dye. In contrast, the more hypsochromic *trans* isomer, known as isoviolanthrone (23), has been little used commercially.

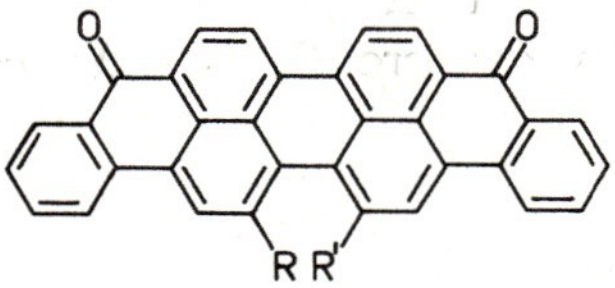

	R	R′	λ_{max}^{DMF}	ε_{max}
(9)	H	H	600	60,260
(10)	OCH_3	OCH_3	636	41,700
(22a)	H	OCH_3	598	38,900
(22b)	CH_3	CH_3	575	—
(22c)	$O{-}CH_2{-}$	$CH_2{-}O$	618	—

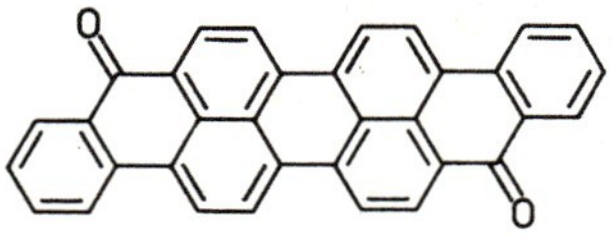

(22)

λ_{max}^{DMF} 588 nm ε_{max} 41,700

(23)

Despite intensive study, a wholly convincing explanation for the green colour of Caledon Jade Green (10) compared to the blue colour of violanthrone (9) has not been forthcoming. Until recently the bathochromic effect was ascribed to steric interactions between the methoxy groups leading to a non-planar molecule. This explanation, however, has been doubted since, theoretically, the assumed weakening of essentially single bonds in the ground state should cause a hypsochromic rather than a bathochromic shift (see Sect. 3.5.9). The doubt is supported by the fact that a substantial *hypsochromic* shift (25 nm) is observed in passing from the parent compound, violanthrone, to the dimethyl derivative (22b). In this derivative the steric interactions are at least as great as in Caledon Jade Green.

The bathochromic and hyperchromic[5] shift in going from the monomethoxy derivative (22a) to the dimethoxy derivative (10, *i.e.* Caledon Jade Green) suggests that electronic effects rather than steric effects are primarily responsible for the enhanced bathochromicity of Caledon Jade Green. An ethylene bridge between the oxygen atoms (22c) results in a hypsochromic shift relative to Caledon Jade Green. This is ascribed to a reduction in overlap between the $2p_z$ orbitals on oxygen and the $p\pi$ orbitals of the ring system because of the non-planarity of the 8-membered ring involved.

We shall now consider that other type of vat dye, the indigoid vat dyes.

5.3 Indigoid Dyes

5.3.1 Introduction

Indigoid dyes are applied to the fibre by a vatting process (see Sect. 6.3.4). They are discussed separately from other vat dyes because of their importance as representatives of a discrete class of coloured compounds quite apart from the majority of vat dyes.

Indigoid dyes are important historically. Indigo itself is one of the oldest known dyes and the 6,6′-dibromo derivative was the principal constituent of Tyrian Purple (see Sect. 1.2.4). Nowadays Indigo is the only natural dye (see Sect. 1.2.5) which is still important as a textile dye[6] (although it is now produced synthetically of course). Hundreds of derivatives of Indigo have been synthesised and evaluated as dyes over the last 130 years. Therefore, it is perhaps rather surprising to find that Indigo itself is still the most important member of the class.

Indigo imparts an attractive blue colour to fabrics. Being a vat dye, the wet fastness properties are excellent; however, its light fastness is only moderate by modern standards. A great advantage of Indigo is that it fades without changing colour, *i.e.* on tone, and it is this feature which has made Indigo dyed fabrics so popular. In fact, Indigo dyed jeans and denims, which gradually fade to paler shades of blue, are held in high esteem by today's younger generation.

5.3.2 Structure and Unusual Features of Indigo

It took many years to establish the structure of Indigo (see Sect. 1.3.6). The key worker was von Baeyer who, after 13 years of painstaking research (1870–1883), assigned the structure (24) to Indigo. However, it was not until 1926 that X-ray crystallography[7] conclusively proved the Baeyer structure to be incorrect: Indigo exists as the *trans* isomer (1) not the *cis* isomer (24), both in the solid state and in solution. X-ray crystallography also shows that in the solid state Indigo forms a

[5] Increase in tinctorial strength, (*i.e.* ε_{max}).
[6] The natural red dye Cochineal (see Sect. 1.2.3) is also still used to some extent but only as a food dye.
[7] Posner, T.: Chem. Ber. *59*, 1799 (1926).

hydrogen-bonded polymer in which each Indigo molecule is linked to four other molecules of Indigo. The Indigo molecule is almost planar with both the central carbon-carbon double bond *a* and the carbonyl carbon-oxygen bonds *b* slightly longer than expected — Fig. 5.2. The carbon-nitrogen bonds *c* are shorter than expected whilst the carbon-carbon bond lengths in the fused benzene rings are approximately equivalent to those in benzene. These observations suggest some contribution from resonance forms such as (25) to the ground state of the Indigo molecule.

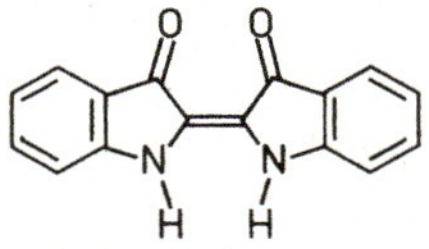

(24)

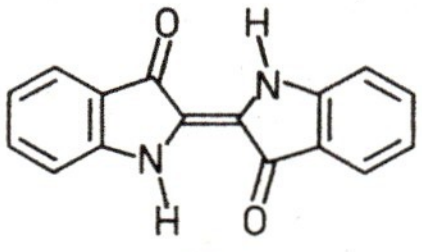

(1)

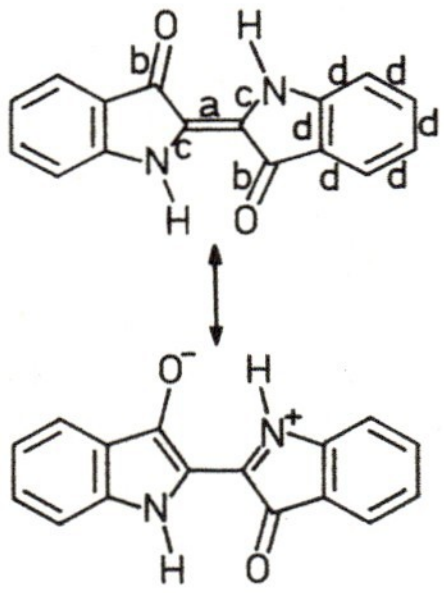

(25)

	Bond lengths (in Å)		
	Observed	Expected	
a	1.37	olefinic C=C	1.34
b	1.25*	carbonyl C=O	1.22
c	1.35*	pyrrole C—N	1.42
		(thiazole C=N	1.37)
d	1.36–1.38	benzene C=C	1.39

* Average of 2 reported values

(+1 other equivalent structure)
Fig. 5.2. Bond lengths of the Indigo molecule

The polymeric and highly polar nature of Indigo are the cause of both its poor solubility and its unusually high melting point (390–392 °C). However, Indigo can be recrystallised from polar, high boiling solvents such as nitrobenzene and aniline to give iridescent copper-red prisms.

The colour of Indigo depends crucially upon its environment — it ranges from red in the vapour phase, to violet in a non-polar solvent and to blue in the solid state, or when dissolved in polar solvents — Table 5.2. Indeed, as shown by Fig. 5.3, even the nature of the solid phase is important. Thus, crystalline Indigo is significantly more bathochromic than amorphous Indigo.

NN′-Dimethylindigo — Table 5.2 — and other indigoid dyes, *e.g.* Thioindigo (Sect. 5.3.4), do not exhibit the same sensitivity to such changes in their environment. Consequently, hydrogen-bonding is thought to be the cause of the marked solvatochromism of Indigo. Infra-red studies have clearly shown the existence of intermolecular hydrogen-bonding in Indigo, *e.g.* (26), and most of its derivatives, although intramolecular hydrogen-bonding, *e.g.* (27), cannot be discounted. Even in the vapour phase, Indigo is still associated to some extent.

Table 5.2. Influence of solvent on absorption[a]

Compound	Solvent	Dielectric constant (20 °C)	λ_{max} (nm)
Indigo	Vapour	—	540
	Carbon tetrachloride	2.2	588
	Xylene	2.3–2.6	591
	Ethanol	24.3	606
	Dimethylsulphoxide	46.3 (25 °C)	620
	Solid, in KBr	—	660
NN′-dimethylindigo	Carbon tetrachloride	2.2	640
	Benzene	2.3	644
	Chloroform	4.8	653
	Ethanol	24.3	656
	Ethanol/H_2O = 1:2	24.3–80.4	672
	Solid, in KBr	—	672

[a] Owing to the poor solubility of the compounds the ε_{max} values obtained are very inaccurate and therefore are not quoted.

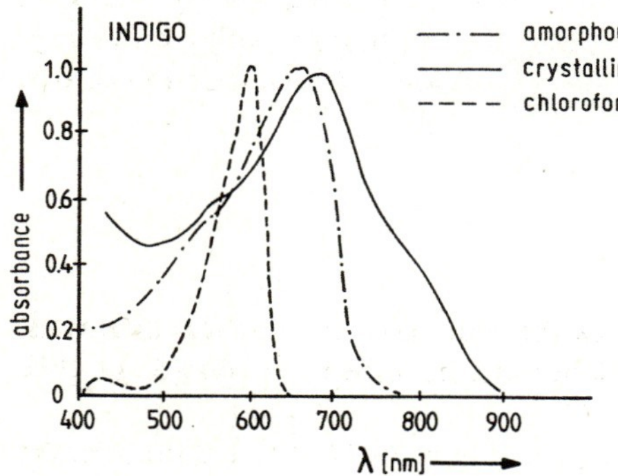

Fig. 5.3. Spectral absorption curves of Indigo in chloroform and in the amorphous and crystalline solid state

Unlike NN′-dimethylindigo and Thioindigo that are incapable of intramolecular hydrogen-bonding, *trans* Indigo cannot easily be converted into *cis* Indigo. This is strong evidence for intramolecular hydrogen-bonding, which would be expected to stabilise significantly the *trans* isomer (27). This lack of photochromism of Indigo is one of the reasons for its commercial importance (see Sect. 5.3.6). In contrast, *trans* Thioindigo may be converted readily to *cis* Thioindigo: since the *cis* isomer absorbs at considerably shorter wavelengths than the *trans* isomer, a striking colour change from violet ($\lambda_{max}^{CHCl_3}$ 546 nm) to yellowish-red ($\lambda_{max}^{CHCl_3}$ 490 nm) accompanies the photochromic change. The phenomenon of photochromism is discussed in Sect. 6.5.

The most striking feature of the indigoid dyes, however, is their unexpectedly bathochromic colour. That such small molecules should be blue[8] is truly remarkable:

[8] For all practical purposes, *e.g.* colour on the fibre, indigoid dyes are blue.

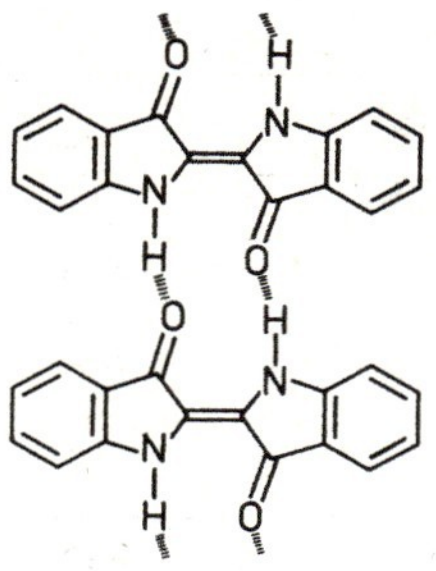

(26)

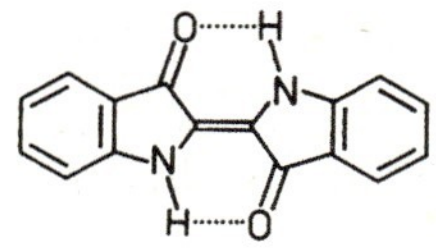

(27)

blue dyes in the azo and anthraquinone classes usually contain a multiplicity of donor and acceptor groups in the optimum orientation for maximum bathochromicity (see Sects. 3.5.4 and 4.6.2). We shall now consider the colour and constitution of the indigoid dyes in more detail.

5.3.3 Colour and Constitution of Indigoid Dyes

The fact that molecules as chemically simple as the indigoids are blue has intrigued colour chemists for many years. Several explanations have been proposed for this phenomenon.

Application of VB theory resulted in two conflicting schools of thought as to the origin of the blue colour of Indigo. Essentially, these boiled down to the necessity or otherwise of the benzene rings.

Arndt[9] suggested that the two benzene rings were not an essential part of the chromogen. He argued that the colour originated from the interaction of the two donor NH groups with the two carbonyl acceptor groups via the ethylene bridge common to both systems. Thus, according to Arndt, the most significant resonance forms are the two dipolar structures (25) and (28). In these extreme ionic structures both benzene rings retain fully their aromaticity. This concept has the advantage of relatively stable charge separated structures contributing to the excited state (hence a small energy difference between the ground state and the first excited state) but has the disadvantage that only a short conjugated path is involved. In Arndt's model, polarisation is parallel to the molecular axis.

[9] See; Knott, E. B.: J. Soc. Dyers and Colourists *67*, 302 (1951) and references therein for a discussion of the application of VB theory to the colour of Indigo.

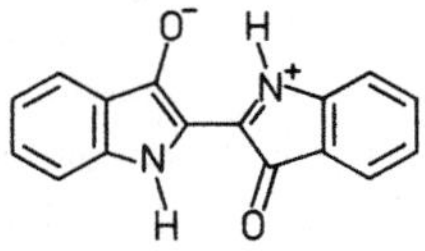

(25) 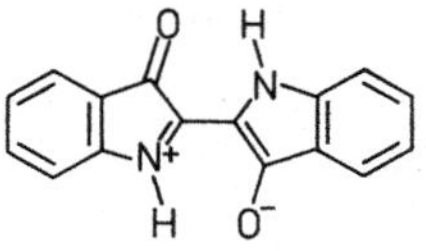 (28)

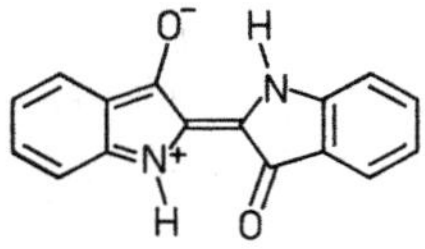

(28A) 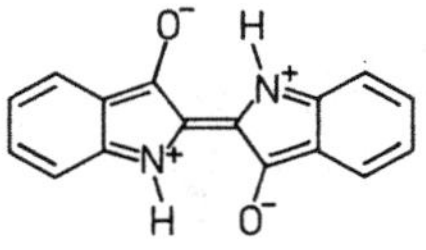 (28B)

Arndt's explanation was challenged first by Kuhn and later by van Alphen and also by Gill and Stonehill. These workers advocated that the benzene rings *were* necessary for the production of a blue colour and suggested that the significant resonance forms were (28A) and (28B). These structures are of higher energy than those proposed by Arndt, *i.e.* (25 and 28), since the RSE[10] of one or both benzene rings has been lost. However, a longer conjugated path is now involved. In contrast to (25) and (28), the polarisation is at right angles to the length of the molecule.

It is only recently that the controversy has been resolved. The observations that the longest wavelength absorption (590 nm) of 6,6′-dibromoindigo is polarised along the length of the molecule and that substituents in the benzene rings have only a minor effect on the colour of Indigo (see later) seemed to favour Arndt's explanation that the benzene rings are not necessary for the blue colour of Indigo. However, it is now known that when Indigo approaches the *monomolecular* state, *viz.* in the vapour phase, it is *red*, not blue (Table 5.2). In other words, the Indigo chromogen is red.

The structural unit responsible for this red colour has been elucidated most convincingly by Klessinger and Lüttke.[11] These workers synthesised model compounds devoid of benzene rings and compared their spectral properties with those of Indigo. Thus, the two compounds (29) and (30) were prepared. If the benzene rings are essential for the blue colour of Indigo, then (29) should absorb at much longer wavelengths than (30) since the latter compound is incapable of assuming structures analogous to (28A) or (28B). Conversely, if the benzene rings are not an essential part of the chromogen, then (29) and (30) should absorb at similar wavelengths, allowing, of course, for a small bathochromic shift in the case of (29) because of the extra conjugation. Experimentally, (29) absorbs at 528 nm and (30) at 480 nm (both in ethanol). This small difference in colour (one is orange, the other red) strongly suggests that the structural unit (31) is the primary chromophore in Indigo. The related merocyanine compound (32), with only three atoms less, absorbs in the ultraviolet at about 280 nm and is completely colourless! Indeed, the basic structure

[10] Resonance Stabilisation Energy — see Sect. 3.3.1.

[11] Lüttke, W., Klessinger, M.: Chem. Ber. *97*, 2342 (1964); Lüttke, W., Hermann, H., Klessinger, M.: Angew. Chem. Intern. Ed. *5*, 598 (1966).

(31) may be regarded as being composed of two such merocyanine units (32) that have been 'crossed' via a common ethylenic bridge. The effect of this 'crossing' or 'crossed conjugation' is obviously very dramatic and in recognition of the special properties of this type of system Klessinger and Lüttke[12] proposed that they be designated 'H-chromophores' (based on the shape of the basic unit (31)). Although other H-chromophores are known, the indigoids are by far the most important.

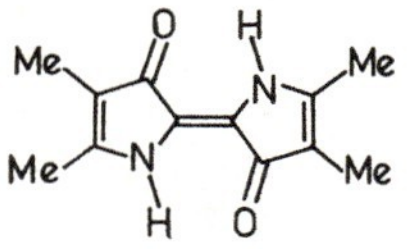

(29)

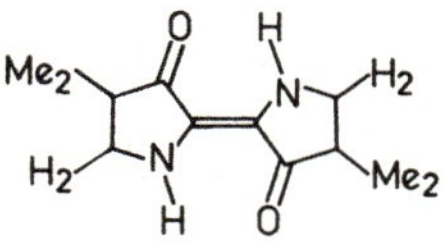

(30)

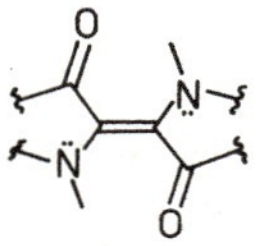

(31)

(32)

PPP MO calculations[13] also agree with the H-chromophore explanation. Thus, both Indigo and the model compounds (29) and (30) are calculated to be red. In the vapour phase, which is what the calculations refer to, Indigo is indeed red (see Table 5.2).

It therefore transpires that neither Arndt nor his critics were correct, although Arndt did deduce the primary chromophoric unit. However, it is red, not blue. From the above evidence, it is likely that the blue colour of Indigo in the solid state and in polar solvents arises from intermolecular hydrogen-bonding effects.

Effect of Substituents

As discussed above, the benzene rings in Indigo play only a secondary role as far as colour is concerned. Consequently, substituents in these rings have only a minor effect on the colour.[14] Thus, for a wide range of substituents the absorption maxima of the derived dyes fall within a relatively narrow range of 570–645 nm (reddish-blue to greenish-blue) — Table 5.3. However, both the type and particularly the position of the substituent in the ring are important for determining its effect on colour. Thus, electron-donating groups *ortho* (7 or 7′ positions) and especially *para* (5 or 5′) to the donor NH groups generally case a bathochromic shift whereas electron-withdrawing groups in these positions cause a hypsochromic shift. In contrast, electron-donating groups *ortho* (4 or 4′) or *para* (6 or 6′) to the electron-accepting carbonyl groups cause a hypsochromic shift whereas electron-withdrawing groups cause a bathochromic shift. Occasionally, anomalous effects are observed for substituents at

[12] Tetrahedron *19*, (Suppl. 2) 315 (1963).
[13] Klessinger, M.: Tetrahedron *19*, 3355 (1966).
[14] However in other indigoid dyes, *e.g.* Thioindigo, substituents can markedly affect the colour.

the 4 or 4′ and 7 or 7′ positions but these are attributable to steric interactions between the substituent and the adjacent carbonyl or imino groups respectively.

Table 5.3. Longest wavelength absorption maxima of substituted indigo dyes and substituted thioindigo dyes

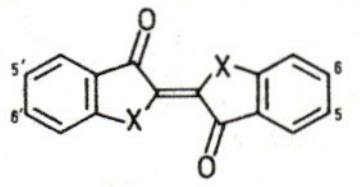

	X=NH		X=S	
	Positions			
	$\lambda_{max}^{C_2H_2Cl_4}$		λ_{max}^{DMF}	
Substituent	5,5′	6,6′	5,5′	6,6′
NO_2	580	635	513	567
H	605	605	543	543
Me	620	595	—	—
F	615	570	—	—
Cl	620	590	556	539
Br	620	590	553	—
I	610	590	—	—
SEt	—	—	573	531
OEt (OMe)	645	570	584	473
NH_2	—	—	638	452

These substituent effects are readily explained by both VB and MO theory. According to VB theory, a donor substituent in the 5 or 7 positions, which are *para* and *ortho* respectively to the NH group, stabilise the first excited state by increasing the electron density in the vicinity of the electron deficient nitrogen atom: thus, a bathochromic shift results, Fig. 5.4a. In contrast, an electron-withdrawing substituent at these positions destabilises the first excited state by increasing the magnitude of the positive charge at an already electron deficient nitrogen atom. Consequently a hypsochromic shift is observed, Fig. 5.4b.

For substituents at the 4 and 6 positions the situation is reversed. Here, the substituents are conjugated to the electron-withdrawing carbonyl group. As depicted in Fig. 5.5a, electron-withdrawing groups at the 4 or 6 positions stabilise the first excited state by sharing the negative charge — hence, a bathochromic shift results. In contrast, electron-donating substituents at the 4 or 6 positions destabilise the first excited state (and also stabilise the ground state) and therefore cause a hypsochromic shift — Fig. 5.5b.

PPP MO calculations support the VB explanation. The former show that the NH groups are the principal donor groups in Indigo and that the carbonyl groups are the principal acceptor groups. The effect of the substituents is also predicted correctly. Thus, the MO calculations show that on excitation the 5,5′ and 7,7′ positions undergo a decrease in electron density whereas the 4,4′ and 6,6′ positions experience

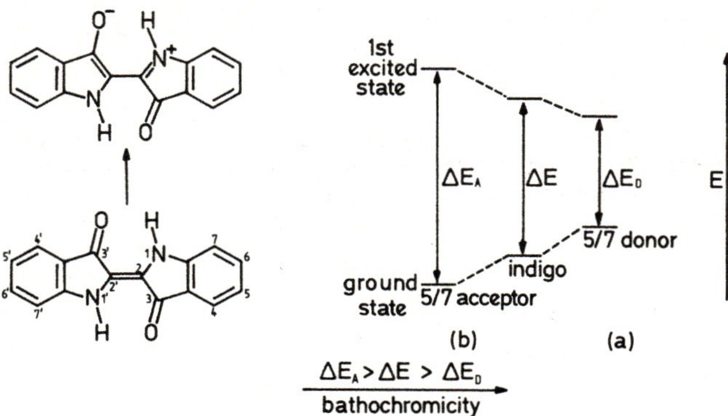

Fig. 5.4. Effect of 5/7 substituents on the colour of Indigo

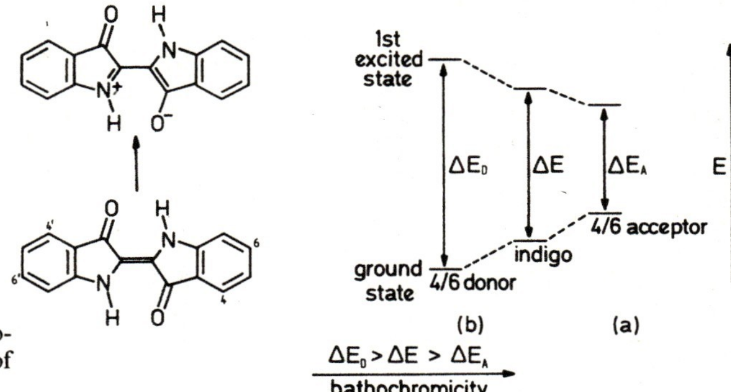

Fig. 5.5. Effect of 4/6 substituents on the colour of Indigo

an increase. Thus, electron-donor groups (at the 5,5′ and 7,7′ positions) are predicted to cause a bathochromic shift and electron-withdrawing groups a hypsochromic shift. The opposite situation is predicted for the 4,4′ and 6,6′ positions.

Only one allowed $\pi \rightarrow \pi^*$ electronic transition is calculated in the visible region of the spectrum — the weaker band at *ca.* 420 nm (see Fig. 5.3) probably arises from a transition forbidden for symmetry reasons.

5.3.4 Other Indigoid Dyes

Other derivatives closely related to Indigo are obtained by replacing one or both NH groups by other heteroatoms possessing lone pair electrons. Thus, Thioindigo (33; X=Y=S) is a useful red dye (see Sect. 5.3.6). The yellow dye, Oxindigo (33; X and Y = O) and the violet dye, Selenoindigo (33; X and Y = Se) are also known, as are various mixed compounds, *e.g.* (33; X=NH, Y=S), but these are far less important commercially. The spectral properties of these chromogens are given in Table 5.4. As expected, the bathochromicity of the compounds increase in line with the electron-donating ability of the *X* and *Y* groups.

Table 5.4. Visible absorption spectra of some indigoid-type systems

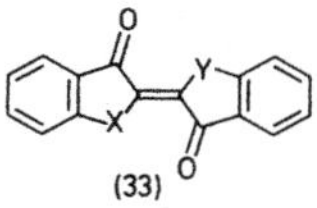

(33)

Structure (33)

X	Y	λ_{max} (nm)	ε_{max}	Solvent
NH	NH	605[a]	16,600	$(CHCl_2)_2$
Se	Se	570[b]	12,020	$CHCl_3$
S	S	546[c]	16,200	$CHCl_3$
O	O	420[b]	12,020	$CHCl_3$
NMe	NMe	650[d]	13,500	$CHCl_3$
NH	NMe	636[e]	—	xylene
NH	S	575[e]	—	xylene

[a] Sadler, P. W.: J. Org. Chem. *21*, 316 (1956)
[b] Pummerer, R., Marondel, G.: Ber. *93*, 2834 (1960)
[c] Wyman, G. M., Brode, W. R.: J. Amer. Chem. Soc. *73*, 1487 (1951)
[d] Pummerer, R., Marondel, G.: Ann. *602*, 228 (1957)
[e] Formanek, J.: Z. Angew. Chem. *41*, 1133 (1928)

Substituents in the benzene rings of these Indigo analogues exert a similar effect to that in Indigo. However, the effect of a powerful electron donor is quite dramatic. Thus, 6,6′-diaminothioindigo absorbs at 452 nm in dimethylformamide whereas the 5,5′ isomer absorbs at 638 nm in the same solvent. Therefore, by merely displacing the amino group by one carbon atom a colour change from orange to bluish-green is obtained: this is almost the full width of the visible spectrum!

Other H-chromophores, which may be regarded as positional isomers of Indigo, are known. These are derived formally by combining any two of the partial structures (34) to (36). This leads to six possible structures, three of which are symmetrical, (*e.g.* Indigo itself), and three unsymmetrical, (*e.g.* Indirubin (37)). The most bathochromic member of the series is Indigo (λ_{max} 642 nm in dimethylsulphoxide) followed by Indirubin (λ_{max} 551 nm in the same solvent). The bathochromicity decreases when the conjugated path from the NH group in one ring to the carbonyl group in the other ring *must* involve a benzene ring. Thus, iso-Indigo (38), in which both benzene rings are involved, is the most hypsochromic, absorbing at 413 nm (yellow) in dimethylsulphoxide.

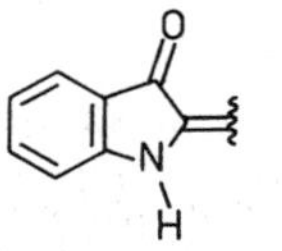

(34) 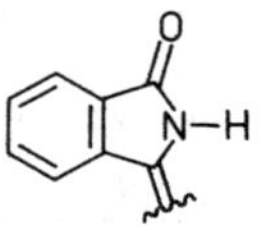 (35) 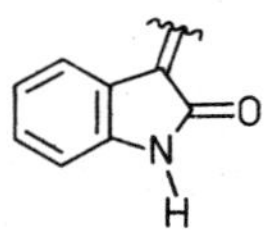 (36)

Indirubin
(37)

iso-Indigo
(38)

A most unusual feature of all the indigoid dyes is that extending the conjugation of the ethylenic link causes a hypsochromic shift! — Table 5.5 (*cf.* the cyanine dyes — Sect. 5.5.4).

Table 5.5. Longest-wavelength absorption maxima of the vinylogous indigo dyes (39) (λ_{max} in nm followed by ε_{max} in brackets)[a]

(39)

X	n = 0	n = 1	n = 2
NH	615 (20,900)[b]	573 (22,400)[c]	
S	554 (14,800)[d]	534 (22,900)[d]	532.5 (36,300)[d]
Se	573 (12,600)[e]	545.5 (17,800)[e]	

[a] in benzene
[b] 5-*i*-$C_{14}H_{29}$-5′-*n*-$C_{16}H_{33}$-derivative
[c] 5,5′-di-*i*-$C_{14}H_{29}$-derivative
[d] 5,5′-di-tert-C_4H_9-derivative
[e] 5,5′-di-C_2H_5-derivative

5.3.5 Protonation and Ionisation

Dilute acids and alkalis have little effect on indigoid dyes. However, strong acids and bases cause protonation and ionisation respectively. Thus, in concentrated sulphuric acid, indigoid dyes undergo a marked bathochromic shift. Presumably protonation takes place at the carbonyl oxygen atom: this increases the electron-withdrawing ability of the carbonyl group — hence the bathochromic shift. Thioindigoid dyes are thought to be doubly protonated (40) in concentrated sulphuric acid.

(40)

λ_{max} 641 nm in conc. H_2SO_4
($\lambda_{max}^{CHCl_3}$ 546 nm for Thioindigo)

Strong bases, (*e.g.* sodium *t*-butoxide), cause monodeprotonation of Indigo. The ionised dye (41) is green due to the greater electron-donating ability of $N^{\ominus}$ compared to NH.

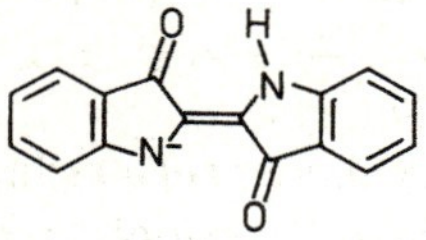

λ_{max} 773 nm in t-BuOH/t-BuO$^-$
(λ_{max}^{EtOH} 606 nm for Indigo)

(41)

5.3.6 Commercial Indigoid Dyes

All the commercial indigoid dyes contain at least two fused benzene rings. There are three reasons for this fact. First, the synthesis of compounds such as Indigo is now relatively simple (see Sect. 2.4.5) whereas the ring-free compounds are extremely difficult to synthesise. Indeed, the parent compound (42) of Indigo has not yet been synthesised! Second, the fused benzene rings confer the necessary affinity to the molecule for the fibre. Third, the simpler H-chromophores are tinctorially weaker (and also more hypsochromic): compare, for example, (29) with Indigo. (This demonstrates that the benzene rings do have some effect on the colour, and strength, of indigoid dyes).

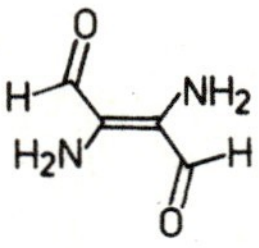

(42)

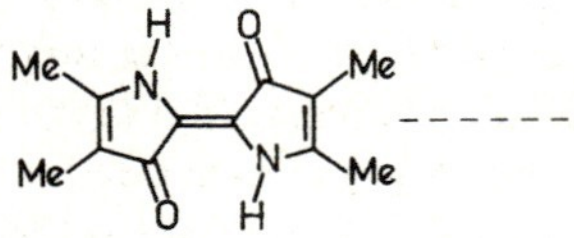

	λ_{max}	ε_{max}
(29)	528 nm	7,080 ($CHCl_3$)
INDIGO	606 nm	17,000 (EtOH)

Indigo itself is the most important indigoid dye. It is a bright blue dye with good all-round fastness properties. It is competitive against the tinctorially strong azo dyes because of its extra brightness and the fact that blue azo dyes usually have complex structures (see Sect. 3.5.4) and are therefore relatively expensive. Indigo is cost effective against the bright, technically excellent, but tinctorially weak, anthraquinone blue dyes (see Sect. 4.6.2). Several halogenated indigoid derivatives are also marketed, *e.g.* CI Vat Blue 41 (43), but these are much less important. They are all blue which is testament to the insensitivity of Indigo to substituent effects (see Sect. 5.3.3).

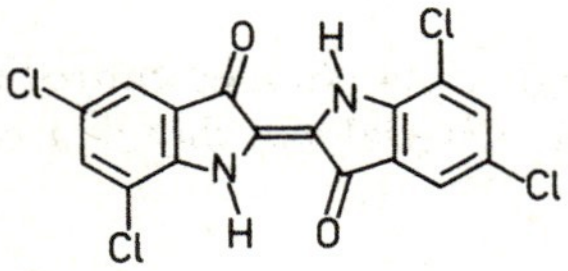

(43)

Although a greater shade gamut is attainable with thioindigoid dyes (orange to turquoise — see Table 5.3), these dyes have not achieved the same commercial importance as Indigo because they are less cost effective. However, Thioindigo itself — CI Vat Red 41 — discovered by Friedlander in 1906, is a useful bright red dye but it is not cost effective against other dye classes, *e.g.* red azo dyes.

Mixed indigoid — thioindigoid dyes provide dull tertiary shades such as browns, greys and blacks. Typical is CI Vat Black 1 (44), which gives bluish-black shades on various fibres.

Indigoid dyes are applied to various fibres such as nylon, silk and wool but the most important fibre is cotton. The use of indigoid dyes, and particularly Indigo, is subject to fashion trends since a major outlet is the dyeing of garments, especially jeans and denims.

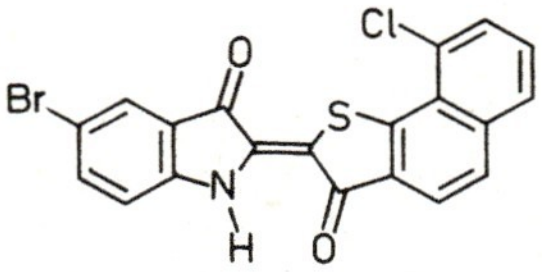

(44)

5.4 The Phthalocyanines

5.4.1 Introduction

Although the number of synthetic organic colourants now exceeds 7,000 (see Sect. 1.1), the vast majority are derived from chromogens discovered in the nineteenth century. The only novel chromogen of major commercial significance discovered since then has been phthalocyanine (45). Obviously, the discovery of new chromogens possessing the demanding properties required of modern commercial colourants is a very rare occurrence. Indeed, the discovery of the phthalocyanines is even more remarkable when one considers how it was made.

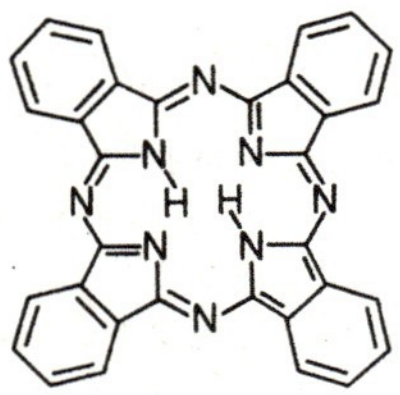

(45)

5.4.2 The Discovery of Phthalocyanines

In 1928, during the routine manufacture of phthalimide from phthalic anhydride and ammonia (Eq. 5.1) it was observed that some of the material was discoloured.

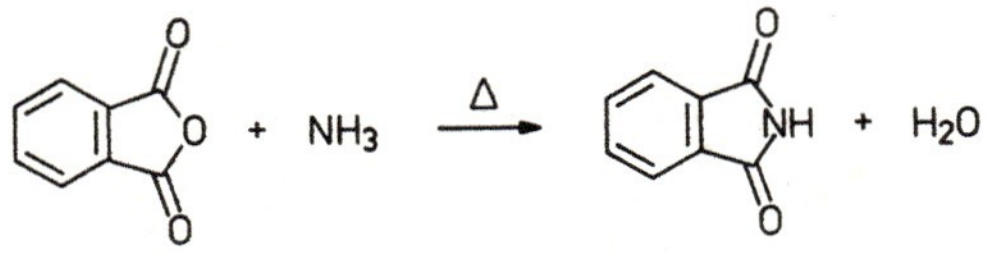

$$\text{phthalic anhydride} + NH_3 \xrightarrow{\Delta} \text{phthalimide} + H_2O \qquad \text{Eq. 5.1}$$

The chemists who investigated the problem, Dandridge, Drescher, Dunworth and Thomas of Scottish Dyes Ltd., (now part of ICI), succeeded in isolating the contaminant — a dark blue, water insoluble, crystalline substance. The material proved to be extremely stable and was shown to contain iron, the latter evidently originating from the reaction vessel. Accordingly, an independent synthesis was undertaken by passing ammonia gas into molten phthalic anhydride containing iron filings and this experiment confirmed the findings. The potential importance of this new chromogen was recognised and a patent application was lodged in the same year.[15]

5.4.3 Elucidation of the Structure of Phthalocyanine

In 1934, Linstead (Imperial College of Science and Technology, London), at the request of ICI, elucidated the structure of the new colourant as the iron complex of (45).[16] He named the new chromogen — phthalocyanine — denoting both its origin and deep blue colour. A year later Robertson[17] confirmed Linstead's structure using X-ray crystallography: in fact, this was one of the earliest applications of X-ray analysis to chemical structures (see Fig. 5.6).

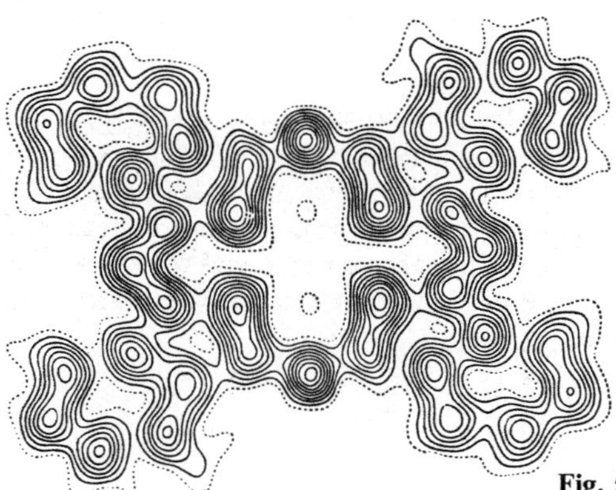

Fig. 5.6. An X-ray shadow picture of a phthalocyanine molecule

Once the methods of preparation and the structure had been established it became apparent that metal-free phthalocyanine had already been prepared by Braun and Tcherniak in 1907[18] (Eq. 5.2) and that de Diesbach and von der Weid[19] had prepared the copper derivative in 1927 (Eq. 5.3). However, these chemists had failed to appreciate the importance of their observations and neither group of workers had bothered to determine the structure of their product. Thus, but for the perception and

[15] GB 322,169 16. 5. 1928 (ICI)
[16] J. Chem. Soc. *1934*, 1016, 1017, 1022, 1027, 1031 and 1033
[17] J. Chem. Soc. *1935*, 615; *1936*, 1195, 1736; *1937*, 219; *1940*, 36
[18] Chem. Ber. *40*, 2709 (1907)
[19] Helv. Chim. Acta *10*, 886 (1927)

perseverance of the ICI and Imperial College chemists, the extremely valuable phthalocyanine colourants could still be locked away in the archives of the older chemical literature!

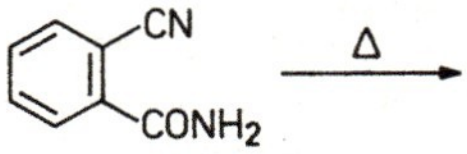

$$\text{o-}C_6H_4(CN)(CONH_2) \xrightarrow{\Delta} (45) \qquad \text{Eq. 5.2}$$

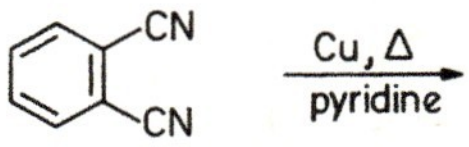

$$\text{o-}C_6H_4(CN)_2 \xrightarrow[\text{pyridine}]{Cu,\ \Delta} \text{CuPc}^{*} \qquad \text{Eq. 5.3}$$

* Shorthand for the copper phthalocyanine molecule

5.4.4 Colour and Constitution of Porphyrins and Phthalocyanines

Linstead found that phthalocyanines were related structurally to the natural porphyrin pigments of which porphin (46) is the parent compound. Since the phthalocyanines are homologues of porphyrins, it is prudent to consider the porphyrins first and then to see how the properties are altered on progressing to the phthalocyanines.

The best known porphyrin derivatives are the biologically important compounds chlorophyll (47) and haemin (48). Chlorophyll is the green pigment, essential for photosynthesis, present in plants and haemin is the red colouring matter essential for oxygen transport in the blood.

X-ray crystallography shows that the porphyrins, like the phthalocyanines, are planar. It also reveals that the cyclic 16-atom pathway represented by the bold line in (49) is the preferred cyclic system for π-electron delocalisation since this pathway exhibits the highest degree of bond equalisation: the outer 1–2, 3–4, 5–6, and 7–8 bonds

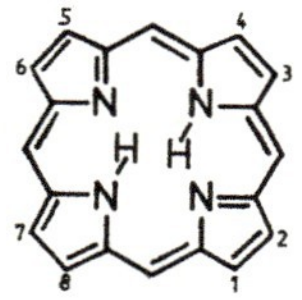

(46)

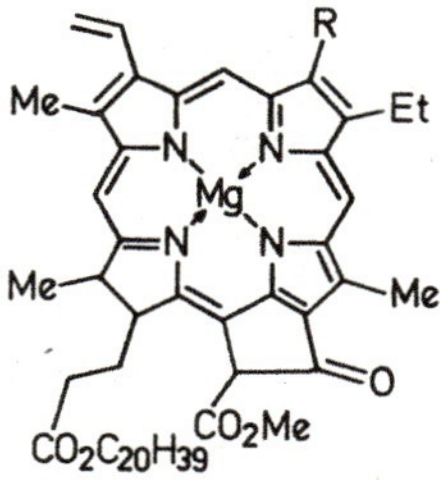

(47)

R = Me (chlorophyll a)
R = CHO (chlorophyll b)

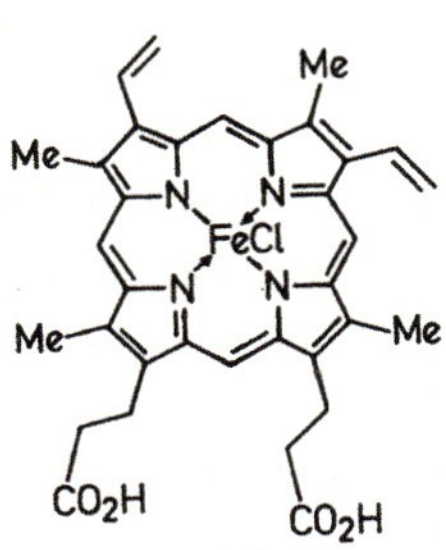

(48)

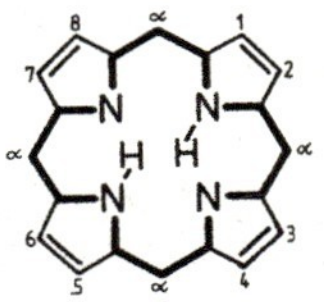

(49)

α = meso position

are essentially pure double bonds. This 16-centre system, which contains 18 π-electrons (the two pyrrolic nitrogen atoms provide 2 electrons each), is the chromogen of the porphyrins. Indeed, this explains a hitherto puzzling feature of the porphyrins, namely that reduction of 1, 2, 3 or even all 4 outer double bonds has little effect on the spectrum despite the large decrease in conjugation. It also explains the chemical stability of the system since, according to the Hückel (4n + 2) rule, it is aromatic, (*i.e.* n = 4 in this case). ^{1}H nmr spectra provide experimental evidence for the aromaticity of the 16-centre ring system since the signals from the meso protons (see 49) are characteristic of aromatic, not vinylic protons, *viz.* nmr shows that the system is capable of sustaining a ring current, a feature of an aromatic system.

Because of their greater symmetry, the spectra of the metallised porphyrins are simpler than those of the metal-free (free base) porphyrins. Thus, the metallised porphyrins display only two principal absorption bands, corresponding to two electronic transitions; one of low intensity ($\varepsilon_{max} \sim 10{,}000$) at *ca.* 550 nm, and the other of much higher intensity ($\varepsilon_{max} \sim 100{,}000$) near 400 nm. These are referred to as the *Q* and *B* bands respectively (the *B* band is also called the Soret band). Fig. 5.7a shows a typical absorption spectrum of a metal porphin and it should be noted that the longer wavelength band (*Q* band) often exhibits vibrational fine structure. The major difference in the absorption spectra of the free-base porphins is that the *Q* band is composed of two separate electronic transitions, Q_x and Q_y (Fig. 5.7b): these are complicated further by vibrational fine structure. Hence, four or more *Q* bands are actually observed. However, the *B* band usually remains a single peak.

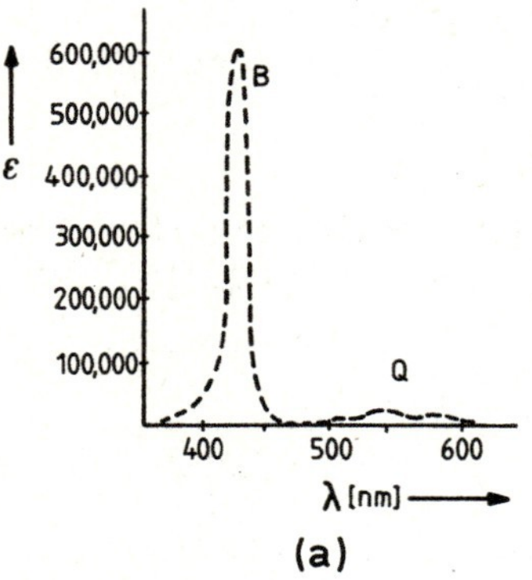

Fig. 5.7.
(**a**) Typical absorption spectrum of a metal porphin

(**b**) Typical absorption spectrum of a free base porphin

Porphyrins in which the *B* band is in the ultra-violet (*i.e.* $\lambda_{max} < 400$ nm) derive their colour from the *Q* band(s) at *ca.* 550 nm and hence they are red. Haemin (48) is a typical example. When the *B* band is present in the visible region (at >400 nm) it imparts a yellow component to the colour. This is the reason why chlorophyll (47), in which the *B* band lies at *ca.* 400 nm (yellow) and the *Q* band is at *ca.* 600 nm (blue) is green, *i.e.* blue + yellow = green.

On progressing from the porphyrins to the phthalocyanines two crucial changes occur in the absorption spectra. The most important is the total reversal of the intensities of the *Q* and *B* bands: the *Q* band, which occurs in the visible region of the electromagnetic spectrum, gains enormously in intensity at the expense of the *B* band: the *Q* band ε_{max} is greater than 100,000, compared to only 10,000 (approx.) in the porphyrins. Thus, phthalocyanines are tinctorially strong. The second factor concerns the position of the absorption maxima of the *Q* and *B* bands. Not only does the *Q* band absorb at longer wavelengths (λ_{max} *ca.* 660 nm) in phthalocyanines than in the parent porphyrins (Table 5.6), but also the *B* band is displaced to shorter wavelengths — at λ_{max} *ca.* 325 nm, it is firmly entrenched in the ultraviolet and so cannot detract from the beautiful turquoise colour of the phthalocyanines. The spectra of phthalocyanine and copper phthalocyanine are shown in Figs. 5.8a and 5.8b respectively to illustrate these effects (*cf.* also Fig. 5.7). As seen from Table 5.6, aza substitution is primarily responsible for the reversal in intensities of the *Q* and *B* bands and the hypsochromic shift of the *B* band, whereas benzannelation causes the bathochromic shift of the *Q* bands.

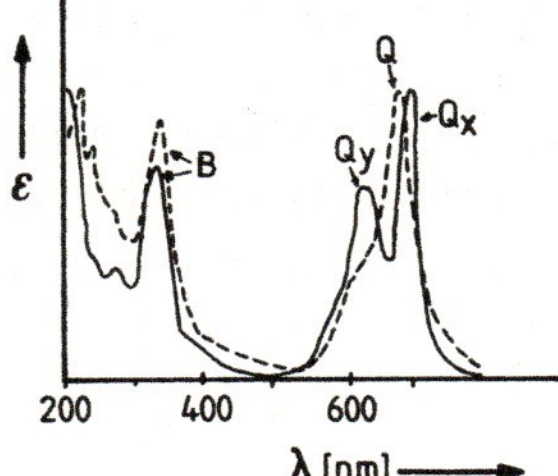

Fig. 5.8. Vapour phase spectrum of: a. ——— Phthalocyanine; b. - - - - Copper phthalocyanine

Table 5.6. Effect of aza-substitution and benzannelation on the spectral properties of porphins

Compound	$\lambda_{max}(\varepsilon_{max})$		
	Qx	Qy	B
Porphin (46)	619 (4,570)	529 (7,586)	398 (112,200)
Tetra-aza porphin	624 (81,300)	556 (46,770)	340 (97,720)
Phthalocyanine (45)	698 (162,200)	665 (151,400)	350 (54,950)
Copper phthalocyanine	678 (218,800)	—	350 (57,540)

Porphin and tetra-aza porphin spectra in pyridine
Pc and CuPc spectra in 1-chloronaphthalene

The brightness of the phthalocyanine colourants is one of the main factors responsible for their commercial success. Phthalocyanines are bright for two reasons. The first and major reason is that the absorption peak in the visible region of the spectrum is relatively narrow: in solution at 25 °C, the half-band width ($\Delta\nu_{1/2}$) is in the range 300→700 cm^{-1} (in the vapour phase at ~500 °C, the peak is broader — $\Delta\nu_{1/2}$ 700→1300 cm^{-1}).[20] For comparison, the half-band width of a typical azo dye is ~5000 cm^{-1} (see Sect. 3.5.6). Second, the phthalocyanines fluoresce (red) and fluorescence nearly always makes a colourant appear brighter (see Sect. 4.6.3).

SCF-CI MO calculations on both the porphyrins and the phthalocyanines are in very good agreement with the experimental observations. Thus, the relative intensities of the *Q* and *B* bands, the splitting of the *Q* band in the free-base compounds (see Figs. 5.7a and 5.8a), and the trends in λ_{max} values, are all calculated successfully. The calculations also show that on excitation from the ground state to the first excited state electrons migrate from the centre of the molecules towards the perimeter. Factors which reduce the electron density at the inner nitrogen atoms are predicted to produce a hypsochromic shift and this is observed experimentally. Indeed, a good linear correlation exists between the colour of metallised porphyrins and the electronegativity of the metal — the more electronegative the metal, the more hypsochromic the complex. In contrast, electron-withdrawing groups at the periphery of the molecule are predicted to produce a bathochromic shift. This is also observed experimentally, *e.g.* the chlorinated copper phthalocyanines are greener than the chlorine-free analogues. PPP MO calculations incorporating Cu II ions[21] indicate that the $d \rightarrow d^*$ and the metal-to-ligand transitions affect the electronic transition energies of the porphin.

X-ray and neutron diffraction studies of phthalocyanines show that the four benzene fragments interact only weakly with the 18π-electron inner ring system (see 49). Thus, substituents in the benzene rings have only a minor effect on the colour of phthalocyanines. Consequently, it appears that the phthalocyanine chromogen is restricted to providing only blue to green colourants.

Commercial Utilisation of Phthalocyanines

The major commercial impact of phthalocyanines is based on three factors; the first is their beautiful bright blue to green shades and high tinctorial strength; the second is their remarkable chemical stability, *e.g.* copper phthalocyanine sublimes unchanged at 580 °C and dissolves in concentrated sulphuric acid without decomposition. Indeed, it is 'purified' by this technique (see later). The third factor is their excellent fastness to light. This combination of properties is extremely difficult, if not impossible, to achieve in other dyestuffs. By comparison, the natural substances chlorophyll and haemin are highly sensitive compounds easily destroyed by light, heat and mild chemical reagents!

Although many metal derivatives of phthalocyanine have been made (all are blue to green in colour), the copper derivative is by far the most important; therefore only copper phthalocyanines are discussed. Copper phthalocyanines are used both as pigments and as dyes. Since the former are more important commercially (the annual world production of copper phthalocyanine pigments is measured in millions of kilos), these are considered first.

[20] Edwards, L., Gouterman, M.: J. Mol. Spectroscopy *33*, 292 (1970)
[21] Roos, B., Sundbom, M.: J. Mol. Spectroscopy *36*, 8 (1970)

Copper Phthalocyanine Pigments

Most phthalocyanine pigments, including copper phthalocyanine itself — CI Pigment Blue 15 — exhibit polymorphism, *i.e.* they can exist in more than one crystalline form. In the case of copper phthalocyanine, the two major polymorphs are the α-form and the greener coloured, more stable, β-form.

All production methods for copper phthalocyanine yield coarse crystals of the β-form which are unsuitable for use as a pigment. These can be converted by various methods (*e.g.* precipitation from sulphuric acid or by grinding with sodium chloride) into the metastable, fine particulate α-form, which is suitable for use as a pigment. It is also possible to obtain the β-polymorph in a fine particulate form suitable for use as a pigment by grinding in the presence of additives (*e.g.* fatty amines) which inhibit the formation of the α-form.

The paint manufacturers prefer the redder shade of the α-form. However, this creates a problem. The metastable α-form, especially in the presence of traces of the more stable β-form, is unstable in paint formulations and gradually reverts to the coarse, crystalline, non-pigmentary β-form — this results in a shade change (greener) and loss in tinctorial strength. Flocculation[22] is also a problem. However, these problems have been overcome by the incorporation of suitable substituents into the copper phthalocyanine molecule: one such method is shown in Eq. 5.4. As an extra safeguard, a small amount of monochloro copper phthalocyanine, which does not readily form the β-polymorph, is intimately ground with the product depicted in Eq. 5.4.

$$[\mathrm{CuPc}]\!-\!(\mathrm{CH_2Cl})_3 \xrightarrow{\mathrm{RNH_2}} [\mathrm{CuPc}]\!-\!(\mathrm{CH_2NHR})_3 \qquad \text{Eq. 5.4}$$

As just mentioned, halogen substituents, especially chlorine and bromine, stabilise the α-form but cause a bathochromic shift so that the halogenated pigments are greenish-blues or greens, depending on the degree of substitution. CI Pigment Green 7 (50), is one of the most important pigments of this type.

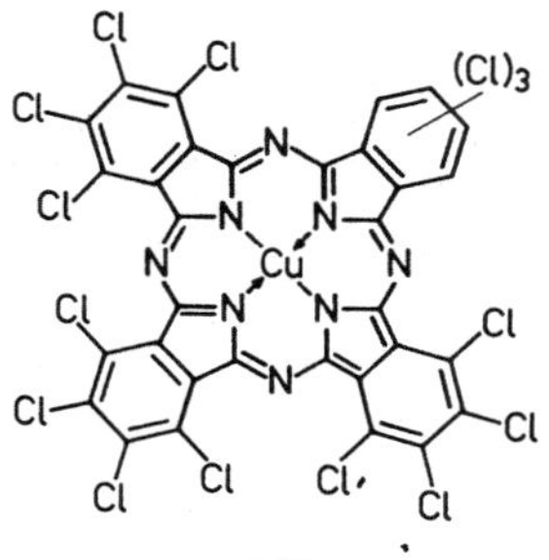

(50)

The greener β-form of copper phthalocyanine finds its major outlet in printing inks. Indeed, this is still one of the cheapest high performance pigments available.

[22] Flocculation is the disintegration of the emulsion and is characterised by the aggregation and precipitation of the pigment.

It is therefore the major coloured pigment employed by the packaging industry. The reader may verify this for himself next time he visits a supermarket by noting the preference for blue and green labelling on products.

Although the paints market is bigger than the printing ink market, blues represent only a fraction of the shade gamut of paints. In contrast, the majority of printing inks are either blue or contain a blue component. Hence, the total sales of the α- and β-forms of copper phthalocyanine are roughly equal.

5.4.5 Copper Phthalocyanine Dyes

The size of the molecule is primarily responsible for the lesser importance of the phthalocyanine dyes compared to phthalocyanine pigments — it is too large to dye the synthetic fibres, polyester and polyacrylonitrile (see Sect. 6.3.3), and is only of limited use for nylon. It is, therefore, used almost exclusively for dyeing the cellulosic substrates, cotton and paper.

A typical phthalocyanine dye is CI Direct Blue 86 (51). This was the first commercial phthalocyanine dye.

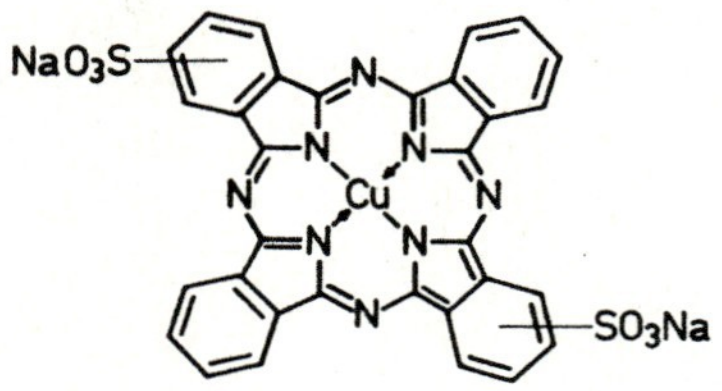

(51)

5.5 Polymethine Dyes

5.5.1 Introduction

The polymethine dyes embrace a wide variety of structural types. They can be sub-divided into three categories: — cationic (52), anionic (53) and neutral dyes (54). Of these, perhaps the best known are the cyanine dyes (52; X=Y=CH).

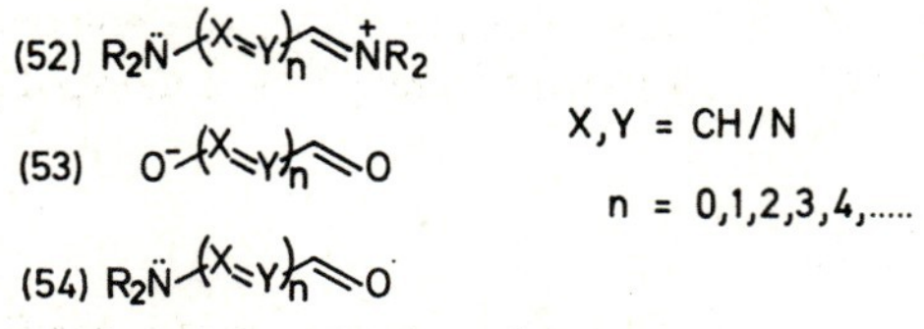

(52) $R_2\ddot{N}\text{-}(X{=}Y)_n\text{-}CH{=}\overset{+}{N}R_2$

(53) $O^{-}\text{-}(X{=}Y)_n\text{-}CH{=}O$

(54) $R_2\ddot{N}\text{-}(X{=}Y)_n\text{-}CH{=}O$

X,Y = CH/N

n = 0,1,2,3,4,.....

The first cyanine dye was discovered by Greville Williams in 1856, the year in which Perkin discovered Mauveine (see Sect. 1.3.2). Williams heated crude quinoline, which fortunately contained some lepidine (4-methylquinoline), with *iso*-pentyl iodide and caustic soda and obtained a blue dye which be named *Cyanine*: its structure was determined later to be (55). Cyanine was useless as a dye for textiles because of its poor fastness properties. Indeed, 'cyanine dyes' could have faded into obscurity but for the important discovery by Vogel in 1875 that cyanine dyes had photosensitising properties. Thus, whereas photographic plates without *Cyanine* were sensitive only to blue light, those treated with *Cyanine* were also sensitive to green light. Unfortunately, the dye caused fogging of the plates but later cyanine dyes such as (56) proved to be of practical value. In 1905, König discovered Pinacyanol (57) and this blue dye was the first sensitiser to red light.

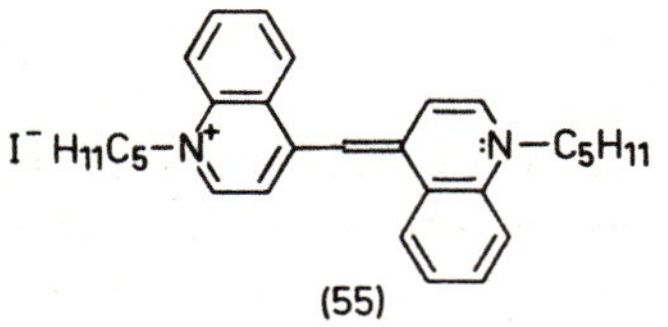

$I^- H_{11}C_5-N^+$ … $N-C_5H_{11}$

(55)

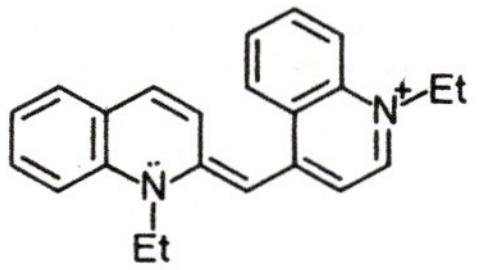

N^+-Et

N

Et

(56)

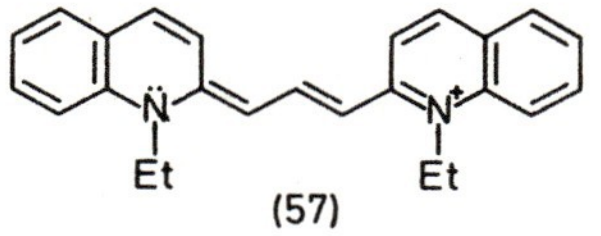

N … N^+

Et Et

(57)

Following these discoveries cyanine dyes have been the subject of extensive research within the photographic industry. However, this section on polymethine dyes is concerned solely with their usefulness as textile dyes. Colour-structure relationships are discussed in some depth.

5.5.2 Oxonols and Merocyanines

Oxonols, of general formula (53), are relatively unstable compounds, the stability decreasing rapidly as n increases. The have not found any practical use and have been little studied.

Although not important as textile dyes, merocyanine (Greek, meros = part) dyes are very important in the photographic industry. However, they are not discussed here. This section, therefore, reduces to a discussion of cyanine dyes, and in particular, their derivatives.

5.5.3 Cyanine Dyes and their Derivatives

Because of poor fastness properties, especially to light, cyanine dyes (58; see Fig. 5.9) have not been used as textile dyes. However, IG (a German dyestuffs combine) discovered that derivatives of cyanine dyes, namely azacarbocyanine (60) and hemicyanine dyes (63) — see Fig. 5.9 — were useful dyes when applied to the first synthetic fibre, cellulose acetate (see Sect. 6.2.3). Their 'Astrazon' range of cationic, water soluble dyes contained many azacarbocyanine and hemicyanine dyes. The azacarbocyanines provided yellow dyes, *e.g.* CI Basic Yellow 11 (65), whereas the hemicyanines provided red to violet dyes, *e.g.* CI Basic Violet 7 (66). However, it was the introduction of polyacrylonitrile fibres containing acidic dye sites (*e.g.* Orlon 42) by du Pont in 1953 (see Sect. 6.2.4) that really heralded the beginning of a new era for cationic dyes. Thus, it was found that azacarbocyanine (60), diazacarbocyanine (61), hemicyanine (63), and particularly diazahemicyanine (64) dyes — see Fig. 5.9 — not only dyed polyacrylonitrile fibres in bright, strong shades, but also that the

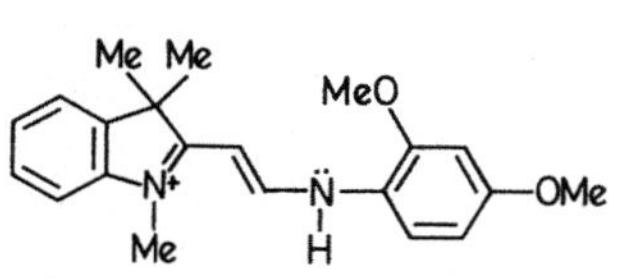

(65)

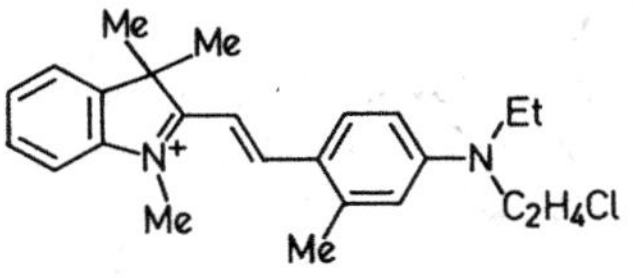

(66)

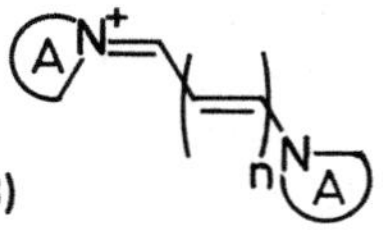

(58)
cyanines

A = heterocyclic residue
B = carbocyclic residue

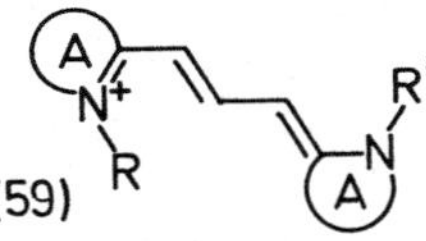

(59)
carbocyanines
(*i.e.* contain 3 carbon atoms between heterocyclic nuclei)

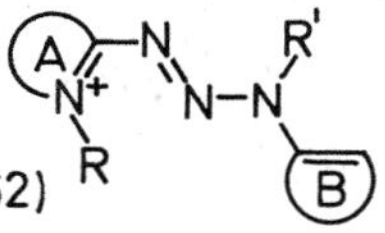

(62)
triazacarbocyanines

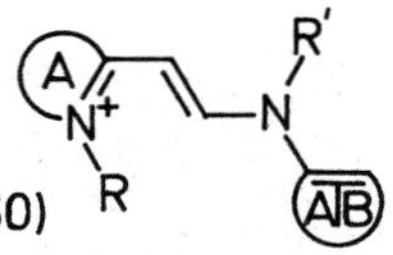

(60)
azacarbocyanines
(*i.e.* 1 CH replaced by N)

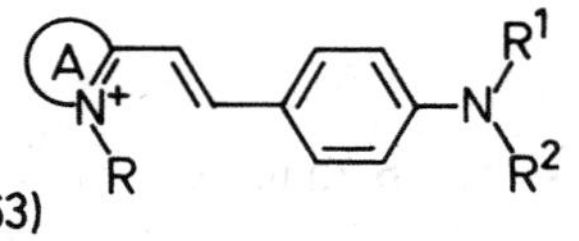

(63)
hemicyanines
(*i.e.* half-cyanines)

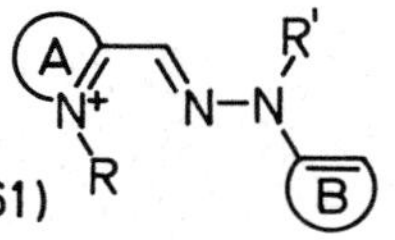

(61)
diazacarbocyanines

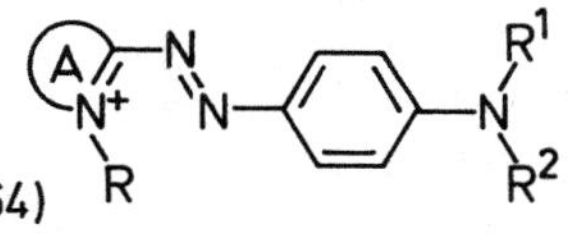

(64)
diazahemicyanines
(*i.e.* 2 CH groups replaced by 2 N atoms)

Fig. 5.9. Cyanine dyes and their derivatives — structural relationships

light fastness of the dyes was much better on this fibre than on any other fibre! (see Sect. 6.4.4). In many cases the light fastness was excellent.

Some dyes, including (65) and (66), were used without modification on polyacrylonitrile. However, it was found that by making minor modifications to the existing Astrazon range of dyes, their properties could be improved greatly. For example, the greenish-yellow analogue of (65) — *viz.* the dye (67)[23] — in which the nitrogen atom is part of a heterocyclic ring, has superior light fastness to (65). A more striking example is provided by the orange dye (68). The reddish-yellow dye (69),[24] which is obtained simply by methylating the diazacarbocyanine dye (68), has vastly superior light fastness (6–7 *vs* <4 at standard depth)!

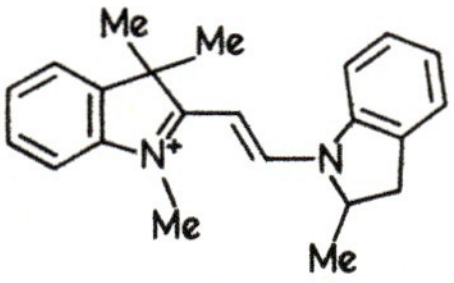

(67)

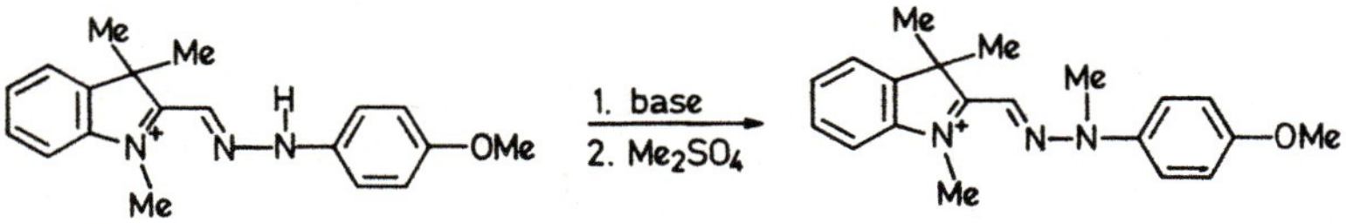

1. base
2. Me_2SO_4

(68)

(69)

Triazacarbocyanine dyes (62) are also used commercially. A typical example is the yellow dye (70).[25]

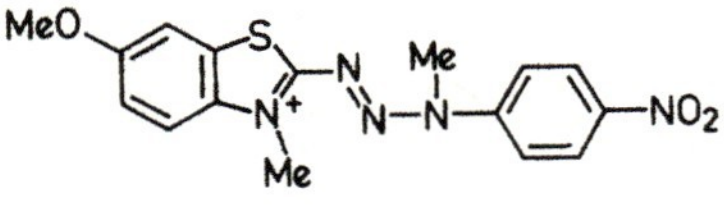

(70)

Diazahemicyanines (64) are perhaps the most important type of dye for polyacrylonitrile fibres. They provide red to blue dyes depending on the heterocyclic diazo component employed (see later).

The more important red dyes are obtained from 3-amino-1,2,4-triazole as diazo component: CI Basic Red 22 (71) is representative of such dyes. An interesting feature of the triazolium diazahemicyanine dyes is that the commercial products are a mixture of two isomers. Three isomers (72, 73 and 74) are theoretically possible by methylation of the dye base (75): in practice, only two are formed — the bluish-red 2,4-isomer (74) ($\lambda_{max}^{H_2O}$ 540 nm) and the yellowish-red 1,4-isomer (73) ($\lambda_{max}^{H_2O}$ 523 nm) in a ~6:1 ratio.[26]

[23] Example 2 of GB 448,936 (IG).
[24] Example 1 of US 3,345,355 (Bayer).
[25] Example 3 of US 3,055,881 (Geigy).
[26] Brierley, D., Gregory, P., Parton, B.: J. Chem. Research (S) *1980*, 174

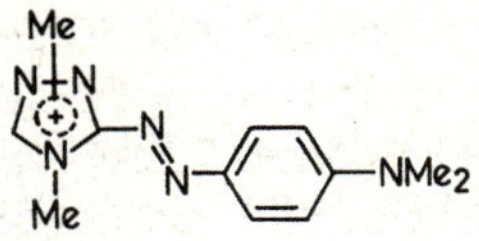

(71)

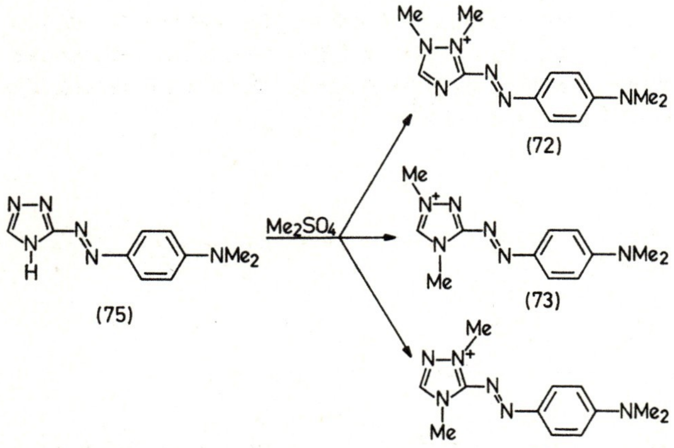

(72)

(75)

(73)

(74)

The bathochromic shade (bluish-red) of the 2,4-triazolium isomer (74) is attributable to the fact that the positive charge is delocalised throughout the molecule, as indicated by the limiting resonance structures (74) and (74A). In the yellowish-red 1,4-isomer (73) such delocalisation is not possible — the positive charge is confined to the triazolium ring, *viz.* (73) and (73A).

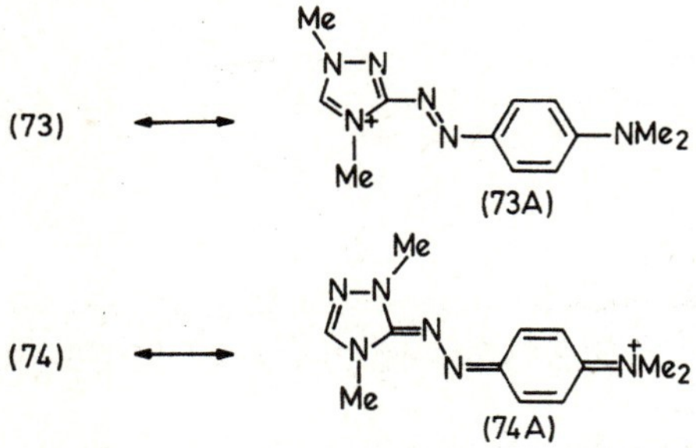

(73) ⟷ (73A)

(74) ⟷ (74A)

Aminoimidazoles (76) and aminopyrazoles (77) are also used in place of aminotriazole to provide red cationic diazahemicyanine dyes. These dyes obviously consist of a single component since only one NN′-dimethyl derivative is possible.

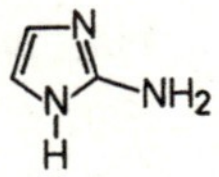

(76) 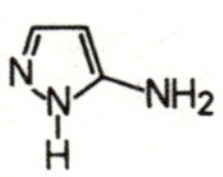 (77)

Blue dyes are obtained from aminothiazole (78) and particularly aminobenzothiazole (79) diazo components. The most important dye of this type is undoubtedly Cl

Basic Blue 41 (80). It is a bright mid-blue dye (λ_{max} 610 nm) with exceptional tinctorial strength (ε_{max} 80,000, $\Delta\upsilon_{1/2}$ 2,800 cm^{-1}) and, apart from a deficiency in heat fastness (see Sect. 6.6), has excellent all-round fastness properties. The corresponding dye without the methoxy group is redder (λ_{max} 600 nm) and has much lower fastness to light (4–5 *vs* 6–7 at standard depth).

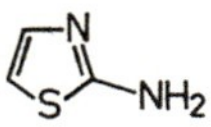

(78)

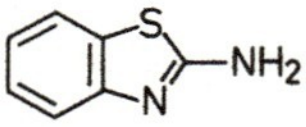

(79)

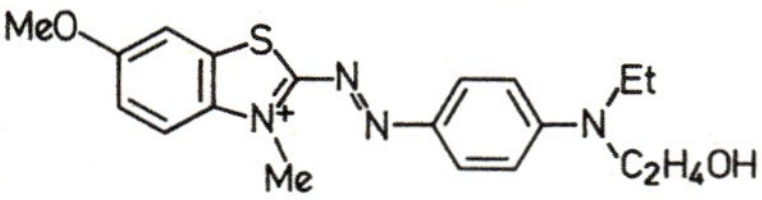

(80)

Representative shades of diazahemicyanine dyes obtained from a range of the more common amino heterocycles are shown in Table 5.7.

Table 5.7. Colours of some typical diazahemicyanine dyes

Hetero ring	Colour	λ_{max}^{MeOH} (nm)
Tetrazole	red	520
Imidazole-2	red	522
1,2,4-Triazole-2	red	531
Indazole-3	rubine	538
Pyridine-2	reddish-violet	553
5-Methyl-1,3,4-thiadiazole-2	reddish-violet	565
Quinoline-2	violet	588
Thiazole-2	violet	588
Benzothiazole-2	reddish-blue	592
3-Methyl-isothiazole-5	reddish-blue	595
β-Naphthothiazole-2	blue	619
Benzoisothiazole-3	green	683[a]

[a] In aqueous acetic acid

5.5.4 Colour and Constitution

The colour-structure relationship of cyanine dyes and their derivatives has been studied thoroughly. There are two reasons for this: one is their importance in the photographic industry, but a second, equally important reason, is that cyanine dyes approximate closest to the 'ideal dye' on which several theoretical models are based. Two of the most familiar of these are the Perturbational MO (PMO) model, from which the well known Dewar's Rules are derived (see Sect. 3.5.5), and the Free Electron MO (FEMO) model developed by Kuhn (the FEMO model is probably more familiar to the reader as the electron-in-a-box model). Both these models assume planar molecules in which there is complete delocalisation of the π-electrons: this leads, of course, to total bond uniformity (bond equalisation) within the molecule. In symmetrical cyanines (81) in particular, these conditions are almost fully satisfied since the two limiting resonance structures (81A and 81B) are exactly equivalent. Hence, both make the same contribution to the ground state of the molecule. Indeed, X-ray crystallographic and nmr spectroscopic investigations on a number of symmetrical cyanine dyes confirm that such dyes are planar (except for the alkyl groups of course) and that all the carbon-carbon bond lengths in the chain are equal. For example, in the dye (82) the mean length of the C—C bonds in the conjugated chain is 1.46 Å. It is interesting that the sulphur atoms are located *cis*, as shown.[27]

R = an alkyl group or forms part of a heterocyclic ring
n = 0, 1, 2, ...

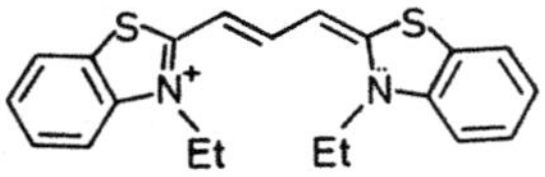

(81A) (81B)

(82)

In complete contrast to the symmetrical cyanine dyes, polyene compounds show a pronounced bond alternation because the limiting resonance structures (83A) and (83B) are far from equivalent. The high energy charge separated structure (83B) makes little contribution to the ground state of the molecule, which is essentially represented by (83A).

R = H, alkyl, aryl
n = 0,1,2,3,.....

(83A) (83B)

More advanced MO models, such as the PPP model (see Sect. 3.5.3), have also been applied to cyanine dyes. In the other extreme, much of the earlier work, notably that by Brooker and his co-workers, relied, quite successfully, on empiricism. These various approaches are now discussed. Electronic and steric effects are considered separately.

[27] Wheatley, P. J.: J. Chem. Soc. *1959*, 3245.

Electronic Effects

The colour of cyanine dyes depends primarily on the terminal groups and the length of the carbon chain.

The effect of a terminal group depends on its electron-donating ability — the more readily it releases electrons, the more bathochromic the dye. The relative electron-donating abilities of the terminal groups are usually determined experimentally, (*i.e.* by making the dyes and comparing their colours), but other methods have been used, *e.g.* correlations with redox potentials. From the symmetrical cyanine dyes shown in Table 5.8, it can be seen that:

(i) heterocyclic termini give more bathochromic dyes than acyclic termini (*cf.* (84) with the rest),

(ii) dyes having unsaturated terminal groups are more bathochromic than corresponding dyes with saturated terminal groups (*cf.* (86) and (87)),

and

(iii) that additional conjugation at the terminal groups can also give a further bathochromic shift (*cf.* (87), (89) and (90)).

Table 5.8. Long wavelength absorption bands of some symmetrical cyanines

	n	0	1	2	3	4	5
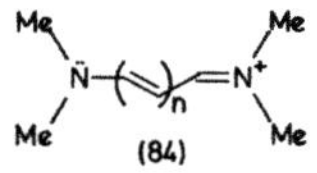 (84)	$\lambda_{max}^{CH_2Cl_2}$	224	312.5	416	519	625	734.5
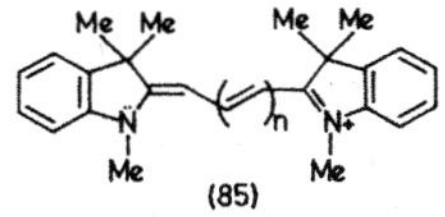 (85)	λ_{max}^{EtOH}	—	434	545	636	—	—
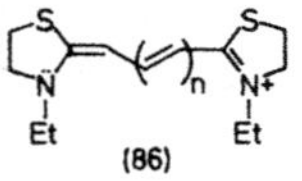 (86)	λ_{max}^{HOAc}	335	442	540	—	—	—
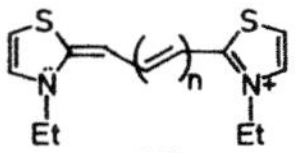 (87)	λ_{max}^{MeOH}	409	556	640	—	—	—
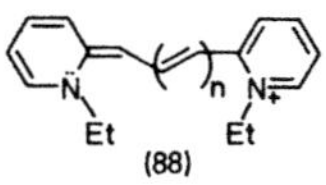 (88)	λ_{max}^{EtOH}	490	558	—	—	—	—
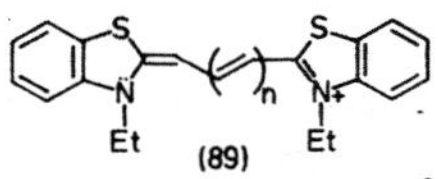 (89)	λ_{max}^{MeOH}	—	557	—	—	—	—
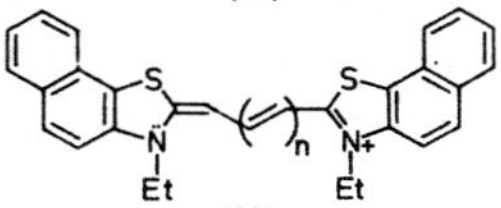 (90)	λ_{max}^{MeOH}	—	597	—	—	—	—

Table 5.8 also shows that the longest wavelength absorption of symmetrical cyanine dyes initially increases regularly with increasing chain length[28]: each additional vinyl unit (—CH=CH—) increases the λ_{max} by ~100 nm. In the case of the more hypsochromic dyes, *e.g.* (84), the long wavelength absorption band progresses from the ultra-violet (n = 0,1) — colourless compounds — through the visible (n = 2, yellow; n = 3, red; n = 4, blue) to the infra-red (n = 5,6). However, the latter two compounds are not colourless since other absorption bands occur in the visible region of the spectrum.

Cyanine dyes in which the terminal groups are different are termed unsymmetrical. If the basicities of the two terminal heterocycles are similar, then the absorption maximum of the unsymmetrical dye is approximately the mean of the two related symmetrical dyes. This is the case for (91) — its λ_{max} approximates closely to the average of (92) and (93).

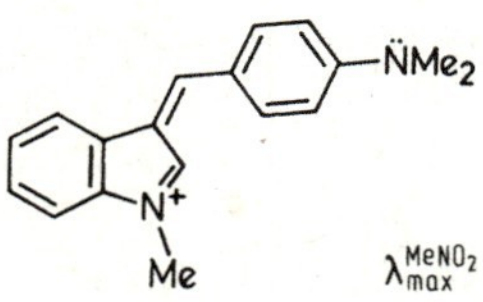

$\lambda_{max}^{MeNO_2}$ 553 nm

(91)

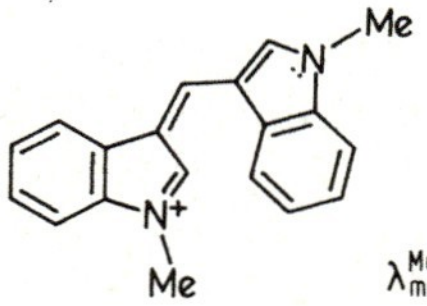

$\lambda_{max}^{MeNO_2}$ 490nm

(92)

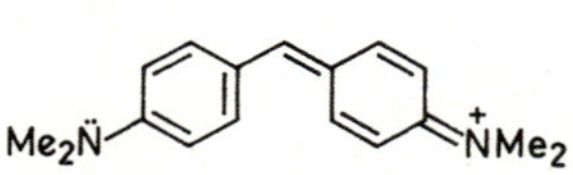

$\lambda_{max}^{MeNO_2}$ 610 nm

(93)

In contrast, if the terminal heterocycles in the unsymmetrical dyes have widely differing basicities, then the dye absorbs at unexpectedly short wavelengths. The 'deviation' from the mean λ_{max}, measured in wavelength units, is known as the 'Brooker deviation'. The unsymmetrical dye (94), derived from a strongly basic benzimidazole heterocycle and a weakly basic aniline derivative, exhibits a large Brooker deviation of 141 nm.

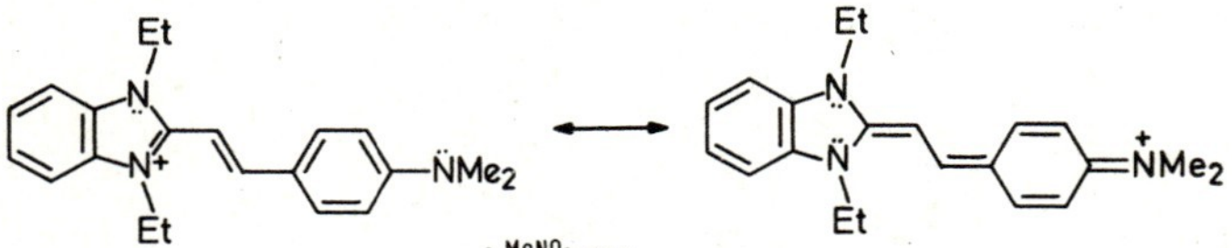

$\lambda_{max}^{MeNO_2}$ 414 nm

(94A) (94B)

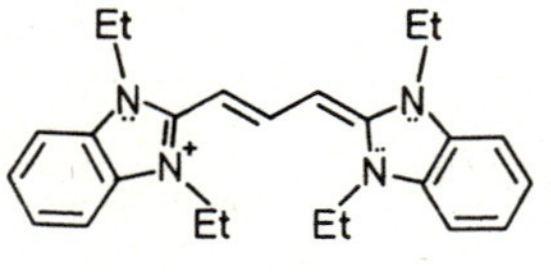

$\lambda_{max}^{MeNO_2}$ 499 nm

(95)

[28] Generally, this only applies up to n = 4 or 5 (see later).

As discussed earlier, symmetrical cyanine dyes show bond uniformity because the contribution from the two equivalent, limiting resonance structures (81A) and (81B) are equal. In unsymmetrical cyanines, the contribution of the limiting resonance structures are not equal — the positive charge resides mainly on the more basic heterocycle. Thus, for (94), structure (94A) makes by far the major contribution to the ground state of the molecule. This results in a degree of bond alternation which increases as the difference in basicity of the heterocycles widens. Hence, unsymmetrical cyanine dyes in which there are large Brooker deviations begin to resemble polyene compounds. Consequently, the bathochromic shift produced by each additional vinyl group diminishes rapidly, *i.e.* the dyes, like the polyenes, form a convergent series.

Cyanine dyes display a highly allowed HOMO → LUMO transition which is polarised along the long molecular axis — shorter wavelength absorptions are polarised both parallel to and perpendicular to the molecular axis. The extinction coefficients are high (ε_{max} 50,000–250,000) and initially increase with increasing chain length (Fig. 5.10). However, as the chain becomes longer, (*e.g.* n = 4 and particularly n = 5 in Fig. 5.10), the ε_{max} decreases dramatically and this is accompanied by a flattening of the absorption curve. The cause of this is thought to be *trans*→*cis* isomerisation, which becomes more facile as the chain length increases. It is known from studies on rigid molecules that the *cis* isomer (*e.g.* 97) absorbs at shorter wavelengths and has a lower ε_{max} value than the *trans* isomer (*e.g.* 96).

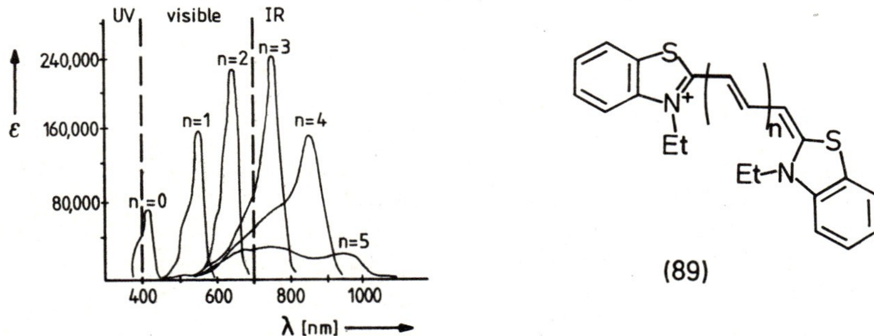

Fig. 5.10. Absorption spectra for the symmetrical cyanine dyes (89)

(96) λ_{max}^{EtOH} 571 nm ε_{max} 63,100

(97) λ_{max}^{EtOH} 561 nm ε_{max} 49,000

Another important feature of cyanine dyes is apparent from Fig. 5.10, namely the narrowness of the absorption bands. Typically, their half-band widths are about 25 nm ($\sim 1000\ cm^{-1}$) compared to values of over 100 nm ($> 5000\ cm^{-1}$) for typical azo dyes and $\sim 3500\ cm^{-1}$ for typical anthraquinone dyes (see Sects. 3.5.6 and 4.6.3 respectively). Thus, cyanine dyes are exceptionally bright. The width of the absorption band depends on how closely the structure (geometry) of the molecule in the first excited state resembles that in the ground state. For cyanine dyes, these geometries are obviously very similar. The geometries become even more similar as the length of

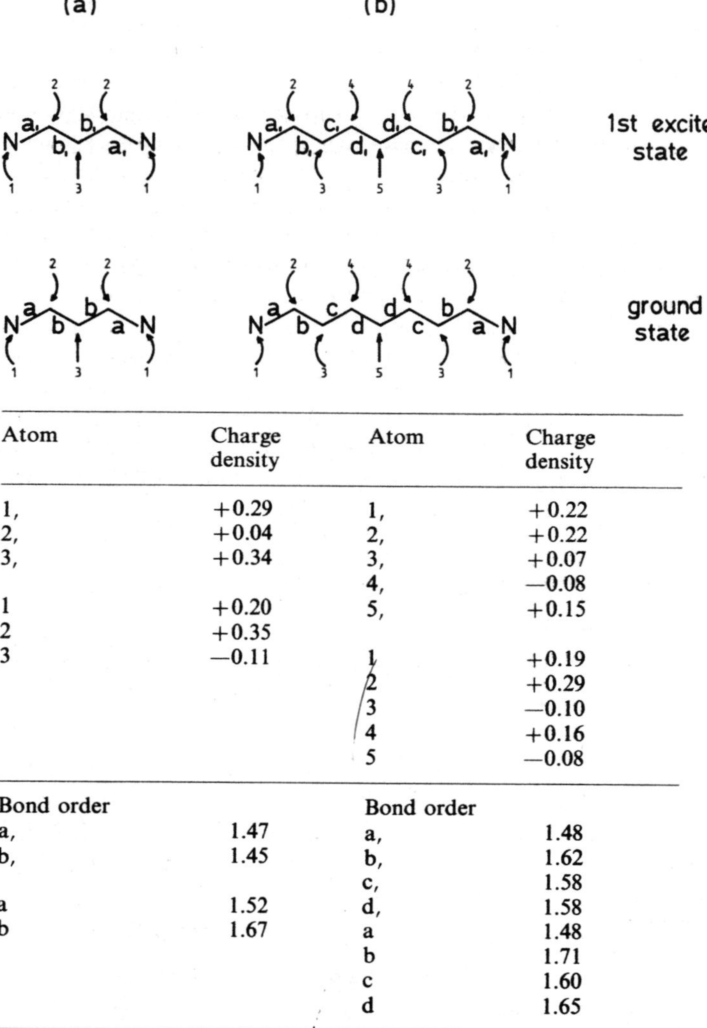

Atom	Charge density	Atom	Charge density
1,	+0.29	1,	+0.22
2,	+0.04	2,	+0.22
3,	+0.34	3,	+0.07
		4,	−0.08
1	+0.20	5,	+0.15
2	+0.35		
3	−0.11	1	+0.19
		2	+0.29
		3	−0.10
		4	+0.16
		5	−0.08
Bond order		**Bond order**	
a,	1.47	a,	1.48
b,	1.45	b,	1.62
		c,	1.58
a	1.52	d,	1.58
b	1.67	a	1.48
		b	1.71
		c	1.60
		d	1.65

Fig. 5.11. Charge densities and bond orders for the ground and first excited states of two symmetrical cyanines

the conjugated chain increases since this is accompanied by a corresponding decrease in band width.

PPP MO calculations[29] support the above experimental observations. The calculations show that bond uniformity in both the ground state and the first excited state increases with increasing chain length and that no major redistribution of electron density occurs on excitation (Fig. 5.11).

Most cyanine dyes show prominent shoulders on the short wavelength side of the intense long wavelength transition (Fig. 5.10). These shoulders arise from vibrational transitions from $0 \rightarrow 1'$ and $0 \rightarrow 2'$ levels in addition to the main $0 \rightarrow 0'$ transition. As seen from Fig. 5.10 the $0 \rightarrow 1'$ vibrational band intensity increases at the expense of the $0 \rightarrow 0'$ transition as the chain length increases.

Many cyanine dyes are fluorescent (see Sect. 6.4).

Qualitative MO Theory — Dewar's Rules

Although Kendal and Knott formulated sets of rules based on empiricism and resonance theory respectively, the most soundly based rules are those of Dewar. Dewar's Rules are based on PMO theory and are stated in Sect. 3.5.5. As mentioned earlier, they are particularly applicable to polymethine dyes, especially symmetrical cyanines. We shall now apply Dewar's Rules first to carbocyanines and their aza derivatives, and then to hemicyanines and their aza derivatives.

(i) Carbocyanines

If we take the basic carbocyanine skeleton and star alternate atoms (see Sect. 3.5.5), we get the general structure (98). According to Dewar's Rules, if an atom at a starred position is replaced by a more electronegative atom (*e.g.* nitrogen) or if an electron-withdrawing group is attached to an atom at a starred position, then a hypsochromic shift should ensue. Thus, α-azacarbocyanines (99) should be hypsochromic relative to

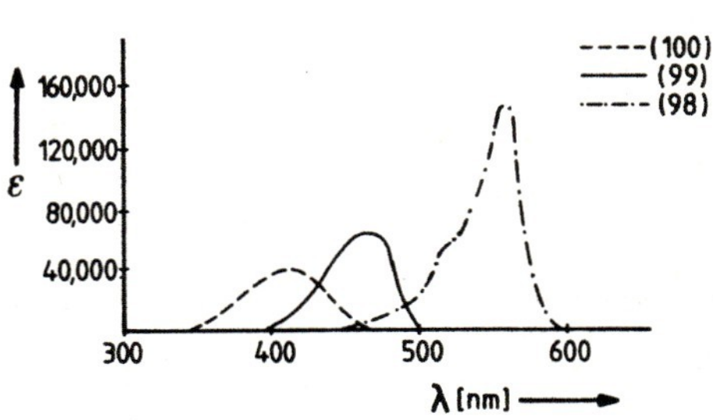

Ref.: Brooker, L. G. S. et al.: J. Amer. Chem. Soc. *73* 1087 (1951)

Fig. 5.12. Effect of aza substitution on carbocyanines (in methanol)

[29] Klessinger, M.: Theoret. Chim. Acta *5*, 251 (1966)

the carbocyanines (98), as should α-nitro and α-cyanocarbocyanines. A greater hypsochromic shift is predicted for α,γ-diazacarbocyanines (100) since carbon atoms at two starred positions have been replaced by nitrogen atoms. These predictions are fully substantiated by experiment (Fig. 5.12). It can also be seen that aza substitution has lowered the ε_{max} value. In contrast, electron-donating groups, *e.g.* alkyl, alkoxy, *etc.*, at a starred position cause a bathochromic shift, as predicted by Dewar's Rules.

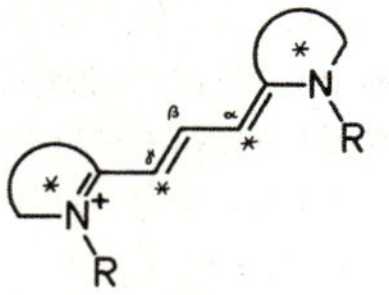

(98)

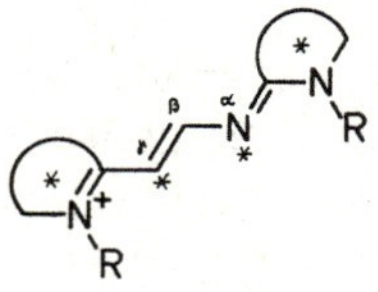

(99)

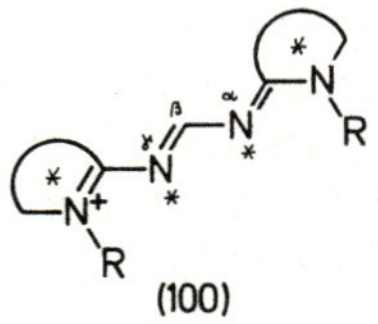

(100)

Dewar's Rules predict that replacing a carbon atom at an unstarred position by a nitrogen atom, or attaching an electron-withdrawing group at such a position, should cause a bathochromic shift. Indeed, this is so. β-Azacarbocyanines, *e.g.* (101), absorb at longer wavelengths than the parent carbocyanines (98), as do carbocyanines substituted at the β-position with electron-withdrawing groups, *e.g.* nitro and cyano. The triazacarbocyanine dye (102) absorbs at a similar wavelength to the monoazacarbocyanine dye (99) (see Fig. 5.12), since the bathochromic shift caused by replacing the CH group at the unstarred β-position by nitrogen approximately offsets the hypsochromic shift of one of the α or γ nitrogen atoms.

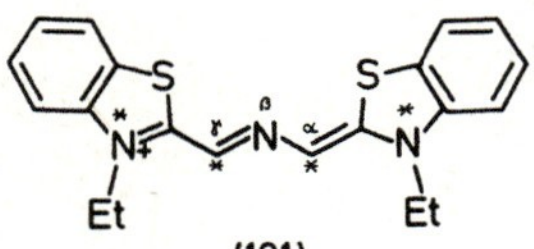

(101) $\lambda_{max} \sim 618$ nm

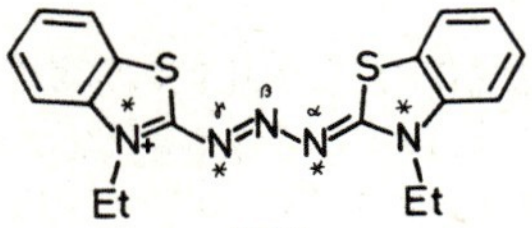

(102) λ_{max}^{MeOH} 487 nm

Dewar's Rules also explain the more bathochromic colour of the commercially important diazacarbocyanines (103), *e.g.* CI Basic Yellow 28 (69), relative to azacarbocyanines (104), *e.g.* CI Basic Yellow 11 (65). Thus, the additional nitrogen atom is at the unstarred β-position and therefore causes a bathochromic shift.

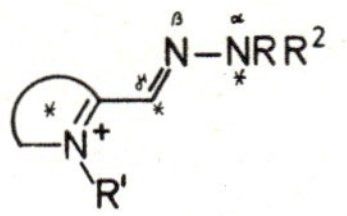

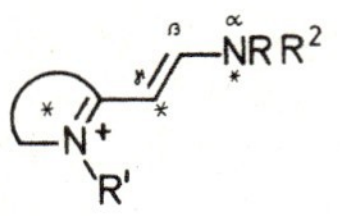

R = H, alkyl
R' = alkyl
(103) R^2 = aryl (104)

(ii) Hemicyanines

The application of Dewar's Rules to hemicyanines and their aza derivatives highlights an important point. Thus, although the bathochromic and hypsochromic shifts respectively of the α-aza- and β-azahemicyanines are predicted correctly, the additional (and large) bathochromic shift observed for the diazahemicyanines is not! (see Table 5.9). Indeed, Dewar's Rules fail when the molecules differ appreciably from the ideal hydrocarbon anion system, as was the case for the neutral azo dyes (see Sect. 3.5.5).

Table 5.9. Effect of aza substitution on the hemicyanines (105)[a]

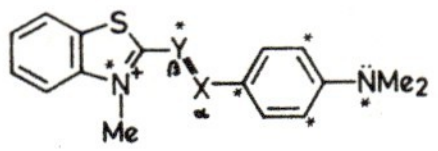

(105)

(105)	$-\overset{\beta}{Y}=\overset{\alpha}{X}-$	$\lambda_{max}^{CH_3NO_2\ b}$ (nm)	ε_{max}
a	–N=CH–	519	79,000
b	–CH=CH–	529	70,000
c	–CH=N–	561	37,000
d	–N=N–	600	88,000

[a] Kiprianov, A. I., Mikhailenko, F. A.: J. Gen. Chem. USSR *33*, 1381 (1963)
[b] Because the β-aza dye is unstable in water and alcohol

In contrast to the diazacarbocyanines (103), the diazahemicyanines (105d) have higher ε_{max} values than the parent hemicyanines (105b). This fact, taken in combination with their excellent fastness properties, is the reason for the commercial importance of diazahemicyanine dyes.

Dewar's Rules can also be used to predict the effect of substituents in diazahemicyanine dyes. As seen from Table 5.10, the following substituent effects are predicted correctly:

(i) substituting methyl groups for hydrogen at the starred terminal amino group causes a bathochromic shift (*cf.* 106a and 106b),

(ii) a methoxy group at the unstarred position in the aniline coupling component gives a hypsochromic shift (*cf.* 106e with 106c),

(iii) a methoxy group at the starred position in the benzothiazolium nucleus gives a bathochromic shift: in contrast, a nitro group at this position gives a hypsochromic shift (dyes 106b, 106c and 106d), and

(iv) a nitro group at the unstarred position in the aniline coupling component gives a bathochromic shift (*cf.* 106f with 106c).

Points (ii) and (iv) suggest that structure (107B), in which the considerable stability of a benzene ring has been lost, unexpectedly makes a major contribution to the ground state of the molecule. This fact has an important bearing on the heat fastness properties of diazahemicyanine dyes (see Sect. 6.6).

Table 5.10. Substituent effects in diazahemicyanines[a] — application of Dewar's Rules

(106)

(106)	R^3	R^2	R	R^1	λ_{max}^{MeOH} (nm)	ε_{max}
a	H	H	H	H	563	—
b	H	H	Me	Me	592	88,000[b]
c	OMe	H	Me	Me	610	55,000[b]
d	NO_2	H	Me	Me	580[b]	77,000[b]
e	OMe	OMe	Me	Me	582	—
f	OMe	NO_2	Me	Me	617	—

[a] Kiprianov, A. I., Mikhailenko, F. A.: J. Gen. Chem. USSR *33*, 1381 (1963); Voltz, J.: Chimia *15*, 168 (1961)
[b] In nitromethane

(107A) (107B)

Free-Electron MO Theory

We shall now consider briefly the most simple MO theory of all, the FEMO theory. In the FEMO model, the transition energy (ΔE) of the first absorption band of a linear, conjugated molecule is given by Eq. 5.5. The value of n is found easily since an N atom system has $N + 1$ π-electrons; halving this gives the number of filled orbitals. If Eq. 5.5 is combined with Planck's equation, (*i.e.* $E = hc/\lambda$), then the wavelength of the first absorption band can be calculated, Eq. 5.6.

In the FEMO model the oscillator strength is given by the simple Eq. 5.7. As seen from Table 5.11, this simple theory calculates both the colour (λ_{max}) and the tinctorial strength (f) of the cyanine dyes remarkably well.

$$\Delta E = \frac{h^2}{8mL^2} \cdot (2n + 1) \qquad \text{Eq. 5.5}$$

where h = Planck constant
m = mass of an electron
L = length of the conjugated chain and
n = the quantum number of the highest occupied orbital

$$\lambda = \frac{8mc}{h} \cdot \frac{L^2}{(2n + 1)} \qquad \text{Eq. 5.6}$$

$$f = 0.134(N + 2) \qquad \text{Eq. 5.7}$$

where N is the number of atoms in the conjugated chain

Table 5.11. Comparison of experimental and calculated absorption spectra of cyanines (84)

n	λ_{max}^{expt} (nm)[a, b]	λ_{max}^{calc} (nm)[c]	ε_{max} (expt.)[a, b]	f (expt.)[d]	f (calc.)[d]
0	224	224	14,500	—	—
1	313	323	64,500	0.87	0.94
2	416	422	119,500	1.12	1.20
3	519	522	207,000	1.32	1.47
4	625	622	295,000	—	—
5	735	722	353,000	—	—
6	848	822	—	—	—

[a] Experimental λ_{max} and ε_{max} values refer to perchlorate salts in CH_2Cl_2 (see Table 5.8)
[b] Malthotra, S. S., Whiting, M. C.: J. Chem. Soc. *1960*, 3812
[c] Calculated from equation 5.6, where the potential well length L is given by (5.82 + 2.46r) Å
[d] Bayliss, N. S.: Quart. Rev. *6*, 326 (1952)

Steric Effects

Brunings and Corwin[30] were the first to examine steric effects in symmetrical cyanine dyes. They discovered that the tetramethyl dye (109) absorbed at longer wavelengths and had a much lower ε_{max} value than the dimethyl dye (108). The dye (108) is essentially planar but (109) is obviously not planar due to the steric interaction of the two N-methyl groups: twisting about the central carbon-carbon bond occurs to relieve this steric crowding.

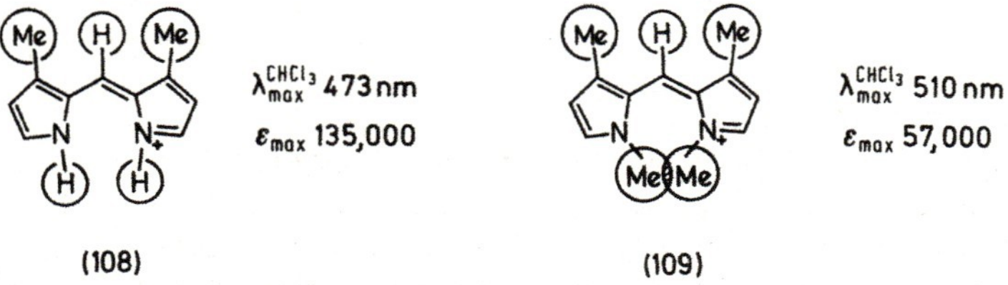

As mentioned in Sect. 3.5.9, steric effects which result in non-planarity *always* cause a reduction in ε_{max} — the large drop in ε_{max} for (109) relative to (108) shows that the deviation from planarity is large. The fact that a bathochromic shift is observed

[30] Brunings, K. J., Corwin, A. H.: J. Amer. Chem. Soc. *64*, 593 (1942).

suggests that the order of the two bonds in the central chain is lower in the first excited state than in the ground state, a postulate supported by PPP MO calculations[31] (Fig. 5.11a). (See Sect. 3.5.9 for an explanation of the theory). Many other examples of the 'Brunings-Corwin' effect have since been found in symmetrical or nearly symmetrical cyanine dyes.

The manifestations of steric effects in highly unsymmetrical dyes, *e.g.* hemicyanines, differ significantly from those observed with symmetrical cyanine dyes. Thus, in addition to the decrease in ε_{max} the absorption shifts to *shorter* wavelengths. This is demonstrated by the hemicyanine dyes (110) and (111). The dye (110) absorbs at markedly shorter wavelengths than (111) and with greatly reduced values of ε_{max}, both effects being more pronounced when R is ethyl. Since structure (110) makes the major contribution to the ground state of the molecule (see earlier), the twisting to relieve the steric crowding occurs about an essentially single bond (α) — hence a hypsochromic shift (see Sect. 3.5.9).

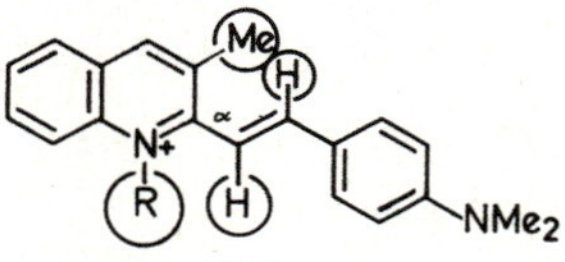

	R	λ_{max}^{MeOH}	ε_{max}
(110)	Me	500	27,000
	Et	493	22,000
(111)	Me	525	60,000
	Et	525	58,000

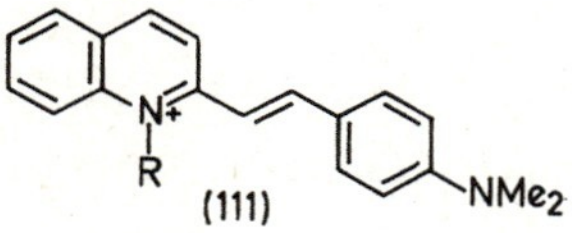

A practical manifestation of steric effects is that non-planar cyanine dyes are non-sensitisers[32] and therefore of no value in the photographic process.

Steric effects in diazahemicyanine dyes are very similar to those in the azonium dyes, which have already been discussed in Sect. 3.5.9.

5.6 Di- and Tri-aryl Carbonium Dyes and their Derivatives

5.6.1 Introduction

Di- and triarylcarbonium dyes and their heterocyclic derivatives comprise the oldest class of synthetic dyes — the majority were discovered in the 19th or early 20th century. Structurally, they are similar to the cyanine dyes just discussed (Sect. 5.5.3) and have rather similar properties, *viz.* exceptional brightness, high tinctorial strength and, unfortunately, low light fastness.

In the following short account, the interrelationships of the dyes within the class, as well as their similarity to cyanine dyes, are highlighted. The discussion is concerned primarily with the cationic amino derivatives (since the important commercial dyes are of this type), but the neutral hydroxy analogues, Phenolphthalein and Fluorescein, are also mentioned.

[31] Falk, H., Hofer, O.: Monatsh. Chem. *106*, 115 (1975).
[32] Sheppard, Lambert, Walker: J. Chem. Phys. *9*, 96 (1941).

5.6.2 Structural Interrelationships

All the important dyes of the di- and triarylcarbonium class are derived from the basic structure (112). Thus, if *X* is a methine group, the resulting dyes are diphenylmethanes (113); if *X* is nitrogen, then azadiphenylmethanes (114) are obtained, whilst if *X* is a C-aryl group, the triphenylmethanes (115) result.

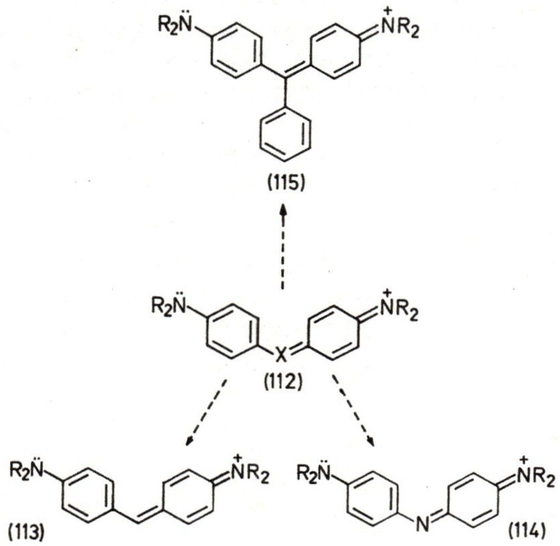

(115)

(112)

(113) (114)

Heterocyclic derivatives are obtained by bridging the diphenylmethanes and azadiphenylmethanes across the *ortho-ortho'* positions with a hetero atom: hence, two series of dyes are possible — Fig. 5.13a and b.

(a) (b)

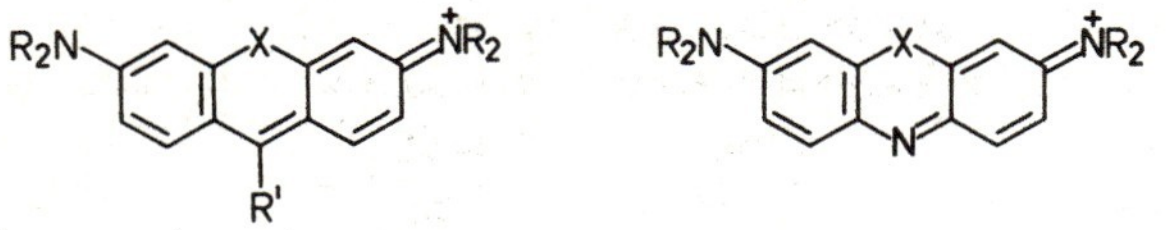

X	Class	X	Class
—O— (known as Xanthenes when R′ = phenyl)	Pyronines	—O—	Oxazines
—S— (known as Thioxanthenes when R′ = phenyl)	Thiopyronines	—S—	Thiazines
—N(R″)—	Acridines	—N(R″)— (known as Safranines when R″ = phenyl)	Azines

Fig. 5.13. Heterocyclic derivatives of (**a**) diphenylmethanes and (**b**) azadiphenylmethanes

5.6.3 General Colour-Structure Properties

The incorporation of an aromatic ring system into the conjugated chain of a cyanine does not alter the characteristic properties of the chromogen. Thus, bond equalisation along the chain (see Sect. 5.5.4) still exists so that Dewar's Rules (see Sects. 3.5.5 and 5.5.4) are eminently suitable for qualitative colour-structure predictions.

The basic diphenylmethane dye, Michler's Hydrol Blue (116), absorbs at 607.5 nm in 98% acetic acid with an ε_{max} of 147,500. Unfortunately, Michler's Hydrol Blue and related bright blue diphenylmethane dyes have poor fastness to light and poor aqueous stability and are therefore of no use commercially.

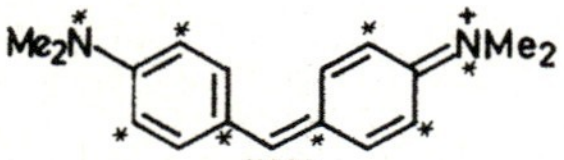

(116)

Attachment of a primary amino group, a powerful electron donor, to the central carbon atom — an unstarred position (see 116) — of Michler's Hydrol Blue produces a tremendous hypsochromic shift. The resulting *yellow* dye, Auramine O (117; CI Basic Yellow 2) is used commercially for several applications, including the dyeing of leather and paper, and in printing inks. Although it is more stable than Michler's Hydrol Blue, Auramine O decomposes in boiling water forming Michler's Ketone (118) and ammonia.

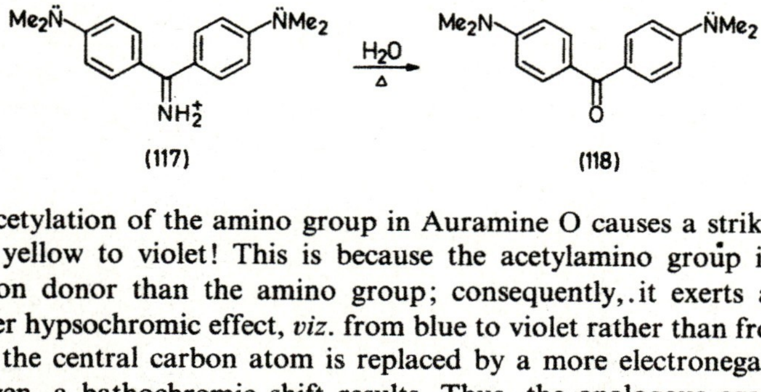

(117) (118)

Acetylation of the amino group in Auramine O causes a striking colour change from yellow to violet! This is because the acetylamino group is a much weaker electron donor than the amino group; consequently, it exerts a correspondingly smaller hypsochromic effect, *viz.* from blue to violet rather than from blue to yellow.

If the central carbon atom is replaced by a more electronegative atom such as nitrogen, a bathochromic shift results. Thus, the analogous aza dye to Michler's Hydrol Blue, namely Bindschedler's Green (119), absorbs at 725 nm in water. However, Bindschedler's Green is even less stable than Michler's Hydrol Blue and therefore it is not used commercially.

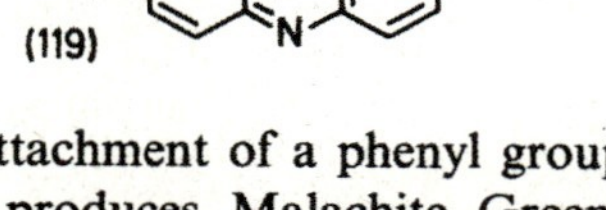

(119)

Attachment of a phenyl group to the central carbon atom of Michler's Hydrol Blue produces Malachite Green (120; CI Basic Green 4), a typical triphenylmethane dye. Triphenylmethane dyes are inherently more stable than diphenylmethane dyes and it is for this reason that the triphenylmethane dyes have been used extensively commercially.

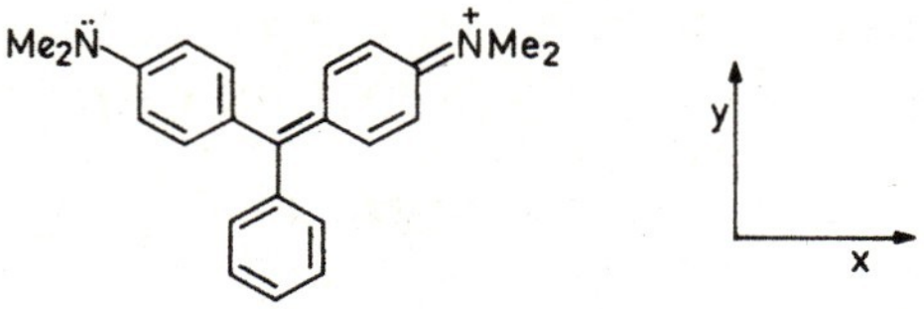

(120)

The long wavelength band of Malachite Green absorbs at longer wavelengths (λ_{max} 621 nm; ε_{max} 104,000) than the long wavelength band of Michler's Hydrol Blue. This is expected for the incorporation of a neutral, conjugating substituent (*viz.* phenyl) at any position in the molecule (see Sect. 3.5.5). However, a new band appears at shorter wavelengths (λ_{max} 427.5 nm; ε_{max} 20,000) which provides a yellow component to the colour (see Fig. 5.14). Hence, the green colour of the dye, *viz.* blue + + yellow = green. The long wavelength band is polarised along the x-axis and the short wavelength band along the y-axis (see 120). Indeed, the two bands are often referred to as the x and y bands respectively.

MO calculations indicate that the x-band arises from a HOMO → LUMO transition associated with the Michler's Hydrol Blue part of the molecule. This produces an excited state with a high electron density on the central carbon atom (hence the bathochromic shift of electronegative atoms, *e.g.* nitrogen, at this position). In contrast, the y-transition is caused by the excitation of an electron from the second highest occupied orbital, *i.e.* the one *below* the HOMO, to the LUMO. Calculations reveal that in the y-transition the electron density in the unsubstituted phenyl ring is reduced.

Substituents in the 'unsubstituted' phenyl ring[33] of Malachite Green affect the major band, the x-band, in a predictable way. Thus, electron-withdrawing groups cause

Table 5.12. Visible absorption spectra of some substituted Malachite Greens in 98% acetic acid

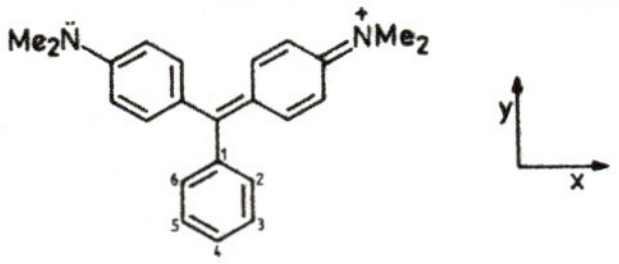

Substituent	x-band		y-band	
	λ_{max} (nm)	ε_{max}	λ_{max} (nm)	ε_{max}
None	621	104,000	427.5	20,000
3-CN	637	89,100	426	15,100
4-CN	643	87,100	429	15,900
4-NO_2	645	83,200	425	17,000
3-MeO	622.5	107,000	435	18,200
4-MeO	608	107,000	465	33,900
4-NMe_2	589	117,000	—	—

[33] *Meta* or *para* substituents only — for *ortho* substituents, steric effects are important (see Sect. 5.6.4).

a bathochromic shift by stabilising the electron rich central carbon atom in the first excited state. In contrast, electron-donating groups cause a hypsochromic shift (Table 5.12). In this case, an interesting effect is observed as the strength of the donor group increases — not only does the *x*-band become progressively more hypsochromic, but also the *y*-band becomes progressively more bathochromic. When all three donor groups are identical, *e.g.* NMe_2 groups, the two bands coalesce into one single absorption band. Thus, Crystal Violet (121) exhibits only one absorption band (λ_{max} 589 nm) in the visible region of the spectrum (Fig. 5.14).

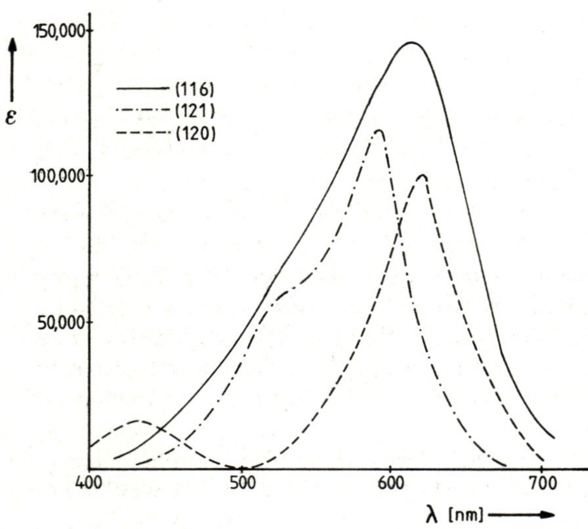

Fig. 5.14. Absorption spectra of Michler's Hydrol Blue (116), Crystal Violet (121) and Malachite Green (120) in 98% acetic acid

Me_2N — $\overset{+}{N}Me_2$ — NMe_2 (121)

The colour of both diphenylmethanes and triphenylmethanes can also be fine-tuned by the donor group. The weaker the donor group, the more hypsochromic the dye. Thus, Doebner's Violet (122) is more hypsochromic than Malachite Green, and Pararosaniline (123) is more hypsochromic than Crystal Violet because a primary amino group is a weaker donor than a dimethylamino group.

$H_2\ddot{N}$ — $\overset{+}{N}H_2$ — Ph (122) λ_{max} 570 nm

$H_2\ddot{N}$ — $\overset{+}{N}H_2$ — $\ddot{N}H_2$ (123) λ_{max}^{AcOH} 538 nm

Extending the conjugation of the phenyl ring produces a bathochromic shift. Indeed, useful blue dyes of the type (124), referred to collectively as the Victoria Blues, have been used commercially for many years.

$Me_2\ddot{N}$ $\overset{+}{N}Me_2$

$\ddot{N}Me_2$ (124)

5.6.4 Steric Effects

Steric hindrance in di- and triphenylmethane dyes can involve twisting about (i) the aryl to central carbon atom bond (*a* in Fig. 5.15) or (ii) the N-aryl bond of the terminal amino groups (*b* in Fig. 5.15). We shall consider type (i) first.

In the Michler's Hydrol Blue series the parent dye appears to be planar[34] even though this implies that the central bond angle (α) must be greater than 120° in order to accommodate the interacting *ortho-ortho'* hydrogen atoms (Fig. 5.15). Bulky substituents adjacent to the central carbon atom (*ortho* position) result in steric interactions. As is evident from Fig. 5.15, one (a methyl group is depicted) or two substituents can be accommodated in the outer *ortho* positions without the planarity of the molecule being greatly affected, but additional *ortho* substituents (which must occupy a more crowded inner position — see Fig. 5.15) cause severe molecular distortions. The relatively large bathochromic shift and massive reduction in ε_{max} of the tetramethyl derivative (Table 5.13) testify to this effect. The bathochromic shift suggests that the central C-aryl bond about which twisting occurs has a higher bond order in the ground state than in the first excited state: this was also the case with cyanine dyes (Sect. 5.5.4).

Unlike the diphenylmethane dyes, the triphenylmethane dyes cannot adopt a planar conformation. X-ray crystallographic studies on both Pararosaniline perchlorate[35] (123) and triphenylmethyl perchlorate[36] (125) show that the three aromatic rings are twisted out of the molecular plane by ~30°. Hence, triphenylmethane dyes are shaped like a 3-bladed propeller. Despite this non-planarity, the steric effects of *ortho* substituents in *symmetrical* triphenylmethane dyes such as Pararosaniline

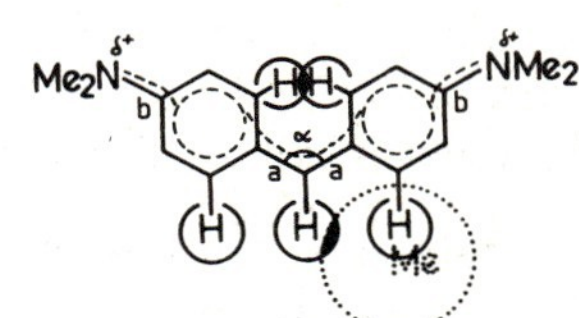

Fig. 5.15. Steric effects in Michler's Hydrol Blue dyes

34 By analogy with a rigid, planar homologue.
35 Eriks, K., Koh, L. L.: Pet. Res. Fund, 230–A1.
36 Eriks, K., et al.: Ann. Rep. Pet. Res. Fund 5, 35 (1960).

Table 5.13. Absorption spectra of *ortho*-methyl derivatives of Michler's Hydrol Blue in 98% acetic acid

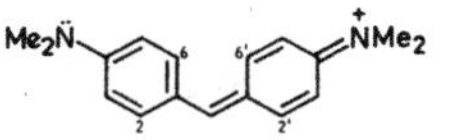

Substituent	λ_{max} (nm)	ε_{max}	$\Delta\lambda$
None	608	147,500	
2-Me	615	130,000	+7
2,2'-Me_2	623	121,000	+15
2,2',6,6'-Me_4	649	55,000	+41

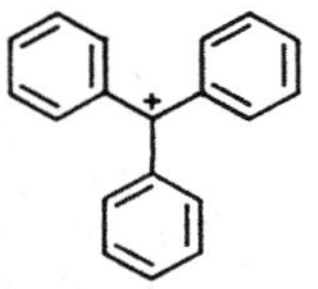

(125)

(123) and Crystal Violet (121) are similar to those observed in the diphenylmethane dyes, *i.e.* a progressive bathochromic shift and reduction in intensity with increasing *ortho* substitution. The uniformity of the shift suggests that the twisting is shared equally between the phenyl rings. However, when two *ortho* substituents are present in one ring, that ring is twisted markedly out of the molecular plane and such dyes, *e.g.* (126), therefore resemble the diphenylmethane dye, Michler's Hydrol Blue, *viz.* they are blue rather than violet.

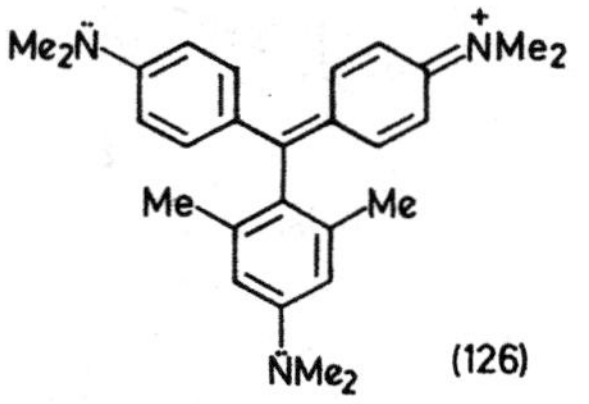

(126)

In unsymmetrical triphenylmethane dyes such as Malachite Green (120) the situation is more complex. For instance, two bands are now involved rather than one. Substitution of methyl groups in the *ortho* positions of the dimethylaminophenyl rings of Malachite Green leads to a reduction in intensity and a regular bathochromic shift of the *x*-band. The *y*-band also undergoes a bathochromic shift and decrease in ε_{max}. In contrast, *ortho* substituents in the unsubstituted ring of Malachite Green cause a hypsochromic shift of the *y*-band (and a reduction in ε_{max}, of course), suggesting a bond order of ~1 for the C-phenyl bond in the ground state. However, the *x*-band undergoes a bathochromic shift and increases in intensity!

Substituents in the *meta* position to the central carbon atom, *i.e.* *ortho* to the terminal dimethylamino groups, leads to complex effects. In the symmetrical dye, Crystal Violet, *meta* methyl groups cause a bathochromic shift which increases as

Table 5.14. Effect of *meta* methyl substituents in Crystal Violet

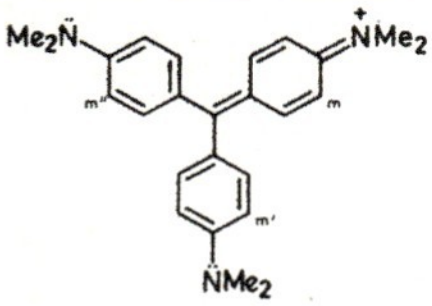

Substituent	$\lambda_{max}^{98\,\%HOAc}$ (nm)	ε_{max}
None	589	117,000
m-Me	599.5	95,000
m,m'-diMe	607.5	78,000
m,m',m"-triMe	615	13,000

the number of *meta* methyl groups increase. The ε_{max}, of course, shows a corresponding decrease (Table 5.14). Shifts are more variable in the Michler's Hydrol Blue and Malachite Green series and the reasons for this are not well understood.

5.6.5 Phenolphthalein

The well-known indicator, Phenolphthalein, is a dihydroxytriphenylmethane. As the lactone (127), it is colourless (even if the hydroxy groups are ionised). Alkali cleaves the lactone ring and the resulting dibasic anion (128) is bluish-red ($\lambda_{max}^{H_2O}$ ~ 550 nm). However, an excess of alkali destroys the colour since the carbinol (129) is produced.

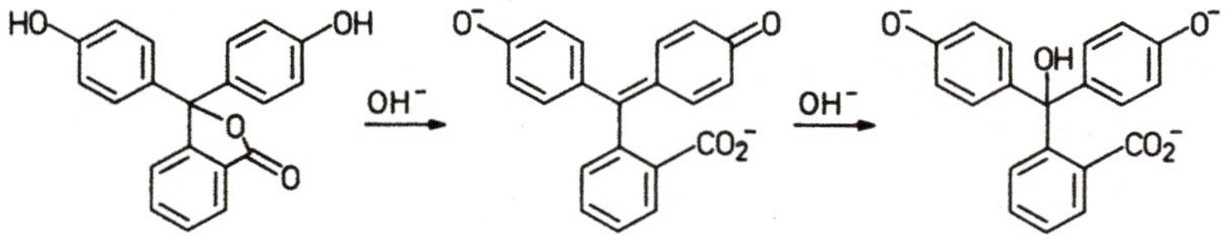

colourless (127) —OH^-→ red (128) —OH^-→ colourless (129)

5.6.6 Heterocyclic Derivatives of Di- and Triphenylmethanes

As mentioned earlier (Sect. 5.6.2), these may be conveniently divided into two series of dyes derived from Michler's Hydrol Blue and Bindschedler's Green respectively (see Fig. 5.13).

The heterocyclic derivatives of both series absorb at shorter wavelengths than their diphenylmethane counterparts; the hypsochromicity increases in the order: — S < O < NH < NR (see Tables 5.15 and 5.16). With the exception of the sulphur atom, this parallels the electron-donating ability of the hetero atom *viz.* the more electron-donating substituents cause the largest hypsochromic shift. Another important feature is that each member of the Bindschedler's Green series (Table 5.16)

Table 5.15. Visible absorption maxima of some heterocyclic analogues of Michler's Hydrol Blue

Structure	Class (Common Name)	λ_{max} (nm)	Colour
	Diphenylmethane (Michler's Hydrol Blue)	607.5	Blue
	Thiopyronine (Thiopyronine)	565	Violet
	Pyronine (Pyronine G)	545	Red
	Acridine (Acridine Orange)	490	Orange
(137)	Acridine (Acriflavine)	460	Yellow

Table 5.16. Visible absorption maxima of some heterocyclic analogues of Bindschedler's Green

Structure	Class (Common Name)	λ_{max} (nm)	Colour
	Aza Diphenylmethane (Bindschedler's Green)	725	Green
(130)	Thiazine (Methylene Blue)	665	Greenish-blue
(133)	Oxazine (—)	645	Greenish-blue
	Azine (—)	567	Magenta

absorbs at longer wavelengths (~100 nm more) than its Michler's Hydrol Blue analogue (Table 5.15).

These experimental facts are explained remarkably well by PMO theory, *e.g.* Dewar's Rules. Thus, the donor group, be it O, NR or S, is attached simultaneously to not one but *two* unstarred positions (see the parent compounds in Tables 5.15 and 5.16). Hence, large hypsochromic shifts are predicted, the magnitude of the shift increasing in proportion to the electron-donating ability of the substituent. Since the central carbon atom also occupies an unstarred position, replacing it with the more electronegative nitrogen atom is predicted to cause a bathochromic shift.

We shall now consider briefly the more important aspects of some of the heterocyclic dyes. More attention is devoted to the derivatives of Bindschedler's Green since these dyes are more important commercially.

Thiazines

The most important thiazine dye, Methylene Blue (130; CI Basic Blue 9 — see Table 5.16) was discovered by Caro in 1876. It is a tinctorially strong (ε_{max}^{EtOH} 92,000) greenish-blue dye with rather poor light fastness (~3 on polyacrylonitrile). In contrast, its nitro derivative, Methylene Green (131; CI Basic Green 5) has superb light fastness! (7–8 on polyacrylonitrile).

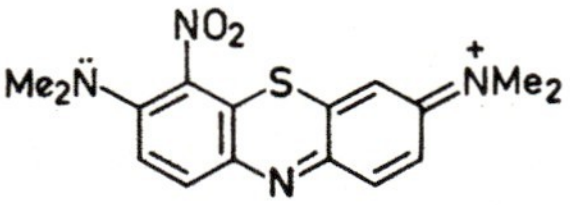

(131)

Thiazine dyes are used on paper (where brilliance and cheapness are required, rather than permanence) and leather. Methylene Blue is also used widely as a stain in bacteriology.

Oxazines

The first oxazine dye was discovered by Meldola in 1879 and is named after him, *i.e.* Meldola's Blue (132; CI Basic Blue 6). The dye is now chiefly applied to leather. However, other oxazine dyes, *e.g.* CI Basic Blue 3 (133; see Table 5.16) are used extensively for dyeing polyacrylonitrile. Bright turquoise shades, which are generally redder than the analogous thiazine dyes, are obtained. The light fastness (~4–5) is better than for the analogous thiazine dyes. Indeed, the light fastness may be improved further (up to 6–7) if arylamino groups are incorporated into the molecule. The blue dye (134)[37] is typical of such dyes.

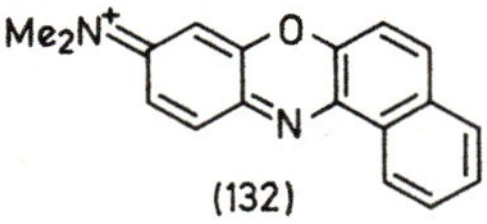

(132)

[37] Example 5 of GB 1,411,479 (BASF).

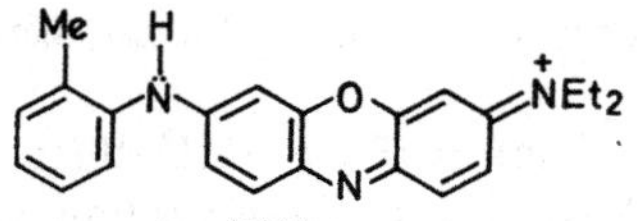

(134)

Azines

The azine dyes are now mainly of historical interest since the first synthetic dye, Mauveine (135) — see Sect. 1.3.2 — belongs to this class. However, a closely related dye, Safranine T (136; CI Basic Red 2), is still used to some extent.

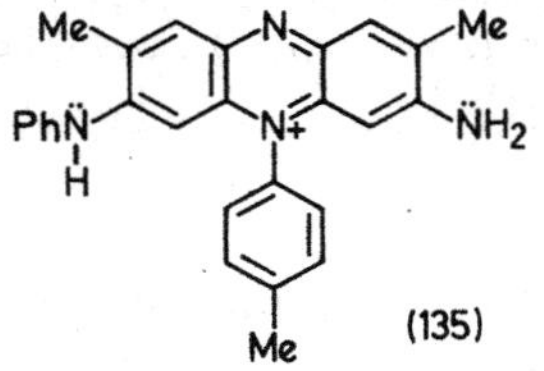

(135)

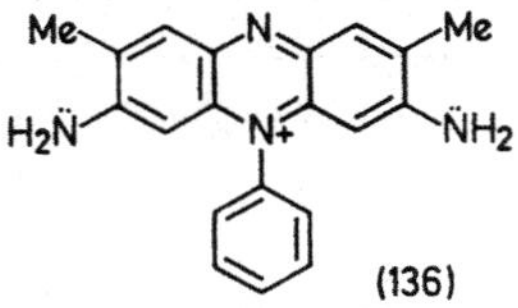

(136)

Acridines

The acridine class provides mainly yellow, orange and brown dyes for leather. The well known antiseptic, Acriflavine (137; see Table 5.15) belongs to this class.

Xanthenes

The outstanding feature of the xanthene dyes, whether they are cationic, *e.g.* Rhodamine 6G (138; CI Basic Red 1), or anionic, *e.g.* Fluorescein (139; CI Acid Yellow 73), is their strong fluorescence. Indeed, Fluorescein's use worldwide as a marker, *e.g.* for life-saving at sea, for tracing the course of underground streams, and for detecting the weak circulation of blood, is due to its strong, green fluorescence in solution, even at very great dilution.

The bright red dye, Eosine (140; CI Acid Red 87), the tetrabromo derivative of Fluorescein, is used widely for colouring paper, printing inks and crayons.

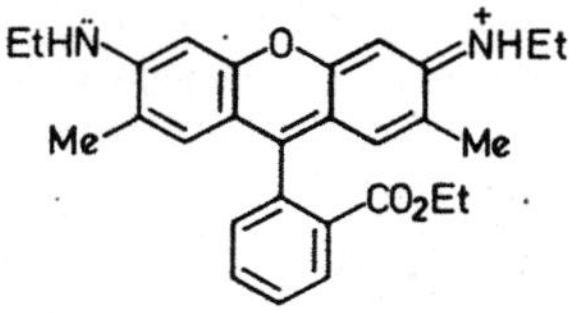

(138)

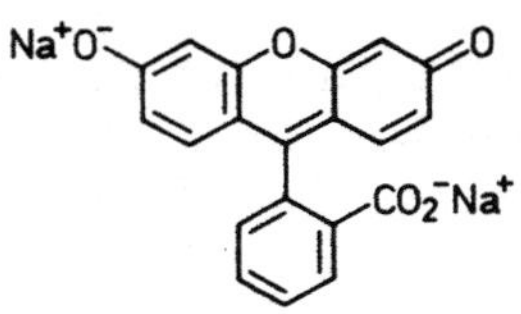

(139)

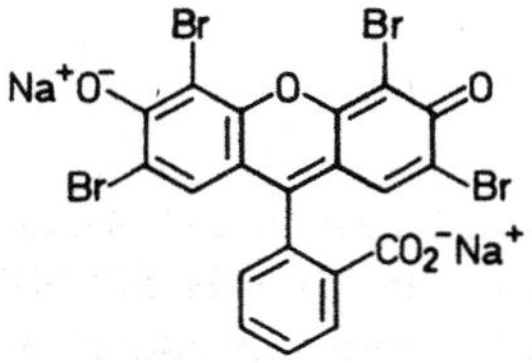

(140)

Because of poor light fastness, xanthene dyes tend not to be used on textiles.

5.7 Nitro (and Nitroso) Dyes

5.7.1 Introduction

Nitro dyes have comparatively simple chemical structures. Typically they are small aromatic molecules (1 to 3 benzene rings) which contain at least one nitro group and one (or more) amino or hydroxy groups. The commercial nitro dyes provide yellow, orange and brown colours. Nitroso dyes are metal complex derivatives of *ortho* nitroso phenols or naphthols. Nitroso dyes are of lesser importance than nitro dyes and consequently less attention is devoted to them.

Nitro dyes are important historically since picric acid (141) was the first synthetic compound to be used commercially for colouring substrates (see Sect. 1.3.1). The 'dye' imparted a greenish-yellow colour to silk. However, picric acid is no longer used because of its toxicity and poor fastness properties.

Martius Yellow (142) — CI Acid Yellow 24 — is a further example of an old nitro dye. Like picric acid, it contains a hydroxy group as the electron donor and is seldom used nowadays.

CI Acid Yellow 1 (143), discovered by Caro in 1879, is the only hydroxynitro dye still in use, even though its fastness properties are poor. Nowadays, all the important commercial nitro dyes are nitrodiphenylamines. Hence, the remainder of this section is devoted to this type of nitro dye.

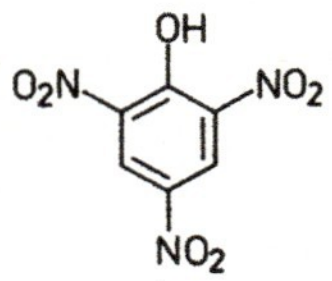

OH, O_2N, NO_2, NO_2

(141)

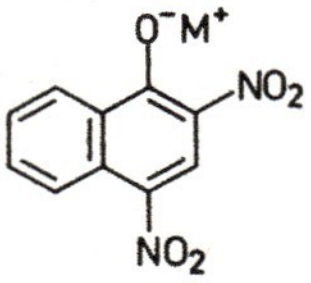

O^-M^+, NO_2, NO_2

$M^+ = NH_4^+, Na^+, Ca^{2+}$

(142)

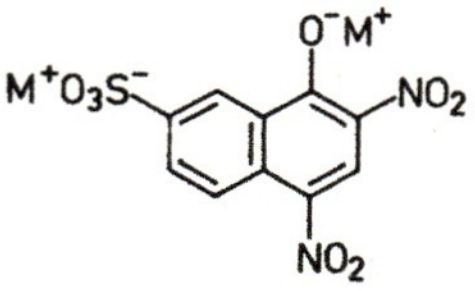

$M^+O_3S^-$, O^-M^+, NO_2, NO_2

$M^+ = Na^+, K^+$

(143)

5.7.2 Nitrodiphenylamine Dyes

Nitrodiphenylamine dyes can be represented by the general structure (144), in which either ring can be further substituted. Typical substituents are Cl, OH, SO_2NR_2 and SO_3H. Most commercial dyes are based on *ortho*-nitrodiphenylamine, *e.g.* CI Disperse Orange 15 (145). The reasons for this are twofold. The first and most important reason is that *ortho* nitrodiphenylamine dyes display a higher order of light fastness than their *para* nitro analogues. Second, *ortho* nitrodiphenylamines are considerably more bathochromic than *para* nitrodiphenylamines and therefore most of the former dyes' absorption curve lies in the visible region of the spectrum, *i.e.* $\lambda_{max} > 400$ nm (see Fig. 5.16 and Table 5.17). In contrast, a substantial portion of the absorption curve of a *para* nitrodiphenylamine dye is in the ultra-violet region of the spectrum (Fig. 5.16 and Table 5.17) and therefore does not contribute to the colour and strength of such dyes (absorption below ~400 nm is not detected by the human eye). However, the ε_{max} values of *para* nitrodiphenylamine dyes are significantly greater than the corresponding ε_{max} values of the *ortho* nitrodiphenylamine dyes (Table 5.17). Consequently, a large number of commercial nitrodiphenylamine dyes contain both an *ortho* and a *para* nitro group in an attempt to increase the tinctorial strength further: the dyes represented by structure (146) are typical.

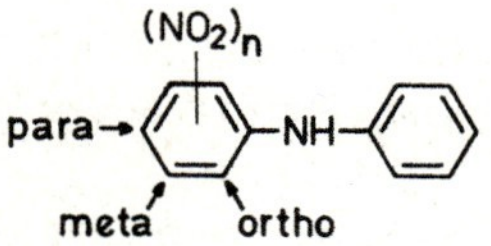

(144) n = 1,2

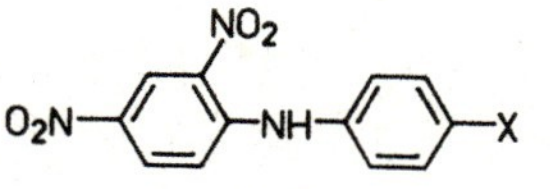

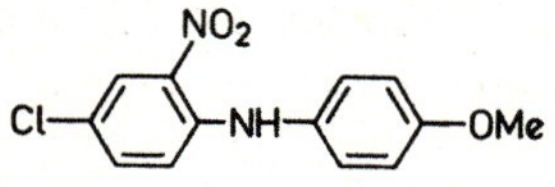

(145)

(146)

X = H CI Disperse Yellow 14
OH CI Disperse Yellow 1
NH_2 CI Disperse Yellow 9

Table 5.17. Absorption data of nitro-substituted diphenylamines and nitroanilines in ethanol

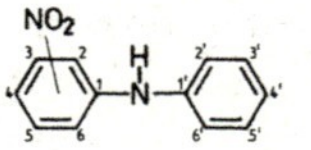

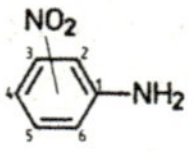

Substituent	λ_{max} in nm (ε_{max}) Diphenylamine		Ref.	Aniline		Ref.
None	289	(19,000)	b	—	—	
2-nitro	425	(8,510)	a	403	(5,890)	a
3-nitro	395	(1,740)	a	373	(1,740)	a
4-nitro	393	(21,900)	a	375	(15,500)	a
2,4-dinitro	354	(20,900)	a	335	(17,000)	a
2,6-dinitro	423	(7,940)	b	—	—	
4,4′-dinitro	402	(38,000)	c	—	—	
2,2′-dinitro	417–423	(9,330)	c	—	—	
2,4′-dinitro	408	(15,140)	a	—	—	
	351	(13,500)				

[a] Asquith, R. S., Bridgeman, I., Peters, A. T.: J. Soc. Dyers and Colourists *81*, 439 (1965)
[b] Bell, M. G. W., Day, M., Peters, A. T.: ibid. *82*, 410 (1966)
[c] Schroeder, W. A., et al.: Anal. Chem. *23*, 1740 (1951)

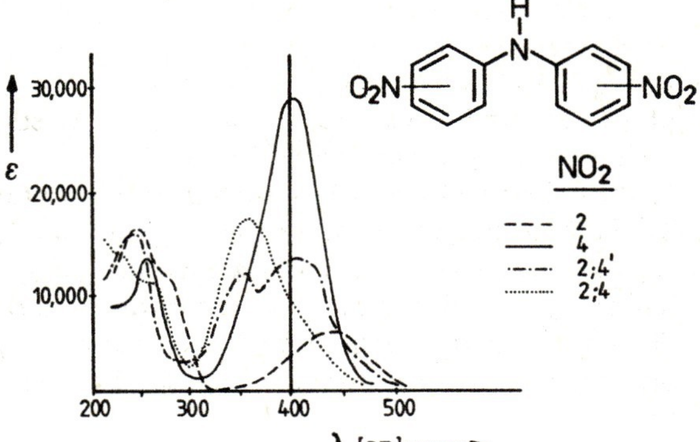

Fig. 5.16. Spectral curves of nitrodiphenylamines in ethanol
Ref. Milliaresi, E. E., Iz'mailskii, B. A.: Zh. Obsch. Khim. *35*, 766 (1965) and references therein.

Dyes containing *meta* nitro groups are not encountered commercially because they are tinctorially weak (see Table 5.17), difficult to prepare,[38] and have poor fastness properties.

The higher light fastness of *ortho* nitrodiphenylamine dyes relative to their *para* analogues is ascribed to intramolecular hydrogen-bond formation, as depicted in (147). This aspect is discussed more fully in Sect. 6.4.4.

(147)

It has been suggested that intramolecular hydrogen-bonding of this type is responsible for the enhanced bathochromic shift of *ortho* relative to *para* nitrodiphenylamines (see Table 5.17). However, although there is no doubt that intramolecular hydrogen-bonding does exist, its influence on the visible absorption band cannot be very pronounced since N-methylation produces no significant change in the λ_{max} value.

The unsubstituted phenyl ring in (144) has little bearing on the colour of nitrodiphenylamines since the corresponding nitroaniline dyes (148) are a similar colour — see Table 5.17. Consequently, substituents in this ring have only a minor effect on the absorption spectrum. In contrast, substituents in the nitrophenyl ring can have a marked effect. In general, electron-withdrawing groups cause a hypsochromic shift and electron-donating groups a bathochromic shift (Table 5.18).

[38] *Ortho* and/or *para* nitrodiphenylamines are readily prepared from the corresponding *ortho* or *para* chloronitrobenzene by reaction with aniline or an aniline derivative, *e.g.*

The chlorine in *m*-chloronitrobenzene is not activated sufficiently by the *meta* nitro group for such nucleophilic reaction to proceed (see Sect. 2.3.1).

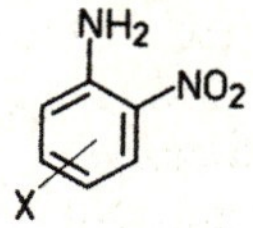

(148)

Table 5.18. Colour bands of nitrodiphenylamines substituted in the nitrated ring

Position of the substituents	λ_{max}^{EtOH} (nm) X CN	CO_2Et	CF_3	H	F	Cl	Br	Me	OMe
2-NO_2,4-X	368	367	366	*393*	385	382	376	396	—
4-NO_2,2-X	412	412	410	*425*	444	437	436	438	465
2,6-di-NO_2,4-X	399	402	407	*423*	—	—	—	—	—

Studies using aniline derivatives as model compounds show that the most bathochromic arrangement of the donor groups is 2,5 relative to the nitro group (*cf.* 149, 150 and 151).[39] This observation, in which the donor group in the 5-position is *meta* to the nitro group, cannot be explained by VB theory. According to VB theory, (149) in which both amino groups are conjugated to the nitro group, should be the most bathochromic isomer (see Sect. 3.5.5). Low tinctorial strength and poor fastness properties have precluded the use of red nitro dyes of this type, *e.g.* (152).

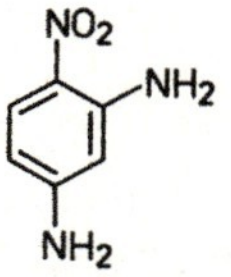

λ_{max} 365 nm (149)

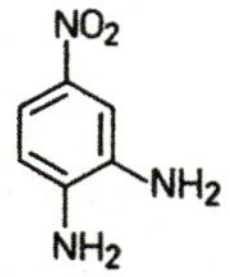

λ_{max} 408 nm (150)

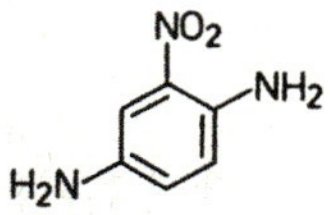

λ_{max} 470 nm (151)

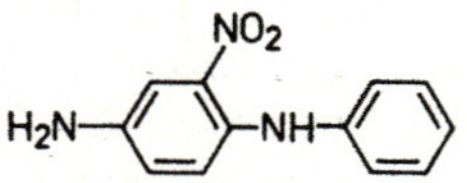

(152) λ_{max} 501 nm ε_{max} 5,260

PPP MO calculations account satisfactorily for the observed colour-structure effects in nitrodiphenylamine dyes. The MO method also shows that on excitation from the ground state to the first excited state there is a decrease of electron density at the amino group and an increase at the nitrophenyl ring, as might be expect-

[39] Dyes of this type, expecially alkylated derivatives, are used extensively as hair dyes since their small molecular size allows ready absorption under mild conditions.

ed. Thus, the amino group is the principal donor group and the nitrophenyl ring the principal acceptor group.

Compared to other dye types, *e.g.* azo dyes, nitro dyes are tinctorially weak. Not only are their ε_{max} values lower (usually $<20{,}000$), but also a significant portion of their absorption curve lies in the ultra-violet ($\lambda < 400$ nm) and therefore does not contribute to the colour/strength (see Fig. 5.16). Nitro dyes are still used for two reasons: — first, they are cheap to prepare, thus compensating for their low tinctorial strength; second, their small molecular size makes them ideal for dyeing tightly packed fibres such as polyester (see Chap. 6).

5.7.3 Nitroso Dyes

Nitroso dyes comprise only a tiny fraction of the commercial synthetic dyes market and so this section will be kept brief.

As mentioned earlier (Sect. 5.7.1), nitroso dyes are metal complex derivatives of *ortho* nitrosophenols or naphthols (they are prepared by nitrosation of the appropriate phenol or naphthol). In the metal free precursors tautomerism is possible between the nitrosohydroxy tautomer (153A) and the quinoneoxime tautomer (153B).[40] The dyes are polygenetic, (*i.e.* form differently coloured complexes with different metals), but only the green iron complexes, because of their good fastness to light, have found application as dyes.

The only nitroso dyes of commercial importance are the iron complexes of 1-nitroso-2-naphthol and its 6-sulpho derivative — CI Pigment Green 8 (154A) and CI Acid Green 1 (154B) respectively. They are relatively cheap colourants and are used chiefly for colouring paper.

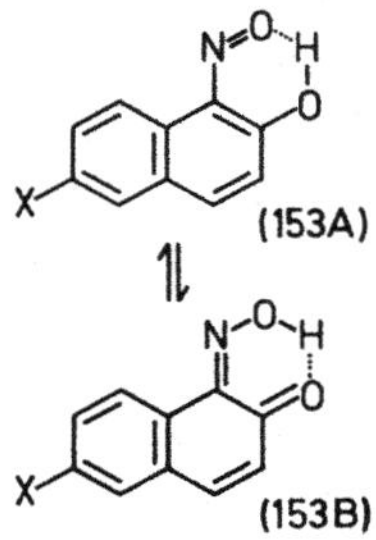

(153A) ⇌ (153B)

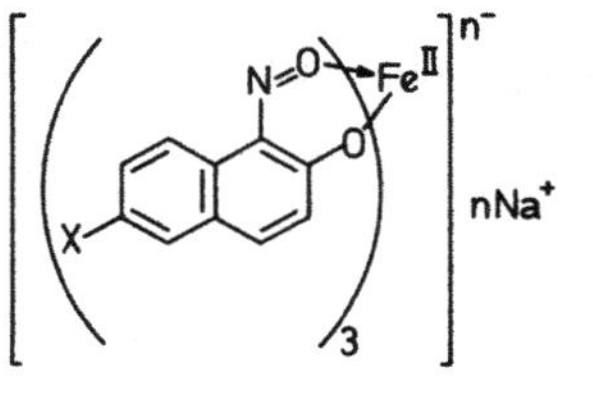

(154) nNa^+

(A) X = H; n = 1

(B) X = SO_3H; n = 4

5.8 Summary

After the azo and anthraquinone dyes, six other important dye types are: —

(i) Vat

(ii) Indigoid

(iii) Phthalocyanine

[40] For a discussion of the tautomerism of nitrosohydroxy compounds (but using *para*-nitrosophenols as model compounds), the reader is referred to: — Hodgson, H. H.: J. Chem. Soc. *1937*, 520 and references therein; — ibid — J. Soc. Dyers and Colourists *40*, 167 (1924).

(iv) Polymethine
(v) Aryl carbonium and
(vi) Nitro and Nitroso dyes.

These eight dye classes cover the majority of dyes used today.

Vat dyes are dyes of any chemical class that are applied to the fibre by a vatting process. In practice, there are two major chemical types of vat dyes — the polycyclic aromatic carbocycles containing one or more carbonyl groups (often called the anthraquinonoid vat dyes), and the indigoid vat dyes. The colours of the anthraquinonoid vat dyes embrace the entire shade gamut but the most important dyes commercially are the blues, greens, browns and blacks. Molecular Orbital calculations indicate that the carbocyclic rings function as the donor groups and that the quinone carbonyl groups are the acceptor groups. The fastness properties of the anthraquinonoid vat dyes are surpassed by no other class of dyes — indeed, vat dyes are renowned for their outstanding fastness properties.

Indigo itself, an ancient natural dye (see Sect. 1.2.5), is the most important indigoid dye. It is unusual in that it is a small molecule yet is blue in colour! Indeed, the controversy concerning the chromophoric unit of Indigo has only recently been resolved. Thus, it has been established that the primary structural unit is the so-called H-chromophore, although hydrogen-bonding and the benzene rings also affect the colour.

The phthalocyanines represent one of the very few new chromophores to have been discovered since the nineteenth century. They are related structurally to the natural porphyrin pigments of which haemin and chlorophyll are well-known examples. However, the phthalocyanine colourants are tinctorially stronger and much more stable. Phthalocyanines are aza [16] annulenes. The chromophoric unit has been established as the 16 atom — 18 π-electron pathway. Consequently, substituents in the fused benzene rings exert only a minor influence on the colour of phthalocyanines. Hence, all phthalocyanine compounds are blue to green in colour. Copper phthalocyanine and its derivatives are by far the most important compounds commercially. They are used mostly as pigments since their large molecular size limits their use as dyestuffs.

Polymethine dyes may be neutral, anionic or cationic but derivatives of the latter category are the most useful as dyes. The best known cationic polymethine dyes are the cyanine dyes. Cyanine dyes have been studied extensively because they approximate closely to the ideal model dye in which there is complete bond equalisation. As a result, qualitative colour-structure rules such as Dewar's Rules, which are derived from the application of PMO theory to such model dye systems, are extremely successful when applied to cyanine dyes. Because of poor fastness properties, especially to light, cyanine dyes are not used as textile dyes. However, they are important in photography. In recent years, certain nitrogen derivatives of cyanine dyes, *e.g.* aza and diazacarbocyanines, and diazahemicyanines, have provided important dyes for polyacrylonitrile fibres.

Diphenylmethane and particularly triphenylmethane dyes are the most important (and familiar) aryl carbonium dyes. Like the cyanine dyes they are bright and strong but suffer from poor fastness properties. The colour-structure properties of triphenylmethane dyes in particular have been thoroughly investigated and these have revealed that complex steric interactions have a crucial bearing on the colour of the dyes. As a result, Dewar's Rules are less successful when applied to triphenylmethane dyes than when they are applied to cyanine dyes. Many heterocyclic dye types are derived from diphenylmethane and azadiphenylmethane dyes simply by bridging with an oxygen, sulphur or nitrogen atom. Thus, pyronines (xanthenes), thiopyronines and acridines are derived from diphenylmethanes whilst the corresponding dyes from azadiphenylmethanes are oxazines, thiazines and azines. Mauveine, the first synthetic dye (see Sect. 1.3.2), belongs to the azine class. The highly fluorescent greenish-yellow dye, Fluorescein, and the well-known indicator dye, Phenolphthalein, are anionic xanthene dyes. Some triphenylmethane dyes and oxazine dyes have found a new lease of life on polyacrylonitrile fibres whilst Fluorescein, because of its strong fluorescence, finds several uses as a marker or tracer dye.

Although nitro and nitroso dyes constitute only a fraction of the total dyes market, they represent some of the simplest dye structures. The most important nitro dyes are nitrodiphenylamines, especially 2-nitrophenylamines. Commercial nitro dyes are yellow to orange in colour and MO calculations indicate that the colour arises because of electron donation from the amino group(s) to the nitro group(s). Nitroso dyes are relatively unimportant commercially. The only commercial nitroso dyes are the iron complexes of 1-nitroso-2-naphthol and its 6-sulphonic acid derivative. These cheap, green colourants are used primarily for colouring paper.

5.9 Bibliography

A. *Vat Dyes*

1. Allen, R. L. M.: Colour chemistry, pp. 162–182. London: Nelson 1971
2. Fox, M. R.: Vat dyestuffs and vat dyeing, pp. 18–29. London: Chapman and Hall 1946
3. Fabian, J., Hartmann, H.: Light absorption of organic colorants. Theoretical treatment and empirical rules, pp. 110–112. Berlin, Heidelberg, New York: Springer-Verlag 1980
4. Venkataraman, K.: The chemistry of synthetic dyes, Vol. II, pp. 861–1002. New York: Academic Press 1952

B. *Indigoid Dyes*

1. Ref. A3.: pp. 115–136
2. Griffiths, J.: Colour and constitution of organic molecules, pp. 195–200. London, New York, San Francisco: Academic Press 1976
3. Rys, P., Zollinger, H.: Fundamentals of the chemistry and application of dyes, pp. 114–118. London, New York, Sydney, Toronto: Wiley — Interscience 1972
4. Ref. A1.: pp. 150–162
5. Sadler, P. W.: J. Org. Chem. *21*, 316 (1956)
6. Sumpter, W. C., Miller, F. M.: Indigo, $\Delta^{2,2'}$-bi-pseudoindoxyl. In: The chemistry of heterocyclic compounds, Weissberger, A. (ed.), Vol. 8, pp. 171–195. New York, London: Interscience 1954
7. Bowen, H. J. M. (comp.), et al.: Tables of interatomic distances and configuration in molecules and ions. Chemical Society Special Publication No. 11, p. M244. London: The Chemical Society 1958

C. *The Phthalocyanines*

1. Lever, A. B. P.: The phthalocyanines. In: Advances in organic chemistry and radiochemistry, Emeleus, H. J., Sharp, A. G. (eds.), Vol. 7, pp. 27–114. New York, London: Academic Press 1965

2. Moser, F. H., Thomas, A. L.: Phthalocyanine compounds. New York: Reinhold Publishing Corporation, London: Chapman and Hall 1963
3. Booth, G.: Phthalocyanines. In: The chemistry of synthetic dyes, Venkataraman, K. (ed.), Vol. V, pp. 241–282. New York: Academic Press 1971
4. Moser, F. H.: Phthalocyanine blue (and green) pigments. In: Pigment handbook, Patton, T. C. (ed.), Vol. I, pp. 679–695. New York, London, Sydney, Toronto: Wiley 1973
5. Ref. A3.: pp. 198–204
6. Ref. B2.: pp. 227–233
7. Ref. A1.: pp. 231–240
8. Cronshaw, C. J. T.: Endeavour *1*, 79 (1942)
9. Ehrich, F. E.: Pigments (organic). In: Encyclopedia of chemical technology, Kirk-Othmer (eds.), 2nd ed., Vol. 15, pp. 580–581. New York, London, Sydney: Interscience 1964
10. Patton, T. C. (ed.): Pigment handbook, Vol. I, pp. 429–434. New York, London, Sydney, Toronto: Wiley 1973
11. Colour Index, 3rd ed., Vol. 4, pp. 4617–4622. The Society of Dyers and Colourists, Bradford; The American Association of Textile Chemists and Colorists, Lowell, Mass., U.S.A.

D. *Polymethines*

1. Ref. A3.: pp. 162–197
2. Sturmer, D. M.: Cyanine dyes. In: Encyclopedia of chemical technology, Kirk-Othmer (eds.), 3rd ed., Vol. 7, pp. 335–358. New York, Chichester, Brisbane, Toronto: Wiley-Interscience 1979
3. Ref. B2.: pp. 240–250
4. Baer, D. R.: Cationic dyes for synthetic fibres. In: The chemistry of synthetic dyes, Venkataraman, K. (ed.), Vol. IV, pp. 161–211. New York, London: Academic Press 1971
5. Ref. A1.: pp. 190–192
6. Brooker, L. G. S. et al.: Chem. Revs. *41*, 325 (1947)
7. Hamer, F. M.: The cyanine dyes and related compounds. In: The chemistry of heterocyclic compounds. Weissberger, A. (ed.), Vol. 18. New York, London: Interscience 1964
8. Ref. A4.: pp. 1143–1186
9. Ficken, G. E.: Cyanine dyes. In: The chemistry of synthetic dyes, Venkataraman, K. (ed.), Vol. IV, pp. 211–340. New York, London: Academic Press 1971

E. *Aryl Carbonium and Related Dyes*

1. Ref. B2.: pp. 250–265
2. Ref. A1.: pp. 103–131
3. Ref. A3.: pp. 137–161
4. Ayyanger, N. R., Iilak, B. D.: Basic dyes. In: The chemistry of synthetic dyes, Venkataraman, K. (ed.), Vol. IV, pp. 103–160. New York, London: Academic Press 1971
5. Ref. A4.: pp. 705–795

F. *Nitro (and Nitroso) Dyes*

1. Ref. A3.: pp. 82–86
2. Ref. B2.: pp. 163–167
3. Ref. A1.: pp. 193–194
4. Ref. B3.: pp. 72–73
5. Venkataraman, K.: The chemistry of synthetic dyes, Vol. I, pp. 401–408. New York: Academic Press 1952
6. Ref. C11.: pp. 4001–4008
7. Krahler, S. E.: Miscellaneous dyes. In: The chemistry of synthetic dyes and pigments, Lubs, H. A. (ed.), pp. 254–259. New York: Reinhold 1955

Chapter 6

Application and Fastness Properties of Dyes

6.1 Introduction

In Chapters 3–5 we have discussed the physico-chemical properties of the more important classes of dyes. We shall now describe how dyes are applied to substrates, mainly fibres, and discuss the fastness properties of the dyed fabric to various agents such as water (laundering) and light. To do this we need to have some idea of both the physical and chemical structure of the substrates to which the dyes are applied for, as will become apparent later, the same dye can exhibit widely differing fastness properties on different fibres.

6.2 Textile Fibres — Types and Structures

6.2.1 Introduction

Although dyes are used to colour substrates other than textile fibres, *e.g.* paper and plastics,[1] textiles are still the principal outlet for dyes. Textile fibres may be classified conveniently into three types: —

(i) Natural
(ii) Semi-synthetic, and
(iii) Synthetic.

As stated in Chapter 1 the early textile industry was based entirely on natural fibres. These were of animal origin (*e.g.* wool, silk, hair, *etc.*) and vegetable origin (*e.g.* cotton, linen, hemp, jute, flax, *etc.*). Even though the natural *dyes* had been totally replaced by superior synthetic dyes by the end of the nineteenth century, natural *fibres* reigned supreme well into the twentieth century.

The first semi-synthetic "fibre" was nitrocellulose, developed by Sir Joseph Swan in England (1883/84) and Chardonnet in France (1884) as a spin-off from the search for a strong carbon filament for lamps. Later, viscose rayon, cuprammonium rayon and cellulose acetate fibres were developed, but it was not until after the First World War that these semi-synthetic fibres made a serious impact on the textile market. In 1919, world production of semi-synthetic fibres was about 11,000 tons; 10 years later, it was 197,000 tons!

[1] Pigments, which may be regarded as water insoluble dyes, are used widely for colouring non-textile substrates.

The three main synthetic fibres, polyamide (nylon), polyester (*e.g.* Terylene) and polyacrylonitrile (acrylic) were discovered, developed and launched as commercial fibres during the period 1930–1950.

The types and structures of the natural, semi-synthetic and synthetic fibres are now discussed.

6.2.2 Natural Fibres

(a) Animal Fibres — Proteins

Several animal fibres have been used as textile materials, *viz.* wool, silk, fur and leather. All these substances are composed of proteins. However, only the structure of wool, the most important and probably the most familiar of these materials, is discussed.

Wool consists of a protein called keratin. Keratin is a polymer of great complexity having as its basic structural unit 18 different amino acids (1). In keratin, the polypeptide (protein) chains (2) are covalently bound at intervals by bridges derived from the amino acid cystine (3). Polymers in which the chains are linked together chemically in this manner are known as cross-linked polymers. Hence, keratin (and therefore wool) is a cross-linked polypeptide (4). As expected, the keratin molecules are very large — the average molecular weight is ~60,000.

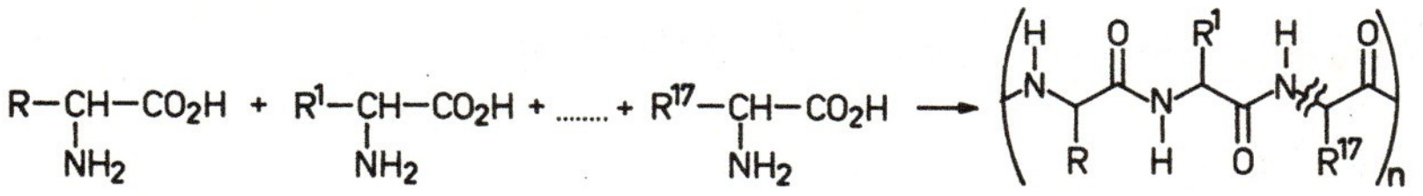

$$R{-}CH(NH_2){-}CO_2H + R^1{-}CH(NH_2){-}CO_2H + \ldots\ldots + R^{17}{-}CH(NH_2){-}CO_2H \longrightarrow$$

(1) (2)

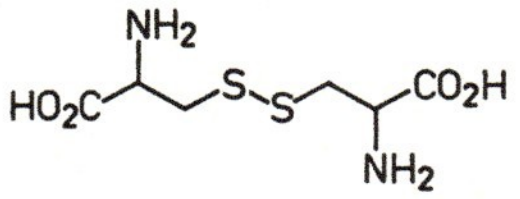

(3)

= polypeptide backbone

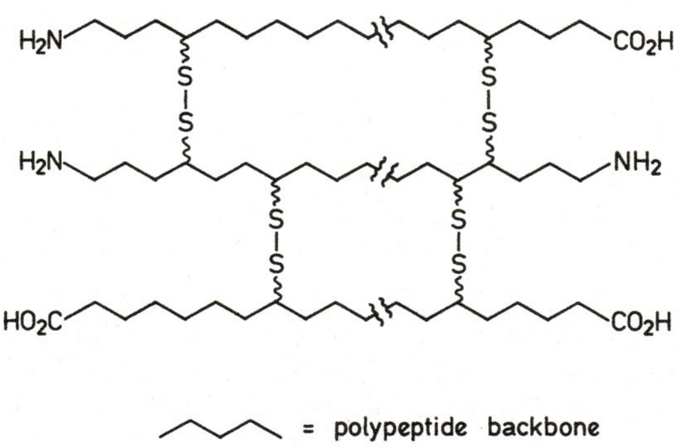

(4)

Strong ionic linkages are also formed within the fibre and these enhance further the binding between the polymer chains. The ionic bonds originate from inter-

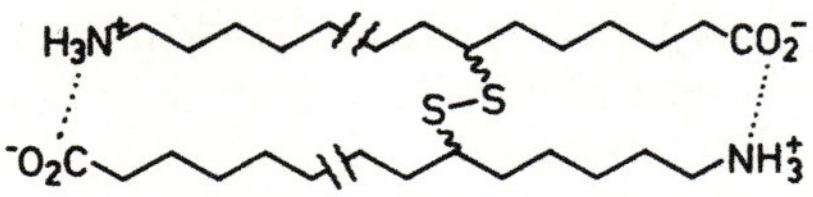

Fig. 6.1. Ionic bonding in wool

actions between the $-NH_3^{\oplus}$ and $-CO_2^{\ominus}$ groups present at the ends of the polymer chains, as shown in Fig. 6.1.

In physical structure the wool fibre is the most complex of all the fibres used for textile purposes. It consists of a central cortex, composed of a bundle of fibrils,[2] surrounded by a scaly covering known as the cuticle. Affinity for dyes is not uniform through the fibre — the cuticle is hydrophobic and highly resistant to penetration. Fortunately for the dyer, the cuticle does not usually extend to the extremities of the fibre.

Under acidic conditions dissociation of the weakly acidic carboxylic acid groups is suppressed and the equilibrium shown in Eq. 6.1 is well to the right. Consequently, wool may be dyed from an acid dyebath by anionic (*i.e.* negatively charged) dyes since the latter will simply displace the colourless counter-ions within the fibre. In practice, weakly acidic conditions are employed since hydrolysis of the peptide linkages occurs under strongly acidic conditions.

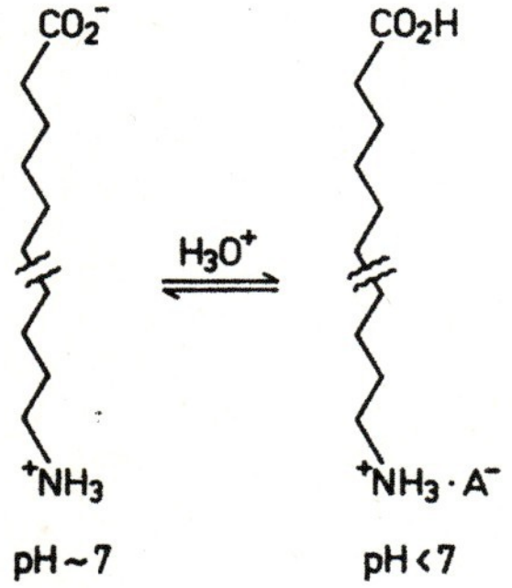

Eq. 6.1

(b) Vegetable Fibres — Cellulose

Cellulose is the most abundant of all organic polymers, natural or otherwise. It occurs in all plants as a skeletal structure. Cotton (from the cotton plant), the most important cellulosic fibre, is almost pure cellulose (up to 95%): other cellulosic fibres, *e.g.* linen, jute, hemp and flax, contain a lower proportion of cellulose. For example, linen contains 82–83% cellulose.

Cellulose is a linear polymer of glucose (5) and is a polysaccharide. X-ray crystallographic data show that the glucose units are linked together in a chain structure in a regular manner. The repeating unit is the cellobiose unit (6) which is two glucose units joined together. As will become evident later (Sect. 6.3.2), the size and configuration of the cellobiose units have a vital bearing on the affinity of certain dyes for cotton.

[2] Small narrow fibres or long cells.

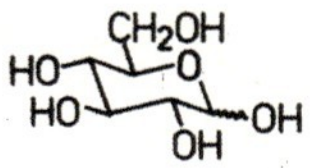

(5)

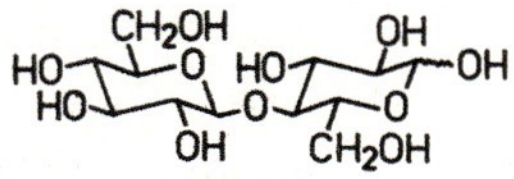

cellobiose unit

(6)

Cotton is a large polymer in which the number of glucose units range from ~1200–3000. Since each glucose unit in the polymer contains 3 hydroxy groups cotton is a hydrophilic (water-liking) fibre. Extensive hydrogen-bonding occurs between the hydroxy groups in cellulose and the more hydrogen-bonding the greater the crystallinity of the fibre. Natural cellulose is highly crystalline, suggesting extensive hydrogen-bonding.

Dry cotton has a density of 1.05 which indicates a fairly open structure (the densities of various fibres are depicted in Table 6.1). In water, it swells laterally rather than longitudinally. In other words, the polymer chains move apart so that the structure becomes even more open: this, of course, facilitates the entry of dye molecules.

Table 6.1. Densities of some important fibres

Fibre	Density
Cotton-dry (wet)	1.05 (1.55)
Nylon	1.14
Polyacrylonitrile	1.17
Polyester	1.78

6.2.3 Semi-Synthetic Fibres

The semi-synthetic fibres are natural fibres that have been modified either by a chemical treatment, *e.g.* viscose rayon and cuprammonium rayon, or by the incorporation of actual chemical groups, *e.g.* nitrocellulose, cellulose acetate and cellulose triacetate. All these semi-synthetic fibres are therefore derivatives of cellulose.

Nitrocellulose was the first semi-synthetic fibre. As mentioned earlier (Sect. 6.2.1), it was developed by Sir Joseph Swan and Chardonnet in 1883/84. Nowadays, nitrocellulose is not used as a textile fibre.

The two most important semi-synthetic fibres are viscose rayon and cellulose acetate. Viscose rayon is a regenerated cellulosic fibre in which up to 15% of the hydroxy groups have been replaced by ether linkages in a cross-linking operation. It is manufactured by dissolving cellulose (*e.g.* wood pulp, straw, *etc.*) in sodium hydroxide solution: carbon disulphide is added and the resultant cellulose xanthate (7) is dissolved in dilute sodium hydroxide solution to give the so-called *viscose* solution. This is then extruded through very narrow holes (spinnarets) into a sulphuric acid bath which neutralises the sodium hydroxide and hydrolyses the xanthate ester, thus regenerating the cellulose as a continuous filament. The chain length of the viscose rayon so obtained is about 600 glucose units, compared to about 1200–3000 units for cotton (see Sect. 6.2.2).

$$R{-}OH \xrightarrow{NaOH} R{-}O^- \xrightarrow{CS_2} R{-}O{-}\overset{\overset{S}{\|}}{C}{-}S^- \xrightarrow[\text{2. extrude into } H_2SO_4]{\text{1. dil. NaOH}} \text{Viscose Rayon}$$

(7)

R = Cellulose residue

In a related process, cellulose is dissolved in a cuprammonium solution ($CuO + NH_3 + H_2O$). The resulting fibre, cuprammonium rayon, is similar chemically to viscose rayon but gives a finer yarn. Thus, it commands a premium price and is used in sheer fabrics.

Unlike viscose rayon and cuprammonium rayon, cellulose acetate (also referred to as secondary acetate or simply acetate) is an actual chemical derivative of cellulose. Indeed, cellulose acetate is often regarded as the first *commercial* synthetic fibre. It was introduced by the British Celanese Corporation soon after the 1914–18 war, utilising a product manufactured during the war as a preservative dope for aircraft fabrics.

Cellulose acetate is made by the partial hydrolysis of cellulose triacetate, obtained by the action of equimolar amounts of acetic anhydride and glacial acetic acid on cotton — Eq. 6.2. On average, cellulose acetate contains 2.3 acetyl groups per glucose unit.

$$[\text{cellulose: } CH_2OH, HO, OH]_n \xrightarrow[3n\ Ac_2O]{AcOH\ +} [\text{triacetate: } CH_2OAc, AcO, AcO]_n \xrightarrow[AcOH]{H_2SO_4/H_2O} \text{Cellulose Acetate} \quad \text{Eq. 6.2}$$

(8)

The removal of more than $^2/_3$ of the hydroxy groups deprives cellulose acetate of affinity for conventional cotton dyes such as the direct dyes. Hence, new dyes were developed for dyeing cellulose acetate (see Sect. 6.3.3).

Cellulose Triacetate

As is evident from Eq. 6.2, cellulose triacetate (8) is a precursor to cellulose acetate. However, processing problems delayed the commercialisation of this fibre. Even though these have now been overcome, the fibre commands only a small share of the total market. The main virtue of cellulose triacetate is its ability to be heat set to give 'wash and wear' fabrics.

As stated earlier, it was not until after the First World War that semi-synthetic fibres made a serious impact on the textile market. The market growth of both natural fibres and semi-synthetic fibres continued until after the Second World War, when the advent of totally synthetic fibres posed a new threat.

6.2.4 Synthetic Fibres

As the name implies the synthetic fibres are those prepared entirely independently of the natural polymers. There are three important synthetic (man-made) fibres — polyamide (nylon), polyester (*e.g.* Terylene) and polyacrylonitrile (acrylic). In each case the polymer is built up from a manufactured monomer or monomers.

(i) Polyamides

The work leading to the first commercial synthetic fibre was started by Carothers (du Pont) in the late 1920's. It culminated in the discovery that hexamethylene diamine adipate could be drawn into a resilient textile fibre. These fibres are now known under the generic name nylon.

Polyamide fibres are made by reacting a diamine with a dicarboxylic acid. They are named according to the number of carbon atoms in the intermediates; for example, nylon from hexamethylene diamine and adipic acid is called nylon 66 (9). Similarly, the nylon from caprolactam (10) is called nylon 6[3] (11). Nylon 66 is a stronger, more durable fibre than nylon 6 and has a higher melting point (265 v 220 °C). Its density is 1.14. Nylon 66 is the dominant nylon fibre in Britain and America whereas nylon 6 dominates the market in Europe and Japan. Although several other types of nylon are manufactured, nylon 66 and nylon 6 remain the most important textile fibres (the other types are used for ropes, canvas, bristles, *etc.*).

$$HO_2C{-}(CH_2)_4{-}CO_2H + H_2N{-}(CH_2)_6{-}NH_2 \longrightarrow [{-}CO(CH_2)_4CONH(CH_2)_6NH{-}]_n$$

(9)

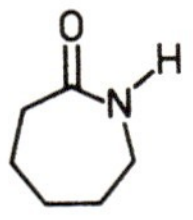

(10) $\longrightarrow$ $-CO[NH(CH_2)_5CO]_nNH-$ (11)

X-ray diffraction studies show that nylon 66 has a highly oriented structure due to intermolecular hydrogen-bonding between the CO and NH groups and Van der Waals forces between the chains (see Fig. 6.2).

Nylon fibres are a chemically simpler version of protein fibres. However, they lack the moisture absorbency and softness of, for example, wool. On the other hand they are hard wearing and make up into garments which are drip dry. As a result, they are used extensively, either alone or in blends with other fibres, as in carpets.

(ii) Polyester

As well as his work on polyamide fibres Carothers also investigated aliphatic polyesters (*e.g.* 12) but no polymer emerged that was useful for producing synthetic fibres of commercial value. However, in 1941 Dickson and Whinfield (Courtaulds) discovered

[3] Tradename — Perlon.

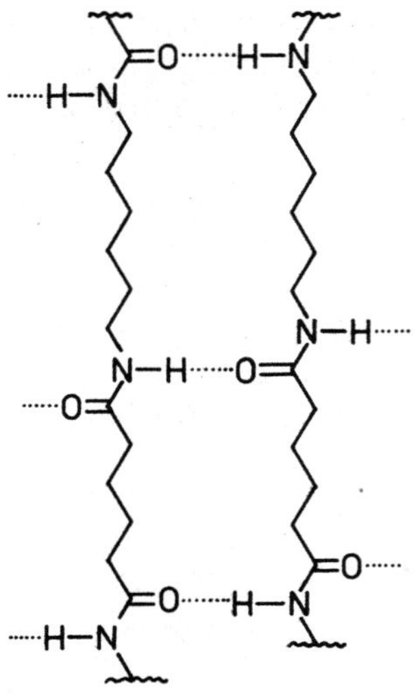

Fig. 6.2. Intermolecular hydrogen-bonding in nylon 66

that the reaction of terephthalic acid (an *aromatic* dicarboxylic acid) with ethylene glycol (an aliphatic diol) gave a synthetic polymer capable of forming a fibre with useful properties. Both ICI and du Pont carried out further development and the first polyester fibres were introduced in 1947.

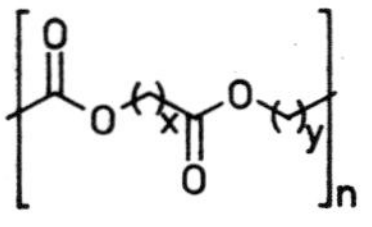

(12)

Polyethylene glycol terephthalate is still the major commercial polyester textile fibre. It is a condensation polymer (*cf.* nylon) made by reaction of ethylene glycol with either terephthalic acid or its dimethyl ester (Scheme 6.1).

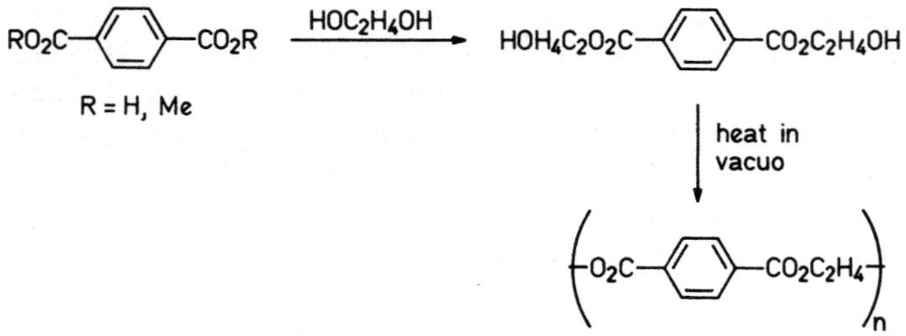

$RO_2C-C_6H_4-CO_2R \xrightarrow{HOC_2H_4OH} HOH_4C_2O_2C-C_6H_4-CO_2C_2H_4OH$

R = H, Me

heat in vacuo

$\left(-O_2C-C_6H_4-CO_2C_2H_4-\right)_n$

Scheme 6.1

In the polymer the chains, which are almost fully extended, are packed tightly side by side (13). They are held together by Van der Waals forces, which are strong because of the close proximity of the molecules. Polyester fibre is therefore remarkably free of voids and this is reflected in its high density (1.78) compared to other fibres — see Table 6.1. As a result, polyester fibre is inaccessible even to small molecules, as demonstrated by its very low water imbibition figure — Table 6.2. Consequently, it is a difficult fibre to dye.

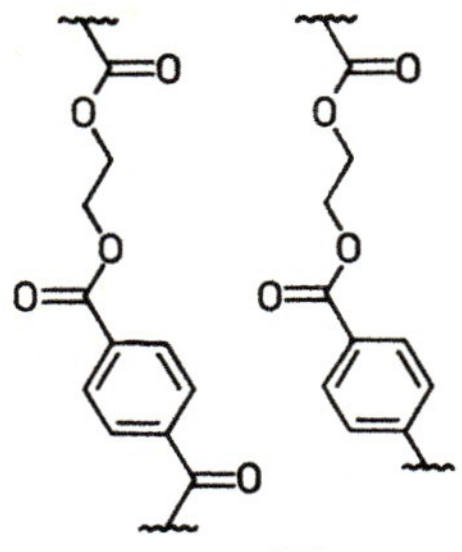

(13)

Table 6.2. Water imbibition of some common fibres

	Water Imbibition %
Viscose	100
Acetate	25
Triacetate	10
Polyamide	11–13
Polyacrylic	8–10
Polyester	3
Polypropylene	0

Polyester fibres are used mainly as textile materials, either alone or in blends with cotton, *viz.* polyester-cotton blends. In recent years the polyester-cotton blends have gained in popularity since they have a better moisture absorbency than pure polyester and have better 'wash and wear' properties than pure cotton.

The high weight/unit area ratio of pure polyester places it at an economic disadvantage relative to less dense fibres (see Table 6.1). Since cotton is a much lighter fibre than polyester, it follows that polyester-cotton blends are less dense than pure polyester and this is a further reason for the increasing importance of such blends.

A secondary use for polyester is in the clear film, Melinex. It is used both as a food wrapping and to make clear plastic bags.

(iii) Polyacrylonitrile

Acrylonitrile came into prominence during the Second World War when it was used as a copolymer with butadiene to produce 'nitrile rubber' to augment the depleted supplies of natural rubber. As a result, acrylonitrile was in plentiful supply in the USA in the 1940's. This no doubt prompted chemists at du Pont to investigate the properties of polymers derived from this available and inexpensive monomer and in 1950 du Pont launched the first polyacrylonitrile fibre — Orlon.

In contrast to polyamide and polyester fibres, which are condensation polymers, polyacrylonitrile is an addition polymer. It is made by the polymerisation of acrylonitrile (Eq. 6.3).

n CH₂=CH–CN → –[CH₂–CH(CN)]ₙ– Eq. 6.3

Although experimental data on the structure of polyacrylonitrile is scanty, X-ray studies indicate a high degree of lateral order but little or no longitudinal regularity of the polymer molecules. The polymer chains are packed together in a hexagonal manner and are held together by Van der Waals forces and polar forces between the nitrile groups. The density of polyacrylonitrile (1.17) is lower than polyester and nylon (Table 6.1) — hence, it is concluded that there are voids in the structure.

Anionic centres are present in the major brands of polyacrylonitrile fibres : —

Orlon	(du Pont)	
Dralon	(Bayer)	SO_3H groups
Acrilan	(Chemstrand)	
Courtelle	(Courtaulds)	CO_2H groups

These arise either from the addition of polymerisation inhibitors or from small amounts of copolymer added deliberately to introduce acidic sites. As will be seen later, these acidic sites allow the fibre to be dyed with cationic dyes.

More recently the so-called modacrylic fibres have been introduced. These are fibres in which acrylonitrile is copolymerised with 20% or more of a comonomer such as vinyl acetate or methyl methacrylate. Modacrylic fibres are usually easier to dye than polyacrylonitrile itself since the presence of a second monomer disrupts the regular chain packing and thereby improves the entry of dye molecules into the fibre. Indeed, it is a general feature that there is more versatility in the synthesis of addition polymers than in condensation polymers: hence, they can be better 'tailored' to optimise the desired fibre properties.

In appearance and handling, polyacrylonitrile fibres are similar to wool; however, they are considerably cheaper. Hence, polyacrylonitrile fibres have made a significant impact on the cheaper end of the wool market.

6.3 Application and Wet Fastness of Dyes

6.3.1 Introduction

It may seem puzzling that such a vast number (~7000) of dyes[4] is required to colour such a small number of different fibres. As we shall see, part of the answer lies in the constraints imposed by a particular fibre upon the physical and chemical structure of the dye. Thus, different dye types are required for different substrates. An equally important reason lies in the enormous diversity of end uses of the finished articles. At one extreme there are dyes for paper where the overriding consideration is cheapness since the fastness requirements of dyes for toilet paper and wrapping paper are obviously minimal. At the other extreme are dyes for high class furnishings such as carpets and curtains: such dyes need to possess excellent properties because the products are expected to last for many years. There are many outlets in between these two extremes where the finished article needs to withstand some particular treatment. For example, dyes for certain types of apparel (clothing) need to have good wet fastness since they are laundered often whereas dyes for swimwear must be resistant to bleaching by chlorine, and so on. However, the dye has first to be applied to the fibre by a dyeing process and since the wet fastness properties of a dye are inextricably linked to the application process we shall consider these two aspects together.

There are five broad types of dye — fibre interaction: —

1. Physical adsorption 2. Solid solution 3. Insoluble aggregates within the fibre	Physical interactions
4. Ionic bonds 5. Covalent bonds	Chemically bound

We shall consider each type separately. For brevity, only the more important examples of each type are considered.

6.3.2 Physical Adsorption

The best example of this type of interaction is provided by the direct dyes for cotton. Unlike wool and nylon, cotton does not contain any cationic or anionic centres capable of forming an ionic linkage with a suitably charged dye. Thus, whereas all anionic dyes can be applied directly to nylon and wool (see later) only certain types have affinity for cotton, *viz.* the direct dyes. Thus cotton proved to be an extremely difficult fibre to dye directly and various pre-treatments or after-treatments were required to 'fix' the dye on the fibre, *e.g.* mordanting (see Sect. 3.4.1). As mentioned earlier cotton is a very hydrophilic fibre and also has an open structure which permits the easy entry of dye molecules into the fibre; the difficult part is keeping them there!

The first dye which was shown to have affinity for cotton by simply boiling the fibre in water containing the dye (and common salt) was Congo Red (14). This discovery prompted workers to explore why this dye had affinity[5] for cotton when all the other dyes examined previously had no affinity. It was subsequently found

[4] This figure includes organic pigments.

[5] Often called substantivity.

that long, planar, polyazo dyes having relatively low water solubility were required. Furthermore, the dyes with the highest substantivity were those in which the distance between the azo groups was *ca.* 10.8 Å.

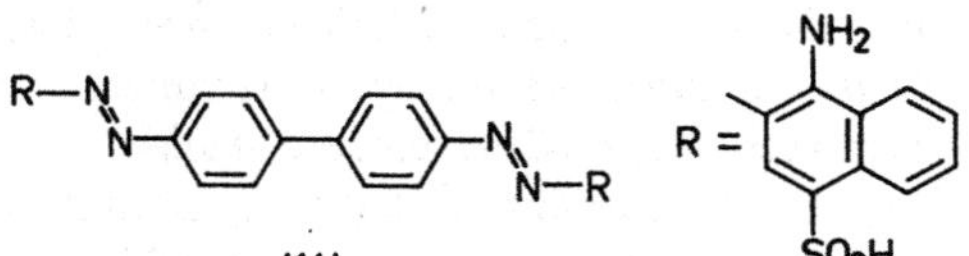

(14)

Azo dyes from benzidine[6] (15), *e.g.* (17), and 1-naphthylamine[6] (16), *e.g.* (18), fit the above requirements very well and therefore most direct dyes are of these types. As expected, non-planar benzidine dyes, *e.g.* (19), have little affinity for cotton.

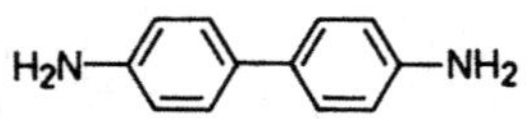

(15)

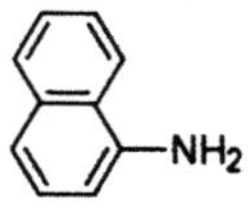

(16)

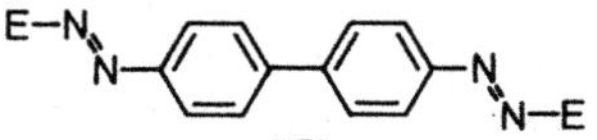

(17)

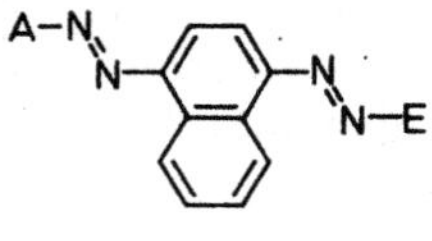

(18)

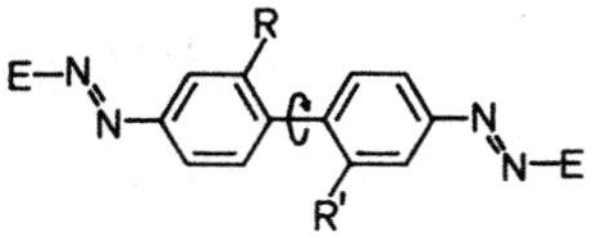

R,R' = Me, Cl, etc.

(19)

A = Diazo component
E = Coupling component

The most likely explanation of the above facts is that such dye molecules are of the correct shape to fit closely on the cellulose polymer. This is supported by X-ray measurements on the cellobiose unit (see structure 6) which show it to have a length of 10.3 Å corresponding closely to the distance between the azo groups. The close proximity of the dye to the cellulose maximises the effect of both intermolecular hydrogen-bonding and Van der Waals forces[7] and the dye is therefore 'anchored' to the fibre. Since Van der Waals forces and hydrogen-bonding interactions are relatively weak forces, the dyes are not firmly bound to the fibre. Consequently, direct dyes on cotton have the poorest wet fastness properties of all the dyes in this section. However, because they are cheap, they are used widely for colouring paper.

[6] Benzidine and 2-naphthylamine, often an impurity in 1-naphthylamine, are human carcinogens. Therefore, several dye manufacturers, including ICI, have ceased to make dyes using these intermediates.

[7] Some interaction between the π-electrons of the dye and the cellulose has also been proposed but the poor wet fastness properties of the dyes militate against this idea.

6.3.3 Solid Solutions

In contrast to cotton, polyester fibres have a tightly packed structure and even small molecules such as water have difficulty in penetrating the fibre under normal conditions (see Table 6.2). Special dyeing techniques have been developed therefore to allow the entry of dye molecules into the fibre. These techniques were devised to make the fibre *temporarily* accessible to the dye molecules by loosening or opening its structure; after dyeing, the fibre reverts back to its closely packed state thus trapping the dye molecules within the polymer structure. Even using these techniques only small dye molecules will dye polyester: large dye molecules such as copper phthalocyanine (see Sect. 5.4) do not. The monoazo dye (20) and the anthraquinone dye (21) are typical dyes for polyester.

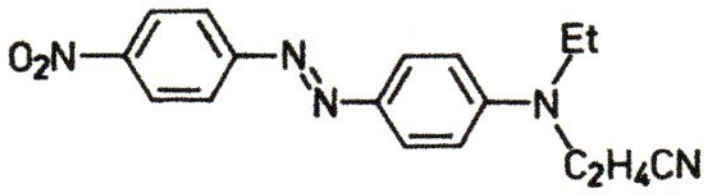

O_2N—C₆H₄—N=N—C₆H₄—N(Et)(C_2H_4CN)

(20)

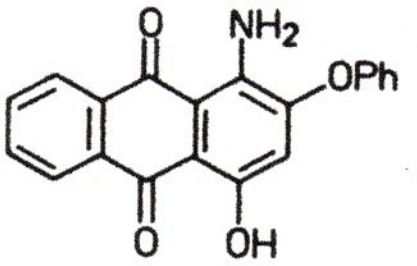

(21)

An additional requirement of dyes for polyester is that they should have only a very low (*i.e.* $\ll 1\%$) solubility in water. Thus, the dyes are devoid of water solubilising groups such as SO_3H, CO_2H and $-NR_3^{\oplus}$. They are normally supplied (and used) as aqueous dispersions — hence the name, disperse dyes.

There are three main techniques which have been developed for temporarily opening up the fibre structure:

(i) dyeing at 95–100 °C with the addition to the dyebath of chemicals called 'carriers',

(ii) dyeing without 'carriers' at temperatures of 120–130 °C: this necessitates the use of pressurised dyeing vessels, and

(iii) pad-dry heat methods involving the use of temperatures of 180–220 °C for short periods of time.

(i) Carriers

Typical carrier molecules are (22–24). At the boil an equilibrium is attained between the fibre, carrier, dye and water. The carrier molecule enters the fibre along with some water molecules thus opening up the fibre structure.[8] This allows the larger dye molecules dissolved in the water to enter the fibre — as they do, more dye dissolves in the water to maintain the equilibrium until eventually almost all the dye passes into the fibre (Fig. 6.3). On cooling to normal temperatures the fibre reverts back to its tightly packed structure — hence the dyes are now trapped within the fibre.

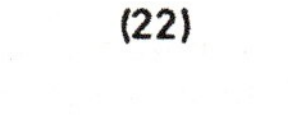

(22)  (23) 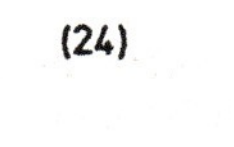 (24)

[8] The mechanism of carrier dyeing is not yet fully understood. Although swelling is accepted as a possibility, the formation of a 'transfer' layer on the fibre surface by the carrier is also a mechanism generally favoured.

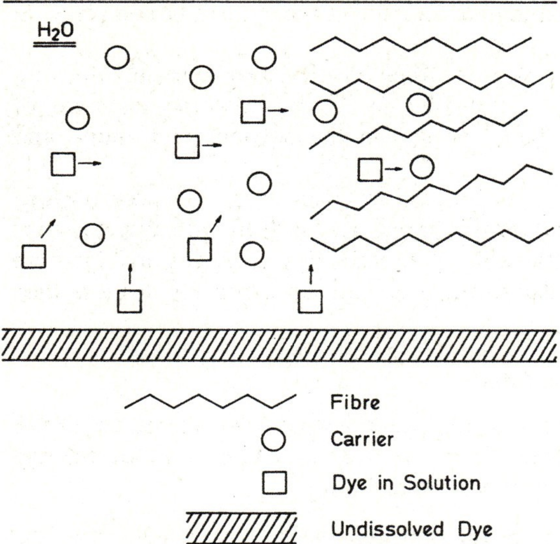

Fig. 6.3. Carrier dyeing of polyester

The structures of the dyes and carriers are closely related and therefore the carriers can also be retained in the fibre after dyeing; this can cause problems. For instance, the light fastness of the dyed fibre may be lowered and the material may retain the odour of the carrier.

(ii) High Temperature (HT) Dyeing

Where possible, it is preferable to avoid the use of carriers. It has been shown that polyester fibres are capable of adsorbing disperse dyes to a large extent at 100 °C providing the dyeing is excessively long (*i.e.* many hours). For deeper shades in reasonable times (*i.e.* $\leqq 1$ hour) the rate of diffusion must be increased. The polymer structure can be loosened by heating it to a temperature sufficiently high to cause a reduction of inter-chain bonding by thermal agitation of the polymer molecules: the energy of the dye molecules is simultaneously increased. Temperatures of 120–130 °C are required to cause this effect. Since this temperature is higher than the boiling point of water under normal atmospheric pressure it is necessary to enclose the dyeing vessel completely and raise the internal pressure above atmospheric. This technique is now used extensively where the polyester material can withstand this treatment.

(iii) Pad-Dry Heat

The use of even higher temperatures raises the rate of diffusion of the dye into the fibre even more thus enabling very short dyeing times to be achieved. The pad-dry heat technique — the Thermosol process — was introduced by du Pont for Dacron[9] in 1949 and equipment has since been developed for heating at 180–220 °C. At this temperature the thermal agitation of both polymer and dye molecules is increased

[9] Du Pont's tradename for their polyester fibre.

enormously; diffusion is therefore rapid and 'fixation' of the dye can be completed in times of the order of 60 seconds.

The fixation of disperse dyes in polyester fibres is believed to be mainly one of a solution of the dye in the fibre, *i.e.* a solid solution. There is also the possibility of hydrogen-bonding between the carbonyl groups in the polymer and amino and hydroxy groups in the dye.

Disperse dyes are also used to dye the semi-synthetic fibre, cellulose acetate. Also, both polyamide and polyacrylonitrile fibres can be dyed with disperse dyes but only pale depths of shade are attainable. The mechanism of dyeing and dye-fibre interactions are believed to be similar to those already described for disperse dyes on polyester fibres.

6.3.4 Insoluble Aggregates within the Fibre

All these dyes work on the principle of producing water insoluble aggregates of dye which are larger than the fibre pores within the fibre structure — hence the dye is firmly trapped within the fibre structure (Scheme 6.2).

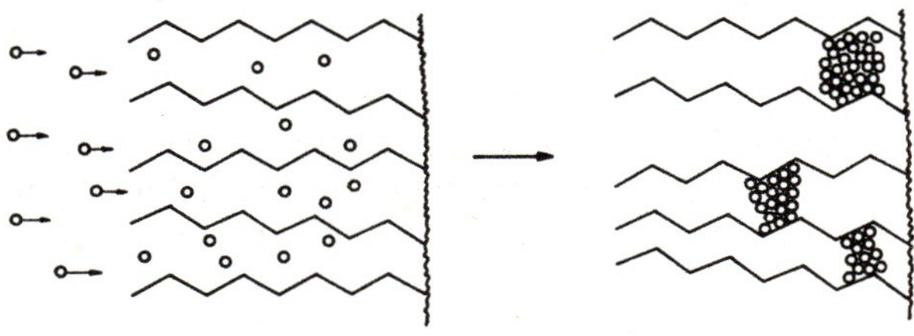

o = Dye molecules

Scheme 6.2

The most familiar examples of this type are the vat dyes (see Sects. 5.2 and 5.3). Vat dyes are water insoluble dyes that have no affinity for cotton but which can be reduced to a water soluble *leuco* form which does have affinity for cotton. The *leuco* form therefore "dyes" the fibre and then atmospheric oxidation regenerates the original water insoluble vat dye within the cotton — aggregation is usually effected by a soaping treatment.

The chemical reaction involved in converting a vat dye into its *leuco* form is simply the reduction of one or more of the carbonyl groups to phenolic groups. The practical reducing agent is alkaline sodium hydrosulphite (hydros). Thus, the water soluble sodium salts of the phenolic groups are formed and it is this anionic species which has high fibre affinity. The simplified vatting process is depicted in Eq. 6.4.

$$\text{C=O} \xrightarrow[\text{NaOH}]{Na_2S_2O_4} \text{C–O}^-Na^+ \xrightarrow{[O]} \boxed{\text{C=O}} \qquad \text{Eq. 6.4}$$

water insoluble vat dye — soluble leuco form — regenerated dye trapped in fibre

Temporarily water solubilised dyes represent another way in which insoluble dye aggregates are formed within the fibre. The principle here is that the dye is rendered water soluble for ease of application to the fibre but that once it is in the fibre the solubilising group is removed in order to generate a pigment within the fibre pores. The Alcian dyes, introduced by ICI in 1948, are based on this principle (Eq. 6.5).

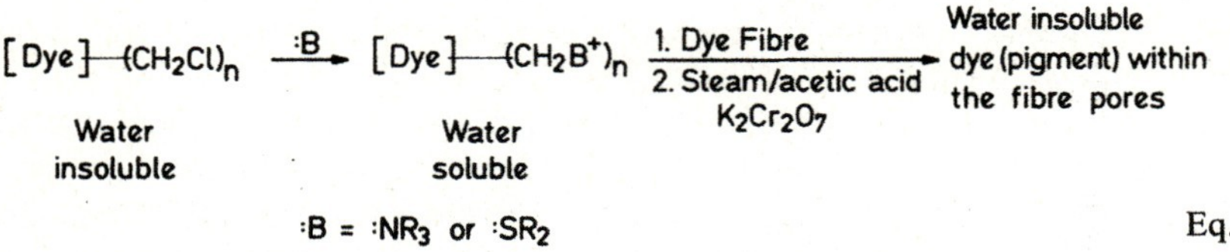

$$\underset{\text{Water insoluble}}{[\text{Dye}]\text{—}(CH_2Cl)_n} \xrightarrow{:B} \underset{\text{Water soluble}}{[\text{Dye}]\text{—}(CH_2B^+)_n} \xrightarrow[\text{2. Steam/acetic acid } K_2Cr_2O_7]{\text{1. Dye Fibre}} \text{Water insoluble dye (pigment) within the fibre pores}$$

$:B = :NR_3$ or $:SR_2$ Eq. 6.5

A third way of forming dye aggregates within the fibre is to actually synthesise the dye within the fibre. The best examples of this type are the azoic dyes for cotton. In this process the fibre is impregnated with a coupling component which has high affinity for cotton — the so-called BON[10] acid arylamides (25) are the supreme examples. A stabilised diazonium salt is then added which reacts with the coupling component already inside the fibre thus producing a water insoluble dye. The simplified process is shown in Scheme 6.3.

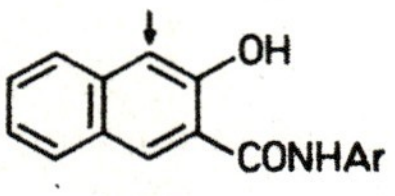

OH
CONHAr
(25)

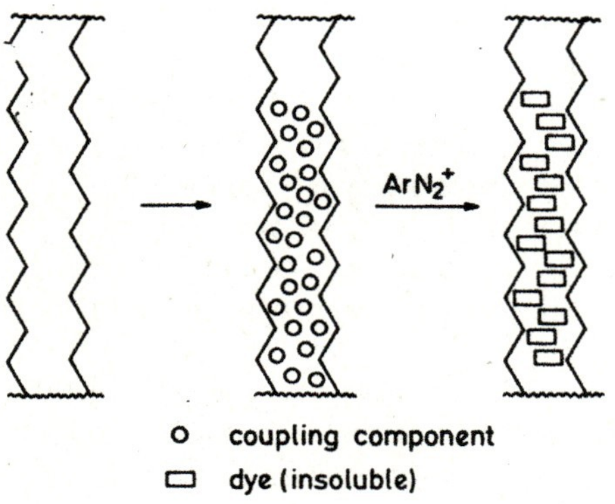

ArN_2^+

o coupling component
▭ dye (insoluble)

Scheme 6.3

Using the same principle phthalocyanine pigments (see Sect. 5.4) can also be formed within the fibre. These 'phthalogen' pigments were introduced by Bayer in 1951 to circumvent the adverse effect that groups such as SO_3H had on wet fastness properties. Here, a precursor in the synthesis of phthalocyanine, 1-amino-3-imino-isoindolenine (26) — see Sect. 2.4.4 — is applied to the fibre in the presence of a copper (or nickel) salt: on developing in acid steam, copper (or nickel) phthalocyanine is formed smoothly within the fibre — Eq. 6.6.

[10] *B*eta-*O*xy-*N*aphthoic.

(26)

Eq. 6.6

The wet fastness properties of all the dyes that form water insoluble aggregates within the fibre pores are excellent. As we shall see later, an additional advantage of aggregation is that it tends to improve light fastness.

6.3.5 Ionic Bonds

There are two distinct types of ionic dye-fibre bonding depending upon whether the dye is anionic or cationic.

Anionic dyes contain negatively charged groups, *e.g.* sulphonic acid groups.[11] Since the sulphonic acid group is strongly acidic, it is virtually fully ionised in water[12] — Eq. 6.7. Obviously, anionic dyes are used to dye fibres which contain cationic centres under the dyeing conditions employed. As mentioned earlier the best examples of this type are nylon and the protein fibres wool and silk. In a weakly acidic dyebath such as dilute aqueous acetic acid the amino groups in the fibre are protonated. During dyeing the negatively charged dye displaces the colourless counter ion and is retained in the fibre by the ionic bonding between the anionic dye and the cationic ammonium group — Eq. 6.8.

$$[\mathrm{Dye}]\!-\!(SO_3H)_n + H_2O \longrightarrow [\mathrm{Dye}]\!-\!(SO_3^-)_n + H_3O^+$$ Eq. 6.7

CO_2H … $NH_3^+ \cdot A^-$ $\xrightarrow{\mathrm{Dye-SO_3^-}}$ CO_2H … $NH_3^+ \cdot {}^-O_3S-\mathrm{Dye}$

Eq. 6.8

In practice the number of sulphonic acid groups per dye molecule is carefully chosen to attain the optimum balance between water solubility and fibre affinity. As a rule, most commercial acid dyes contain one or two sulphonic acid groups.

Cationic dyes are used mainly to dye polyacrylonitrile fibres. As stated earlier, polyacrylonitrile fibres contain anionic centres such as $SO_3^{\ominus}$ or $CO_2^{\ominus}$ groups. It is the

[11] Hence the name, acid dyes.
[12] Most dyes are sold as sodium salts. Obviously these are fully ionised in water.

ionic attraction between these groups and the positively charged dye which binds the dye to the fibre.

As is the case with anionic dyes for nylon and wool, cationic dyes are applied to acrylic fibres from a weakly acid dyebath. It is thought that the dye is first adsorbed on to the fibre surface: heat then causes it to diffuse into the fibre interior where it becomes anchored to an anionic centre (Eq. 6.9).

$$\text{Dye}-X^+ \quad M^+O_3^-S- \longrightarrow O_3^-S- \; X^+-\text{Dye} \qquad \text{Eq. 6.9}$$

Because there is a limited number of negative sites in polyacrylonitrile and of positive sites in nylon and wool, cationic and anionic dyes respectively exhibit a definite saturation value for these fibres, *i.e.* no further dye can be taken up by the fibre once all the sites have been occupied by dye molecules. However, in practice there are sufficient sites available to enable the desired depths of shade to be obtained.

The wet fastness properties of dyes bound by ionic forces are better than those bound by physical adsorption, *e.g.* direct dyes for cotton, but inferior to those which form insoluble aggregates in the fibre, *e.g.* vat dyes.

6.3.6 Covalent Bonds

As seen earlier, the problem fibre for dyeing directly is cellulose. Indeed, specially designed molecules — the direct dyes — are needed. These are polyazo dyes and therefore only dull shades are attainable. Furthermore, direct dyes have only poor to moderate wet fastness since only physical forces are involved in dye-fibre binding. Although good wet fastness properties are obtained from azoic and vat dyes, these also give dull shades and need special application techniques. Hence, there was a tremendous incentive to achieve better dye types for cellulose.

The idea of linking a dye to a fibre by a covalent chemical bond in order to achieve high wet fastness had been perceived for many years. Indeed, numerous attempts were made to attach dyes to cellulose by a covalent chemical bond. Although in some cases such a bond was formed, the conditions employed were so drastic that serious degradation of the fibre occurred.

The first practical success was achieved by Rattee and Stephen (ICI) in 1954 when they discovered that dyes containing a dichlorotriazinyl group (*i.e.* 27) reacted with cellulose under mildly alkaline conditions (no fibre degradation occurred) — Eq. 6.10.

$$\text{Dye}-NH-C_3N_3Cl_2 \xrightarrow{RO^-} \text{Dye}-NH-C_3N_3(Cl)(OR) + Cl^- \qquad \text{Eq. 6.10}$$

(27) RO^- = ionised cellulose

Dyes based on cyanuric chloride (28) were not new — in the 1920's Ciba used it as a chromophoric block to produce green dyes, *e.g.* of type (29), from blue and yellow dyes, with a substantivity for cellulose similar to that of a direct dye.

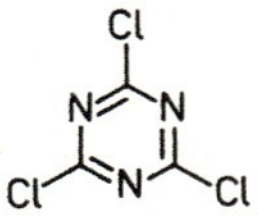

(28) 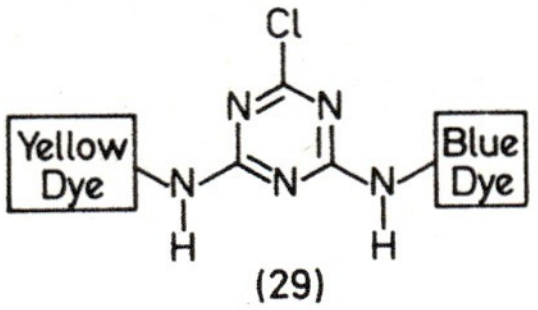 (29)

In 1956, exactly 100 years after Perkin's discovery of Mauveine, ICI introduced the first reactive dyes; a yellow, a red, and a blue. Other dye manufacturers quickly followed suit and nowadays all the major dye manufacturers market a range of reactive dyes for cotton.

The majority of reactive dyes utilise the nucleophilic displacement reaction generalised in Eq. 6.11. Basilene (BASF), Cibacron (Ciba-Geigy), Drimarene P (Bayer) and Procion (ICI) ranges are based on 1,3,5-triazines (30), whereas the principal Drimarene range and the Levafix range (Bayer) are based on pyrimidines (31).

$$\text{Dye—X} + \text{Cell—O}^- \quad \text{Dye—O—Cell} + \text{X}^- \qquad \text{Eq. 6.11}$$

X = a leaving group

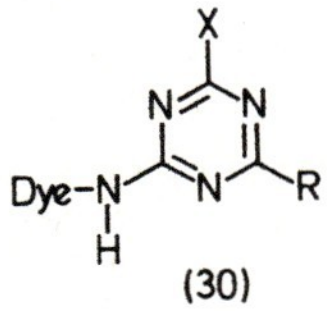

(30) R = Cl, NH_2, NR^1R^2; X = Cl, F

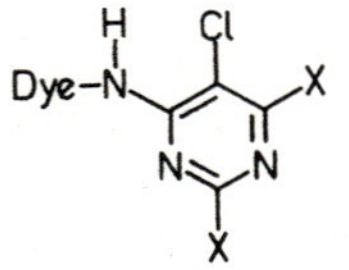

(31) X = Cl, F

Hoechst chemists invented a different type of reactive system, a vinyl sulphone group, which reacts with the cellulose by Michael addition — Eq. 6.12.

All the above reactive dyes need mildly alkaline conditions to achieve chemical reaction with the cotton. In contrast, the Procion T dyes, introduced in 1979, react chemically with cellulose under neutral conditions — Eq. 6.13.

$$\text{Dye—}SO_2CH_2CH_2OSO_3H \xrightarrow{OH^-} \text{Dye—}SO_2CH{=}CH_2 \xrightarrow{\text{Cell—O}^-} \text{Dye—}SO_2CH_2CH_2O\text{—Cell}$$

Eq. 6.12

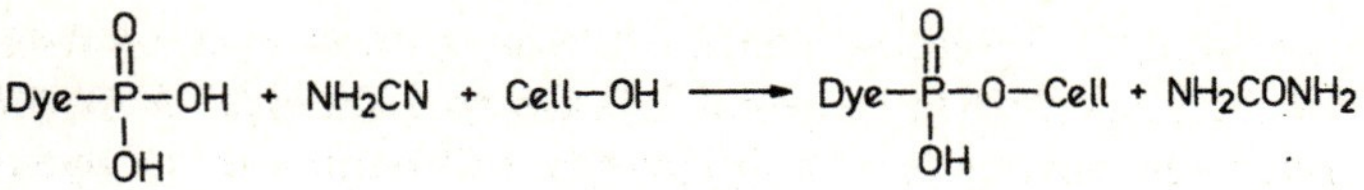

$$\text{Dye}-\overset{\overset{\displaystyle O}{\|}}{\underset{\underset{\displaystyle OH}{|}}{P}}-OH + NH_2CN + \text{Cell}-OH \longrightarrow \text{Dye}-\overset{\overset{\displaystyle O}{\|}}{\underset{\underset{\displaystyle OH}{|}}{P}}-O-\text{Cell} + NH_2CONH_2 \qquad \text{Eq. 6.13}$$

As might be expected in all of the reactive dyes, by no means all of the dye reacts with the cellulose — reaction with water in the dyebath (hydrolysis) is a competing reaction (Scheme 6.4). The level of fixation of the dyes varies enormously (from ~30% to 90%) and consequently considerable effort is still being directed towards achieving 100% fixation.

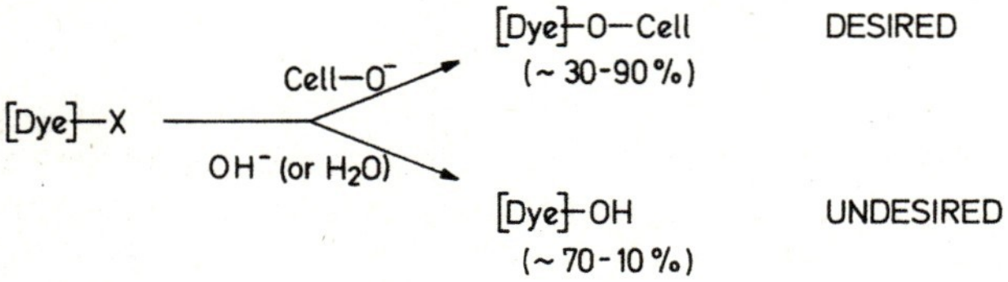

[Dye]–X → (Cell–O⁻) → [Dye]–O–Cell (~30-90%) DESIRED

[Dye]–X → (OH⁻ (or H_2O)) → [Dye]–OH (~70-10%) UNDESIRED

Scheme 6.4

Since virtually any chromogen can be converted to a reactive dye, there is a vast choice of colours and chemical types available. Hence, for the first time bright shades are available on cotton. Monoazo dyes provide the bright yellow to red shades, *e.g.* (32) and (33) respectively, anthraquinone dyes the bright blue shades, *e.g.* (34), and copper phthalocyanine dyes the bright turquoise shades, *e.g.* (35).

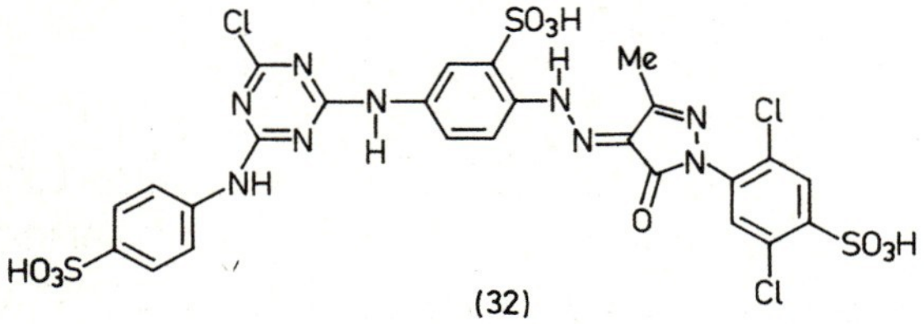

(32)

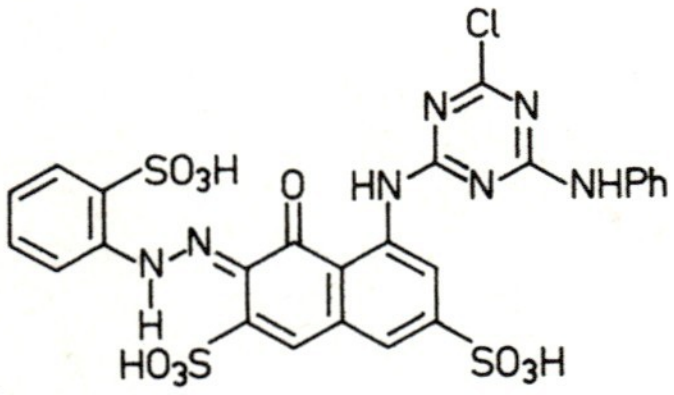

(33)

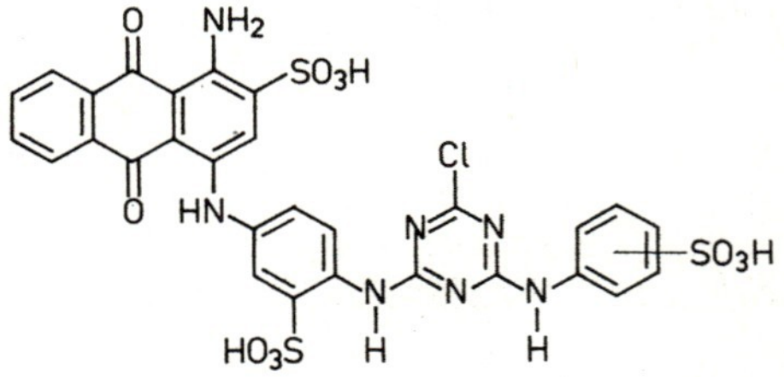

(34)

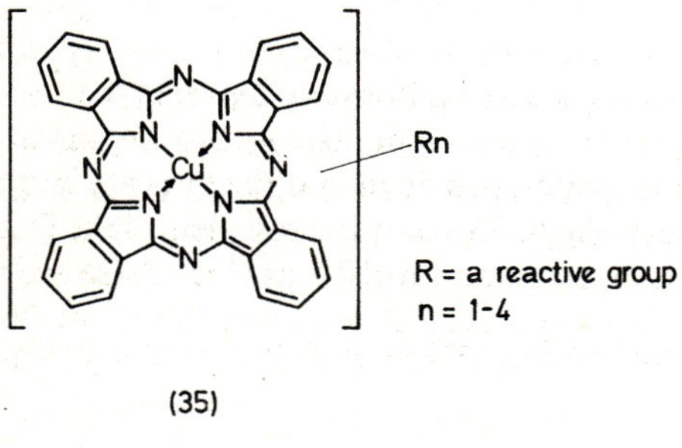

(35)

Because of their simpler application process (Scheme 6.5) reactive dyes have won their share of the market at the expense of vat and azoic dyes. However, the cost effective sulphur and direct dyes, and pigments, still dominate the dyeing and printing of cellulose fibres — Fig. 6.4.

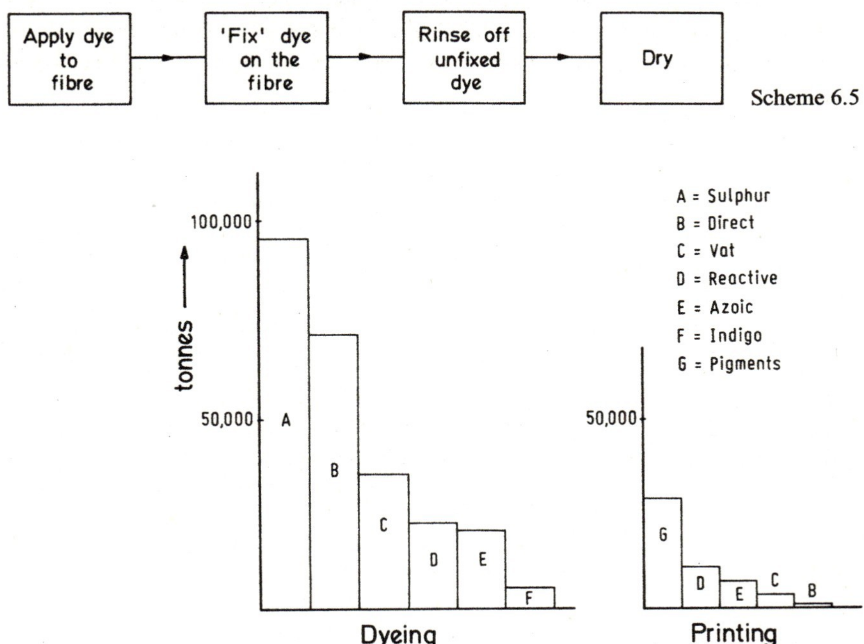

Fig. 6.4. Relative consumption, worldwide, of the major classes of dyes used in colouring cellulosic fibre [1980]

6.4 Light Fastness of Dyes

6.4.1 Introduction

As mentioned earlier a successful dye must satisfy many conditions and light fastness rates amongst the most important of these. Obviously, no consumer will purchase a dyed product he knows will fade within weeks and usually requires the dye to last as long as the purchased article. In general, a low light fastness is acceptable in dyes for paper, *e.g.* wrapping paper and toilet paper, provided that the dyes are inexpensive since the lifetime of such goods is very short. In contrast, excellent light fastness is demanded of articles such as curtains, carpets and furnishings since these are usually expected to last for many years. Dyes for apparel fall in between these two extremes. Consequently, a considerable amount of effort has gone into studying the photo-fading process not only from scientific interest but also from a commercial viewpoint.

Not surprisingly azo dyes have received most attention which is consistent with their enormous commercial importance. As a result, a lot of information has been accumulated and several theories and mechanisms have been proposed to explain the photofading of azo dyes and, to a lesser extent, other dye classes, *e.g.* anthraquinones, triarylmethanes, *etc.* However, the photofading processes are far from being well understood and, as will become evident, further research is still required.

The major reasons for the limited progress in the area is not because of a lack of effort or ingenuity but because of the immense practical problems encountered in studying the fading mechanism on dyed fibre. For instance, a fibre is usually dyed with a relatively small amount of dye;[13] thus, recovering the photodegradation products in quantity without removing impurities that were present in the fibre before dyeing can be extremely difficult. In addition, because the photofading process is relatively slow, even for dyes of very poor light fastness, the primary products from the fading reaction may well react further and even volatilise from the fibre in cases where the photodegradation products are of low molecular weight!

In the following discussion more attention is devoted to the study of dyed polymers than to the photochemical behaviour of dyes in solution. Hopefully, this reflects the greater relevance of the former approach to the dyes industry.

6.4.2 Test Methods

Scientific tests for assessing the light fastness of dyed fibres have been established for approximately three hundred years. Not surprisingly, the first methods used sunlight as the light source and this was probably adequate for the early natural dyes which generally had poor light fastness (see Sect. 1.2.1). However, the duration and intensity of sunlight varies from day to day, especially in countries like Britain whose climate is notoriously variable. As a result, tests on dyes of superior light fastness can take weeks, months and even years! Clearly then, alternative testing procedures were sought in which the dyed fibre could be subjected to a constant light source for twenty-four hours a day, thus shortening the test considerably. The problem with using an alternative light source is that it might not correspond to the light output of the sun. Very little is gained from having a dye which has an excellent light fastness to one particular light source if the same dye fades in sunlight in a matter of hours. As a result, many light sources are quite unsuitable; for example, the low pressure mercury vapour lamp emits very strongly at 253.7 nm but has a very weak emission over the rest of the electromagnetic spectrum. In this case dyed fibres that do not absorb at 253.7 nm would appear to have an excellent light fastness rating; however, such a test is quite meaningless since no light below 290 nm reaches the Earth's surface. Fortunately, the output from a xenon arc lamp, with suitable filtration, corresponds quite closely to sunlight and therefore this source is used widely nowadays.

The light fastness of a dye is usually assessed against a set of standard dyeings. The International Organisation for Standardisation (ISO) has chosen a set of eight blue dyes (36–43) which vary in light fastness (on wool) from very poor (grade 1) to excellent (grade 8). The dyes have been selected carefully so that an increase of one grade is approximately equivalent to doubling the light fastness, *i.e.* a decrease in the

[13] Typically this might be a 1% dyeing, *i.e.* 1 g of dye per 100 g of fibre.

fading by a factor of two; this means that dye (43) fades approximately 250 times slower than dye (36)! In the test, a dyed fibre is irradiated along with the eight standards. At intervals the dye is assessed visually to compare its fading rate with that of the standards. A value for the light fastness of the dyed fibre is obtained when its rate of fading corresponds to one of the standards. For example, if it begins to fade when standard 3 fades then it exhibits a light fastness of 3.

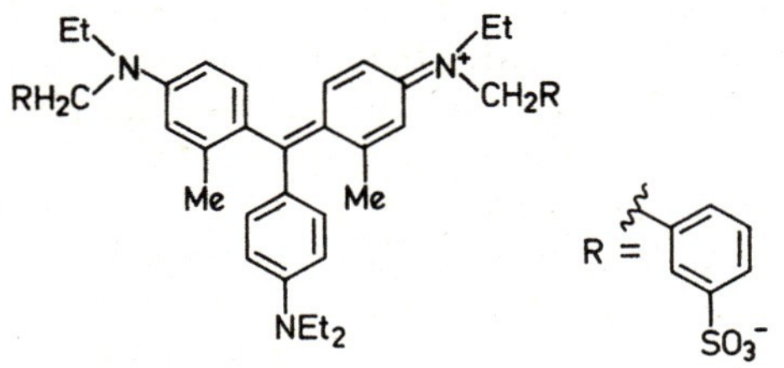

CI Acid Blue 104

(36)

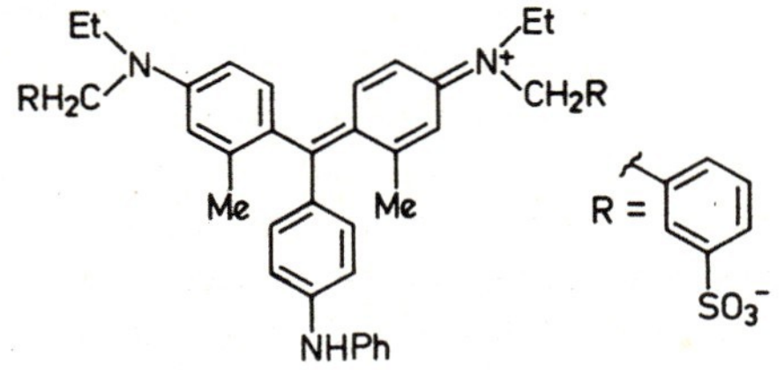

CI Acid Blue 109 (37)

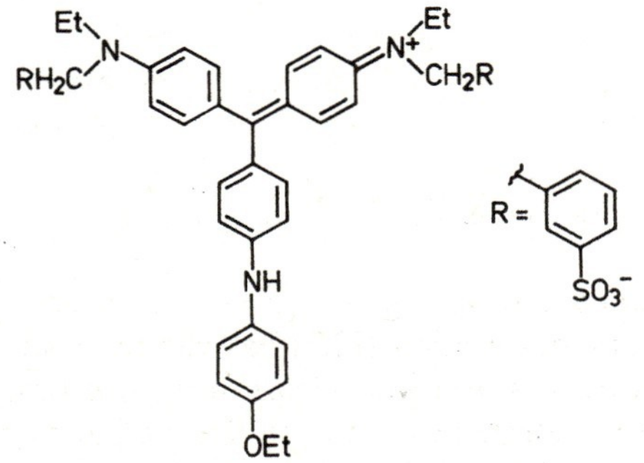

CI Acid Blue 83 (38)

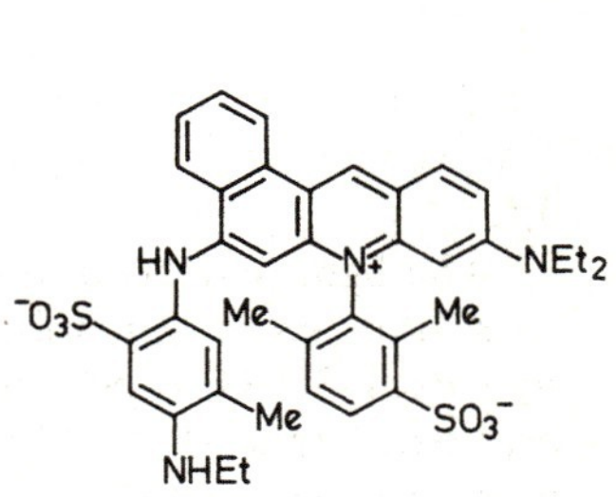

CI Acid Blue 121 (39)

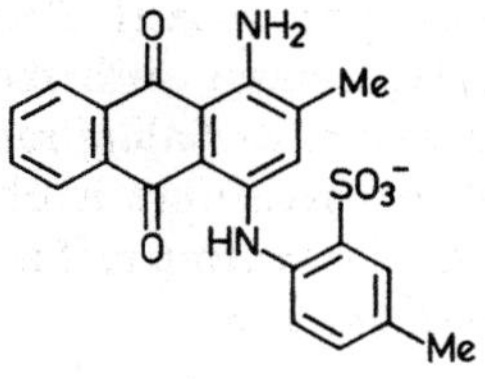

CI Acid Blue 47
(40)

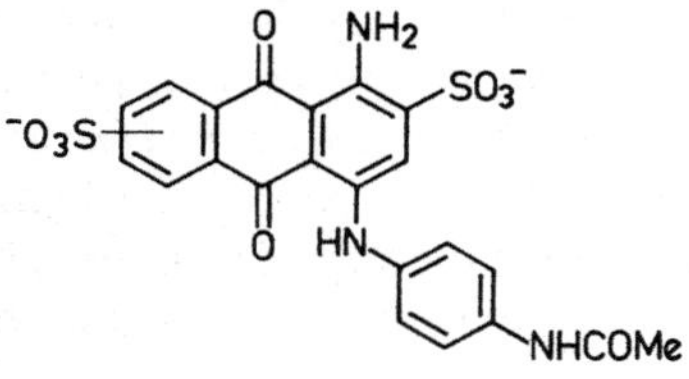

CI Acid Blue 23
(41)

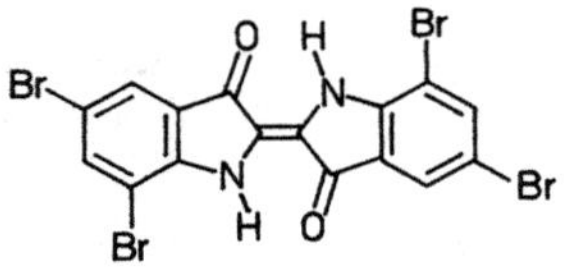

CI Vat Blue 5 (42)

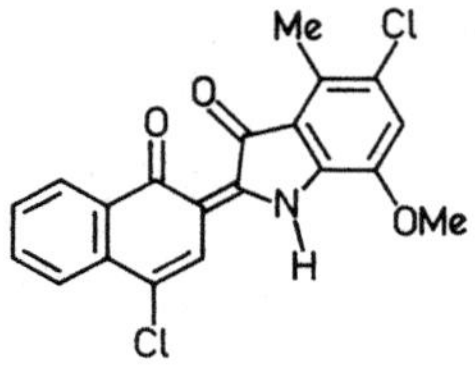

CI Vat Blue 8 (43)

An alternative to the ISO test is the AATC test method. In this method seven standards are used (L2–L8); however, all the standards are made up from just two dyes, (43) and (44). Each standard is prepared by mixing the dye of low light fastness (44) with that of high light fastness (43). Increasing the percentage of (43) obviously increases the light fastness of the mixture.

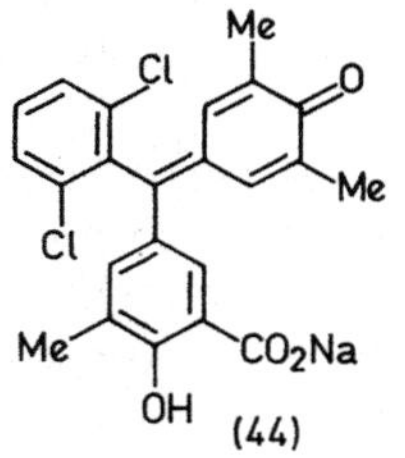

(44)

6.4.3 Basic Photochemical Principles

Once a molecule absorbs a photon of light energy it is raised to an excited level (see Fig. 6.5). The excited molecule can then lose its energy by a variety of processes to return to the ground state. These are shown in Fig. 6.5 (1–4).

Alternatively, the excited molecule might decompose or react with another molecule before it has time to return to the ground state, (*i.e.* pathway 5 — Fig. 6.5); such a reaction is termed a photochemical reaction. The two most important energy states for photochemical reaction are the first excited singlet state and the first excited triplet state; in azo dyes these are the $^1(\pi\pi^*)$ and $^3(\pi\pi^*)$ or the $^1(n\pi^*)$ and $^3(n\pi^*)$ states respectively.

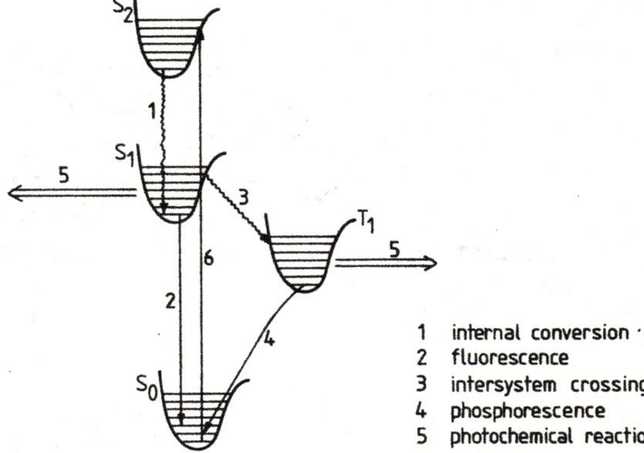

Fig. 6.5. The various processes by which an excited molecule can return to the ground state

The efficiency of a photochemical reaction is measured by the quantum yield Φ, where: —

$$\Phi = \frac{\text{number of molecules reacted}}{\text{number of quanta absorbed}}$$

If $\Phi \gg 1$ then the reaction is a chain reaction and the light energy initiates the first step. If $\Phi = 1$ then each photon absorbed leads to a chemical reaction, whilst $\Phi \ll 1$ indicates a very low efficiency in which other processes are preferred to the chemical reaction. It is usually the latter situation which applies in the photofading process of dyes.

Several factors may affect the ultimate photostability (to reaction) of a dye and amongst these are: the nature of the light source, the absorption characteristics of the dye, the lifetime of the S_1 and T_1 states, the efficiency of intersystem crossing, the nature of the solvent or substrate and the presence or absence of air or moisture. Obviously the longer the molecule spends in the excited states then the more chance it has to react. It is worth mentioning here a widely held misconception regarding fluorescence in dyes. It is often said that fluorescence in dyes causes low light fastness. This is not the case. Although many dyes that are fluorescent do have low light fastness the fluorescence is merely a visible manifestation that the excited singlet state is longer lived — hence, it is more likely to take part in a photochemical reaction. In addition many external factors affect the photoreactivity of the dye, such as the nature of the substrate, temperature and environment.

6.4.4 Mechanism of Fading

As pointed out earlier azo dyes have been studied in great depth in an attempt to gain an insight into the factors which determine light fastness in dyes. Therefore, the following discussion is weighted towards azo dyes. However, it is possible to draw general conclusions from these studies, particularly regarding dye-fibre interrelations.

(a) Azo Dyes

Oxidation and reduction reactions are the two most important pathways for the fading of dyes. There is some doubt as to whether these reactions are direct photoreactions of the dye, *i.e.* from the photoexcited state of the dye, or whether they are in fact indirect and are caused by another photoexcited molecule, *e.g.* oxygen, possibly sensitised by the dye. The problem in differentiating between the direct and indirect mechanisms is exacerbated by the inherently low quantum efficiency for photofading in which Φ is $<10^{-3}$, even for dyes of poor light fastness (rated 1 on the ISO light fastness scale). With such a low quantum yield it is extremely difficult to be certain that trace impurities in the fibre are not participating in the fading reaction.

It has been recognised for some time that many dyes fade much faster in the presence of oxygen. It has also been demonstrated on polypropylene that some dyes that previously faded by an oxidative route faded by a reductive route when oxygen was excluded. However, it is not always necessary to remove oxygen before a reductive mechanism predominates. This conclusion arose out of several studies wherein it was demonstrated that a major influence on the reaction mechanism was the nature of the fibre. These investigations used plots of (log) fading rate *versus* the Hammett σ constants for a variety of substituents in aromatic azo compounds as a means to determine the mechanism of fading. It was reasoned that more electronegative groups (than H) would decrease the propensity with which an azo compound could be oxidised[14] and increase the rate of reduction with respect to the parent azo compound. Thus, the graph should show a negative slope for an oxidative mechanism and a positive slope for a reductive mechanism. This is found experimentally (Fig. 6.6).

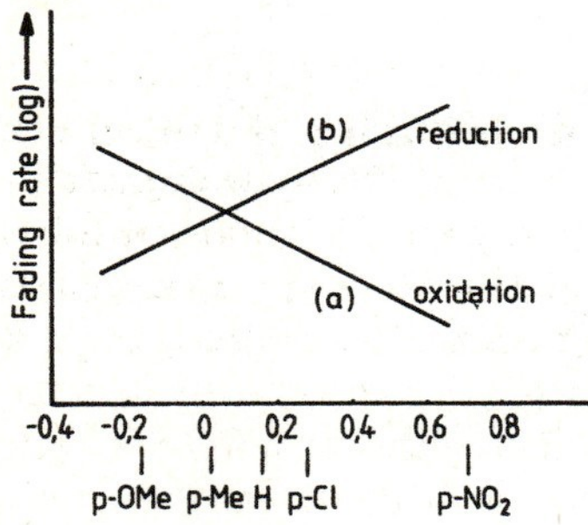

Fig. 6.6. Plot of Hammett σ-values *vs* log fading rate of dyes: an example of the results to be expected for a monoazo dye series by varying the *para*-phenyl substituent on (a) an oxidising substrate (b) a reducing substrate.

On wool the benzeneazo compounds (45) faded by a reductive mechanism since a positive slope was observed. However, on cellulose acetate the dyes (46) showed a negative slope thus indicating an oxidative mechanism. These results have since been supported by further studies. These show that there are two distinct fibre types — proteins and non-proteins — which determine whether a dye is oxidised or reduced. On protein fibres, *e.g.* wool and silk, a reductive mechanism takes place whereas on non-protein fibres, *e.g.* cellulose, polyester and polyacrylonitrile, an oxidative

[14] Oxygen is assumed to function as an electrophile in the oxidation reaction.

mechanism operates. Polyamides, *e.g.* nylon, are somewhat anomalous but tend to fall into the second category; however, as will be discussed later, dyes containing nitro groups are reduced on polyamide fibres.

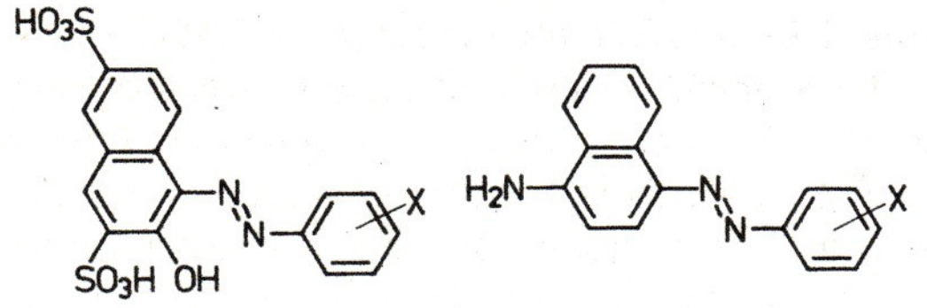

(45) (46)

In the oxidative mechanism the non-protein fibre does not seem to play a major role in the photodegradation of the dye. However, the available evidence strongly suggests that the protein fibre participates at some stage in the reductive pathway of the dye. In particular, the histidine (47) residues in the protein have been directly implicated in the reductive pathway, possibly at some early stage. This theory is supported by the observation that a photoreductive mechanism can take place on non-protein fibres impregnated with histidine.

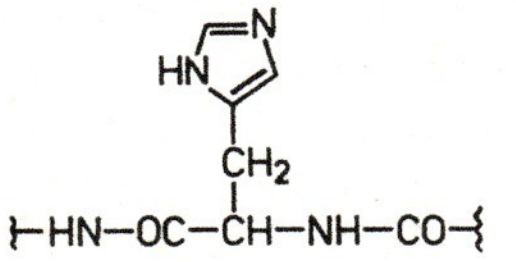

(47)

Oxidative Fading

As mentioned previously oxygen markedly accelerates the fading of dyes on non-protein fibres and the obvious conclusion to be drawn is that the photodegradation pathway involves an oxidation step. It was suggested that the oxygen was involved in the conversion of the azo group to an azoxy group. However, experimental evidence does not support this postulate as a general mechanism. Indeed, only one example of this pathway is definitely known (Eq. 6.14).

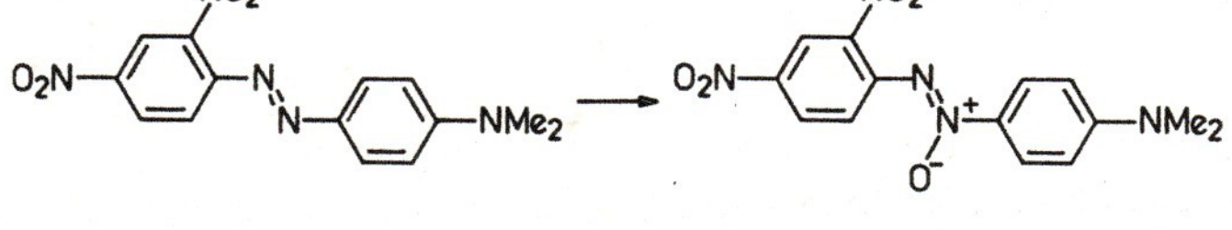

Eq. 6.14

However, more plausible mechanisms that appear to be quite general have been put forward. The first of these concerns azo dyes that are capable of exhibiting azo-hydrazone tautomerism. In these cases it is believed that photodegradation actually takes place from the hydrazone form rather than from the azo form. In the example shown (Scheme 6.6) the postulated mechanism involves an initial attack by singlet oxygen upon the hydrazone form. Here, the singlet oxygen may arise from dye sensitisation of ground state triplet oxygen although other molecules (*e.g.* ketones) could also cause sensitisation. The reactive singlet oxygen molecule then participates in a thermally allowed 6π process, the Ene reaction, to give a peroxide which would then be expected to decompose to the products shown in Scheme 6.6. Experimentally, naphthoquinone and the dye (48) were indeed isolated; (48) was easily rationalised as originating from the coupling reaction between the starting azonaphthol and free aryldiazonium salt formed during the decomposition of the peroxide.

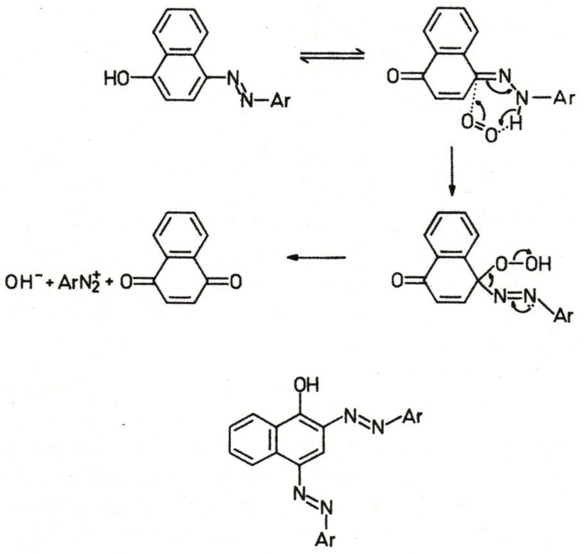

(48) Scheme 6.6

Similarly, 2-phenylazonaphthol (49) gives 1,2-naphthoquinone (50) as well as other oxidation products such as phthalic acid, nitrogen and peroxides, when it is irradiated on cotton.

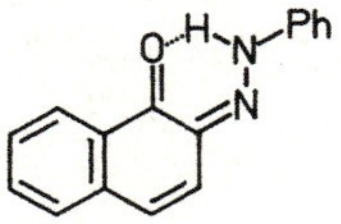

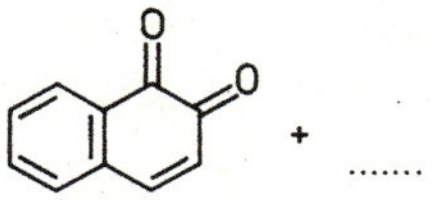

(49) (50)

Further experimental observations have been made which support the general mechanism outlined in Scheme 6.6. For instance:

(i) singlet oxygen produced non-photolytically produces a similar reaction,
(ii) singlet oxygen sensitisers promote fading, and
(iii) quenchers suppress fading.

If the azonaphthol is O-methylated and thereby locked in the azo form the resultant azo dyes appear to be more resistant to attack by singlet oxygen — this is consistent with the proposed mechanism.

An alternative mechanism (see Scheme 6.7) is illustrated for 4-arylazonaphthols which does not require the intermediacy of singlet oxygen. Such a mechanism could operate in those cases where the dye is either resistant to attack by singlet oxygen or where special features exist that promote the mechanism, *e.g.* where radical formation is favoured.

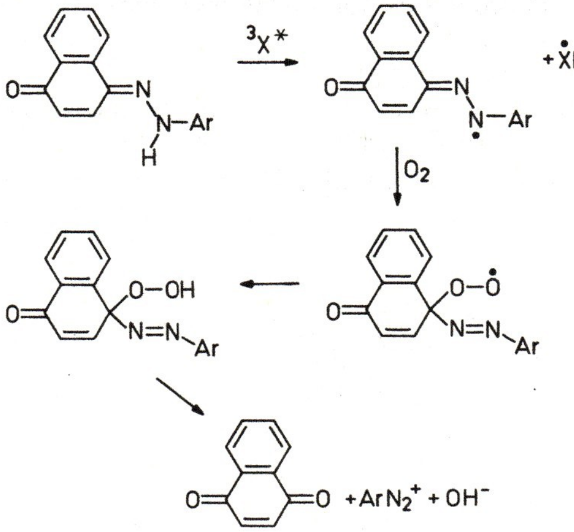

Scheme 6.7

However, not all azo dyes can exist in a hydrazone form; for example, the important class of azo dyes derived from *para*-coupling amines. A plausible mechanism (Scheme 6.8) has also been proposed for the photochemical fading of these dyes. It starts by oxidative cleavage of the C—N bonds with subsequent release of hydrogen peroxide. Once formed, the hydrogen peroxide could react further with the azo chromogen to give, what might be expected to be, a complex reaction mixture. Although this scheme needs further substantiation and experimental proof, it does rationalise known data on the light fastness of azo dyes, *e.g.* that the lower the basicity of the amino nitrogen atom, the greater the light fastness of the dye.

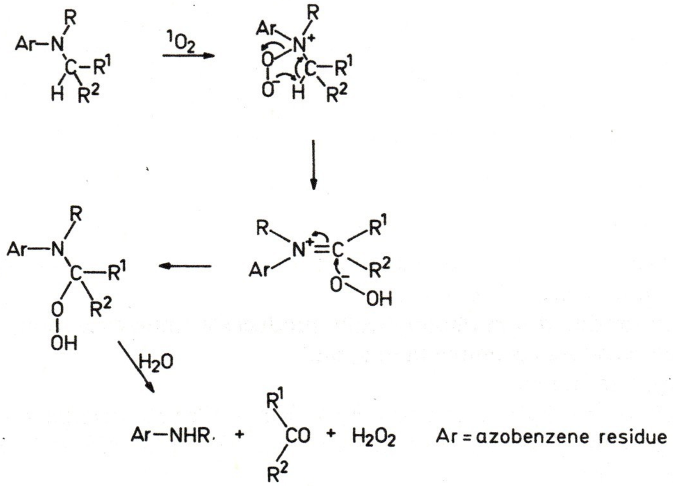

Ar = azobenzene residue

Scheme 6.8

Reductive Mechanisms

As pointed out earlier it is found that the reductive mechanism is predominant in certain dyes on polypropylene once oxygen is excluded. However, even in the presence of oxygen, dyes on protein fibres still fade by a reductive route. Histidine seems to play a vital role in this change of mechanism and it is thought to do so both by quenching and reacting with singlet oxygen. It is possible that the histidine is thus converted to a reactive species capable of promoting reduction of the azo dye although this suggestion needs experimental verification.

The intermediacy of ketyl ($R_2\dot{C}$—OH) and carboxy ($\cdot CO_2H$) radicals have been proposed in the reductive fading mechanism and some model studies support this theory — a plausible mechanism for such a reduction is detailed below (Scheme 6.9). In support of this mechanism evidence for radicals (51) and (52) has been obtained.

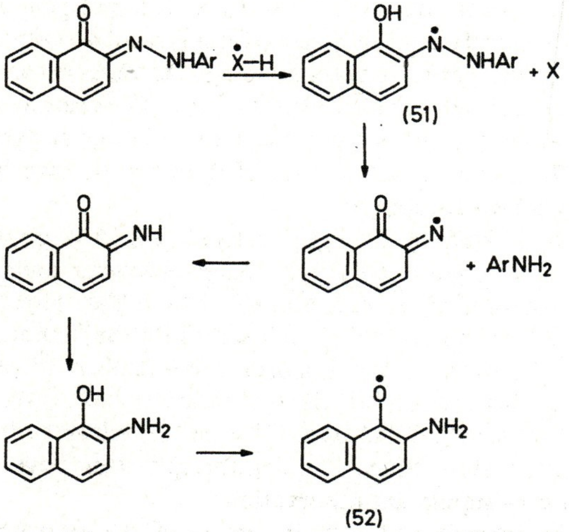

NHAr $\dot{X}$–H → NHAr + X

(51)

NH ← N + ArNH$_2$

OH NH$_2$ → O NH$_2$

(52)

Scheme 6.9

Azo dyes which contain nitro groups have anomalously low light fastness on polyamide fibres, *e.g.* nylon. A reductive mechanism (Scheme 6.10) has been proposed for this anomaly and is consistent with much of the available experimental data. Photoreduction of the nitro group is well precedented and alkylamines, *e.g.* butylamine, promote the reaction. By analogy, the amino end groups on the polyamide chains would also be expected to promote the reduction of the nitro groups.

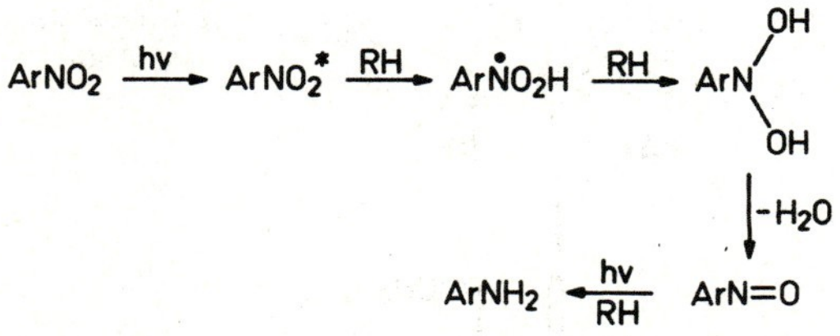

$$ArNO_2 \xrightarrow{h\nu} ArNO_2^{*} \xrightarrow{RH} Ar\dot{N}O_2H \xrightarrow{RH} ArN(OH)_2$$

$$\downarrow -H_2O$$

$$ArNH_2 \xleftarrow[RH]{h\nu} ArN{=}O$$

Scheme 6.10

It might also be anticipated that more complicated products, *e.g.* azoxy compounds, could also be formed, possibly as side reactions. Indeed, model studies support this postulate.

(b) Metallised Azo Dyes

As described in Chapters 2 and 3 metallisation of azo dyes often results in an improvement in properties, particularly light fastness. Two explanations have been propounded: —

(i) that the metal protects, both by steric and electronic effects, the azo-hydrazo nitrogen atoms from attack by, for example, singlet oxygen and

(ii) that the metal promotes aggregation of the dye and thus leads to an improvement in light fastness (see later).

It is, of course, quite possible that both and not just one of these processes operate.

(c) Anthraquinone and Vat Dyes

Although it is possible to predict with some degree of accuracy the light fastness of a dye in a particular class, *e.g.* an azo dye, on a given fibre, it is not possible to extrapolate to comparisons between different dye classes. Some dyes have atrocious light fastness on one type of fibre and yet have acceptable light fastness on another type. Moreover, it does not always follow that if one class of dyes show an improvement in going from one type of fibre to another type, that a second class of dyes will show the same improvement. Thus, each separate class of dyes has its own 'rules' and anthraquinone and vat dyes are no exception.

In general, anthraquinone dyes have good light fastness, *e.g.* 5–6 on wool, 5–7 on acetate fibres. Unfortunately, many of the light fastness studies have been of an empirical nature and suggest general structural modifications that might lead to an increase in light fastness rather than providing an insight into the fundamental mechanism of photodegradation. However, one important mechanism of photodegradation is known to be the dealkylation of alkylaminoanthraquinones; this was first inferred following an observation that the dyed fibre reddened on fading, a fact consistent with a dealkylation step. Since aminoanthraquinones have great commercial importance this is a very significant observation.

The mechanism illustrated in Scheme 6.11 explains many of the experimental observations. Products resulting from hydroxylation of the anthraquinone skeleton and replacement of the methylamino groups by oxygen can all be rationalised on the basis of a reactive intermediate such as (53). However, the mechanism proposed

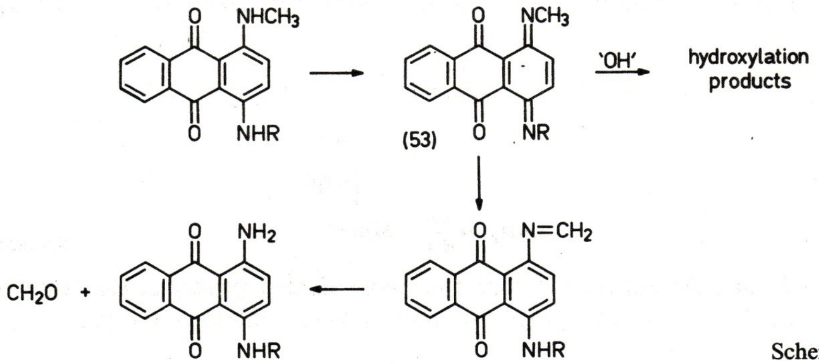

Scheme 6.11

for the dealkylation of aminoazo dyes (see Scheme 6.8) could also operate. Because the mechanism is an oxidative one it might be expected that it should be faster on non-protein fibres than on nylon and there is some support for this supposition since some anthraquinone dyes do appear to fade faster on polyester fibre than on nylon.

Methods to improve the light fastness of aminoanthraquinone dyes nearly always, as a first step, aim at avoiding an alkylamino substituent and this approach does meet with success. For instance, anthraquinone dyes containing primary amino groups, arylamino groups and good hydrogen-bonding groups, *e.g.* an acetylamino group, all have superior light fastness to the corresponding alkylaminoanthraquinone dyes.

The vat dyes have been little studied for their photodegradation mechanisms. Part of the reason could be the excellent light fastness of these dyes (on cotton) which makes it difficult to determine practically their rate of fading on fibres. Their excellent light fastness has been ascribed to the lack of strongly polar substituents such as the donor and acceptor groups found in azo dyes.

(d) Other Dye Classes

In contrast to the anthraquinone and vat dyes the arylcarbonium class of dyes is renowned for poor light fastness, especially on wool, silk, and cotton. However, they have an acceptable light fastness rating on polyacrylonitrile fibres. Within this class of dyes are found some of the strongest and brightest dyes available and several types exhibit very strong fluorescence, *e.g.* xanthene dyes. Indeed, it is probable that these very characteristics are connected with the reasons for their poor light fastness.

Water, and in particular oxygen, have a large effect in lowering the light fastness of arylcarbonium dyes. Studies on triarylcarbonium dyes, *e.g.* Malachite Green (54), reveal that two mechanisms are operative and both may apply to arylcarbonium dyes in general. The first is a dealkylative mechanism rather similar to that found in anthraquinone and aminoazo dyes. However, the second mechanism involves fragmentation of the molecule and products such as 4-dimethylaminobenzophenone, 4-monomethylaminobenzophenone and 4-dimethylaminophenol have been isolated, which support a fragmentation mechanism. Experimental evidence suggests that the fragmentation of Malachite Green takes place from the carbinol base (55), possibly by radical formation, rather than from the dye itself. Thus, substances which absorb light in the same region as the carbinol (55) also protect Malachite Green from fading — this strongly indicates the intermediacy of (55) in the photodegradation reaction.

Me_2N … $\overset{+}{N}Me_2$ $\underset{H_3O^+}{\overset{OH^-}{\rightleftharpoons}}$ Me_2N … OH … NMe_2 $\xrightarrow{h\nu}$

(54) (55)

The pK_a of the dye binding sites may also have a large effect upon the light fastness of arylcarbonium dyes. For example, Malachite Green has a much better light fastness when dyed on sulphoethylated cotton than on carboxyethylated cotton. This is expected since the more acidic sulphonic acid group displaces the equilibrium (54) $\rightleftharpoons$ (55) further to the left than does the weaker carboxylic acid group. However, more careful experimentation must be carried out before this hypothesis can be accepted since the effect might well have little to do with the acidity of the group but rather with the ease with which they homolyse. Thus, carboxylic acid groups are known to form radicals far easier than sulphonic acid groups, and a radical mechanism initiated by carboxy radicals could also explain the difference in light fastness.

Like the arylcarbonium dyes, polymethine and cyanine dyes have poor light fastness on all fibres except polyacrylonitrile type fibres. Comparatively little is known about their fading mechanisms but what little is known suggests that fading proceeds by oxidative routes. For example, in the dyes (56) the light fastness is decreased by incorporating electron-donor groups, *e.g.* —OMe. However, there are many exceptions to the above generalisation and further investigations are required.

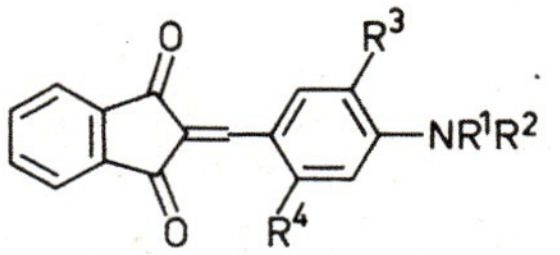

(56)

Once again the oxidative mechanism seems to operate in the fading of indigoid dyes. Evidence for this postulate is that isatin (57) and dibromoisatin (58) are recovered after the irradiation of indigo and tetrabromoindigo respectively on cotton. Recent evidence suggests that singlet oxygen is the species involved in the fading reaction.

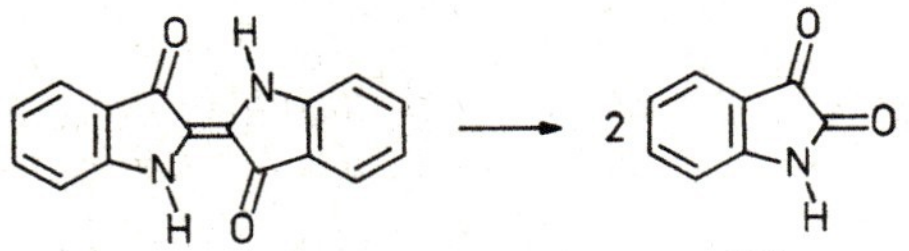

(57)

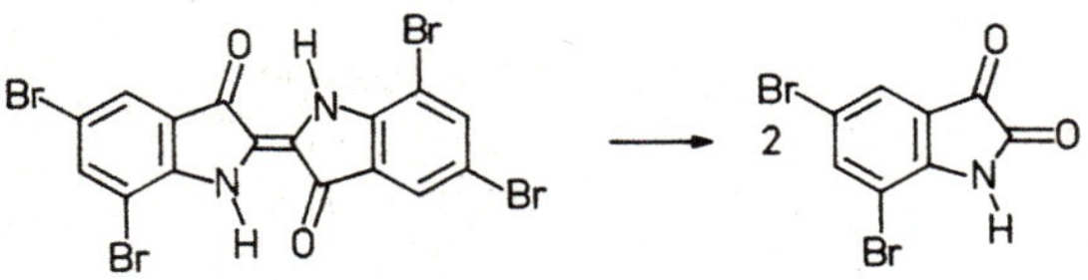

(58)

In some dyes light fastness is increased where hydrogen-bonding is possible. An example of this is found in nitro dyes. 2-Nitrodiphenylamine (59) has been shown to have superior light fastness to 4-nitrodiphenylamine (60). This was rationalised in terms of a hydrogen-bonding effect in the excited state of the 2-nitro dye which increases its photostability and possibly protects the donor-acceptor groups from attack. The introduction of a second nitro group (61) caused a decrease in light fastness due possibly to either steric hindrance, which would destabilise the hydrogen-bond, or alternatively to attack at the unprotected nitro group.

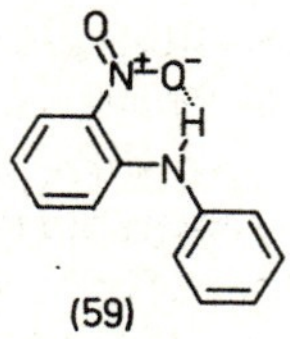

(59)

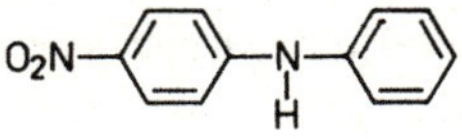

(60)

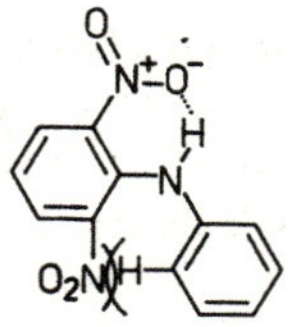

(61)

6.4.5 Effect of Aggregation on Light Fastness

The photostability of a dye and its interaction with the substrate are not the only factors that affect light fastness. The aggregation of the dye also seems to have a crucial bearing upon the rate of fading of dyed fibres. Experimental evidence strongly supports the theory that the more aggregated the dye, the higher its light fastness and conversely the nearer a dye is to the monomolecular state, the lower its light fastness. Thus, in one study the light fastness of a series of dyes was measured in ultrathin films of viscose after the degree of aggregation had been determined. It was found that those dyes which were highly aggregated generally had good light fastness whilst those that were close to the monomolecular state generally had poor light fastness. This observation fits in well with the characteristic tendency of strong dyeings (heavy depth of shade) having superior light fastness to weak dyeings (pale depth of shade); presumably, the strong dyeings contain more highly aggregated dye. It is also known that symmetrical sulphonated dyes generally have a higher order of light fastness than unsymmetrical sulphonated dyes. The most plausible explanation is again based upon aggregation of the dyes since the symmetrical dye molecules would be expected to aggregate better.

An experimental fact that is easy to rationalise in terms of the aggregation of a dye is the effect on light fastness of hydrocarbon chains. In a series of azo dyes having a hydrocarbon chain carrying from eight to sixteen methylene units a noticeable drop in light fastness accompanies the increase in chain length. This result would be anticipated since long alkyl chains have more surface activity than shorter chains and therefore destabilise aggregates.

The reasons for the large effect that aggregation appears to have on light fastness are not fully understood. However, both of the following theories explain the observations: first, highly aggregated dyes should dissipate the energy from the excited state of a dye molecule before it has a chance to react, *i.e.* the lifetime of the excited state is reduced: second, in a highly aggregated dye there is a smaller surface area available for attack by reactive species, *e.g.* singlet oxygen, radicals, hydrogen peroxide, *etc.*

6.4.6 Catalytic Fading

Several factors are responsible for the increase in the rate of fading of dyed fibres and these vary from environmental contaminants to fibre treatments. However, the term 'catalytic fading' is often used to refer to the increased rate of fade of one component in dye mixtures. The effect is particularly pronounced in mixtures containing yellow and blue dyes, *e.g.* greens and greys.

Two mechanisms have been proposed to explain the effect: —

(i) transfer of energy from the excited triplet state of one dye to another dye in the ground state, Eq. 6.15;

(ii) energy transfer from the triplet state of an excited dye to ground state triplet oxygen to generate singlet oxygen (Eq. 6.16) which then attacks the dye molecule.

$$\text{Dye 1}(T_1) + \text{Dye 2}(S_0) \longrightarrow \text{Dye 1}(S_0) + \text{Dye 2}(T_1) \qquad \text{Eq. 6.15}$$

$$\text{Dye 1}(T_1) + O_2(T_0) \longrightarrow \text{Dye 1}(S_0) + O_2(S_1) \qquad \text{Eq. 6.16}$$

The first mechanism is more likely for catalytic fading in bathochromic dyes since energy transfer from the excited state of a hypsochromic dye to the ground state of a bathochromic dye is much more likely to occur than transfer of energy from the excited state of the bathochromic dye to the ground state of the hypsochromic dye. This is because the dye that transfers the energy must have a higher energy triplet excited state than the first excited singlet state of the receiving dye and this situation is met for most yellow→blue dyes (see Fig. 6.7). Obviously the intersystem crossing ($S_1 \rightarrow T_1$) for the sensitising dye must be very fast in order to generate the excited triplet state.

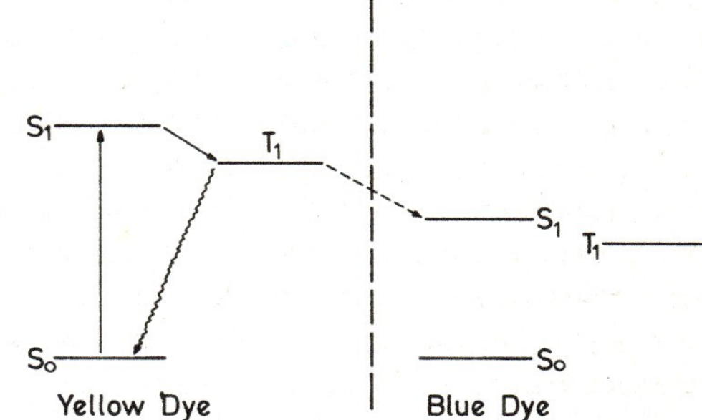

Fig. 6.7. Sensitisation of a blue dye by a yellow dye

The second mechanism has received experimental backing in some studies on mixtures of azo dyes and (more bathochromic) anthraquinone dyes. It was found that singlet oxygen quenchers suppressed markedly the catalytic fading of the azo dyes. The anthraquinone dyes were later shown to be fairly effective singlet oxygen sensitisers and yet more stable to singlet oxygen than azo dyes. Thus, once the singlet oxygen is generated it preferentially attacks the azo dyes — hence the catalytic fading effect. Of course, this mechanism is equally applicable to the catalytic fading of more bathochromic dyes caused by a hypsochromic dye. Obviously, in this case the hypsochromic dye would have to be a singlet oxygen sensitiser. The tendency for fluorescent brightening agents to cause catalytic fading may also be explained by a similar mechanism.

Apart from this narrow definition of catalytic fading the term may be more widely applied to the numerous effects which lower the light fastness of dyes. What causes many of these effects remains to be fully investigated. However, plausible suggestions have been offered in several cases. For example, titanium dioxide, which is used as a delustrant, often decreases the light fastness of disperse dyes. It is thought to do so by an oxidative mechanism, possibly by the catalytic generation of hydrogen peroxide. Similarly, peroxides are thought to be responsible for the adverse effect that crease resistant resins have on dyed fibres. Thus, resins using the easily per-oxidised dimethylol urea (62) cause catalytic fading whereas those using the dimethylol urea (63) show no pronounced catalytic effect.

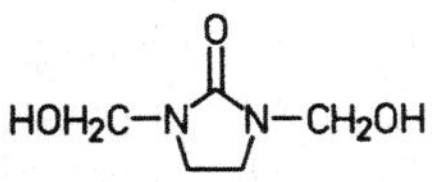

O
HOH_2C-N $N-CH_2OH$
(62)

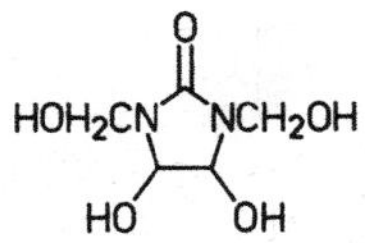

O
HOH_2CN NCH_2OH
HO OH
(63)

However, other reasons such as the different effects that (62) and (63) might have upon dye aggregation cannot be discounted since it is known that resins do decrease the degree of aggregation of dyes on regenerated cellulose.

Humidity also has a vital role to play in the fading rate of dyes. Generally, the higher the humidity the lower the light fastness of dyed fibre. This generalisation has been substantiated by controlled tests which have shown that dyes fade slower in drier atmospheres. This also explains the interesting observation that dyed fibres will sometimes fade slower in direct sunlight than in bright, cloudy conditions. This was later rationalised in terms of the moisture content of the fibre for in direct sunlight the fibre is at a higher temperature than on a cloudy day and therefore has a lower moisture content. The effect that the water has in the fading process is not known for certain but it appears to be primarily a chemical effect, as in the hydrolysis and hydroxylations of anthraquinone dyes. However, water could also play a part in the oxidative fading mechanism since it may facilitate the generation of hydrogen peroxide, which has been shown to be present during many 'photooxidations'.

The situation is complicated further by the known, and also possibly unknown, effects that several environmental pollutants have on the light fastness of dyed fibres. Thus, the oxides of sulphur and nitrogen adversely affect the light fastness of many dyed fibres. The combined effects of these two gases on acetate rayon dyed with certain blue or violet dyes causes a discolouration in the dye known as 'gas fume fading'; the discolouration was first noticed near gas flames from whence it gets its name. More recently, and perhaps not surprising, ozone has been shown to cause an increase in the fading of dyes.

From time to time attempts have been made to make colourless compounds which would improve the light fastness of dyes. Hydroxybenzophenones have been used as ultra-violet filters and these do indeed increase the light fastness of many dyes. Unfortunately, the effect is not substantial enough to offset the extra cost of incorporating these compounds into the fibre. This tends to be the case with most of the light fastness improvers since a relatively large amount of improver has to be used to gain a worthwhile effect — by this time economic constraints have usually ruled it out.

6.4.7 Phototendering of Dyed Fibre

In addition to the catalytic fading effect that one dye may induce upon another, a dye may also cause a change in the fibre substrate; this phenomenon is known as phototendering. The effect usually manifests itself as a change in the physical properties of the fibre such as a reduction in the tensile strength, abrasion resistance, *etc*. The effect is most frequently encountered on cellulosic fibre dyed with vat dyes, particularly yellow→red vat dyes. Azo dyes do not cause phototendering.

Two mechanisms have been proposed to account for phototendering; the first involves attack of singlet oxygen generated from triplet ground state oxygen and excited vat dye. The second mechanism assumes that triplet sensitised dye abstracts hydrogen radicals or electrons from the fibre substrate; phototendering is then caused by ensuing reactions at the radical centres so generated within the fibre.

Of the two, the first mechanism appears to be the more likely on the basis of experimental evidence. Thus, if solutions of yellow and red vat dyes are exposed to light then singlet oxygen can be trapped by tetraphenylcyclopentadienone. Furthermore, phototendering also occurs in undyed fibre when it is placed in very close proximity to, but not touching, dyed fibre. This strongly suggests that a volatile component, such as singlet oxygen or hydrogen peroxide formed from singlet oxygen, causes phototendering.

Less support is available for the second mechanism although this mechanism does explain rather elegantly why yellow vat dyes cause phototendering and blue vat dyes do not.

Fig. 6.8 shows a typical energy level diagram for (a) a yellow vat dye and (b) a blue vat dye. It will be noticed that the relative energies of the $n\pi$ and $\pi\pi^*$ states are reversed in the two dyes. It is thought that hydrogen abstraction will only occur easily from the $n\pi$ state. Hence, only the yellow vat dye causes hydrogen abstraction since the $n\pi$ level is the lower excited state. The only direct evidence for the abstraction

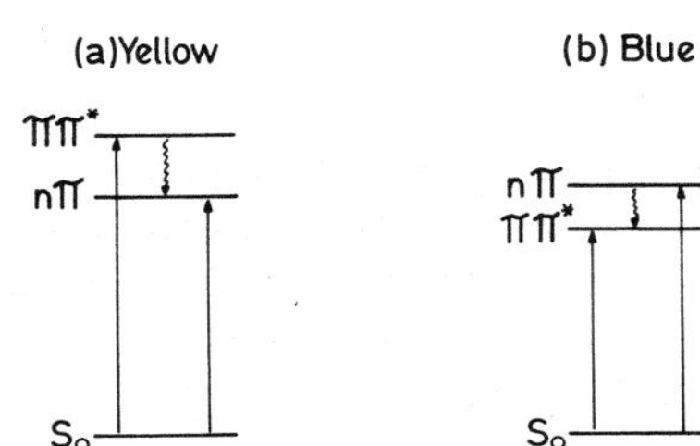

Fig. 6.8. Energy levels for typical yellow and blue vat dyes

of hydrogen radicals is provided by model studies in which the vat dyes were shown to abstract H· from the solvent. However, evidence has been obtained for electron abstraction by photoexcited quinone dyes from the substrate by flash photolysis studies.

6.5 Photochromism

Some dyes change colour reversibly upon exposure to light. This effect is known as photochromism (or phototropism) and is considered a defect in textile dyes.

In azo dyes the change is caused by *trans*→*cis* isomerism about the azo linkage and is related to the isomerism about the carbon-carbon double bond in stilbenes — a well documented process. Photochromism in azo dyes manifests itself as a bathochromic shift in the colour of the dye. The origin of the bathochromic shift is interesting and is demonstrated in Fig. 6.9 for a hypothetical dye. It can be seen that the $\pi \rightarrow \pi^*$ transition undergoes a hypsochromic shift, and weakens, on going from the *trans* azo dye to the *cis* azo dye. On this basis the dye might be expected to undergo a hypsochromic shift. However, another effect is working in precisely the opposite direction — *viz.* the $n \rightarrow \pi^*$ transition in the *cis* dye is stronger and more bathochromic than in the *trans* dye. Overall, this effect tips the balance and the *cis* isomer therefore appears more bathochromic than the *trans* isomer.

Photochromism is less common on natural fibres where the dye-fibre interactions are stronger than in many of the synthetic fibres, *e.g.* nylon, polyacrylonitrile and polyester. Strong dye-fibre interactions probably hinder the *trans*→*cis* isomerism and promote *cis*→*trans* isomerism: therefore, at equilibrium, the *cis* form is present to only a very minor extent.

The presence of strongly polar groups also tends to reduce photochromism since the resonance interaction lowers the activation energy for *cis*→*trans* isomerism, *e.g.* dyes containing nitro groups tend not to suffer from photochromism. Since most red→blue dyes contain at least one nitro (or similar polar) group, photochromism is not a problem in these shade areas. In contrast, it can be serious problem in yellow-orange dyes since these do not usually contain strongly polar groups such as nitro groups. A further ploy to reduce or even eliminate photochromism is to incorporate a hydroxy group *ortho* or *para* to the azo linkage. This is thought to result from the

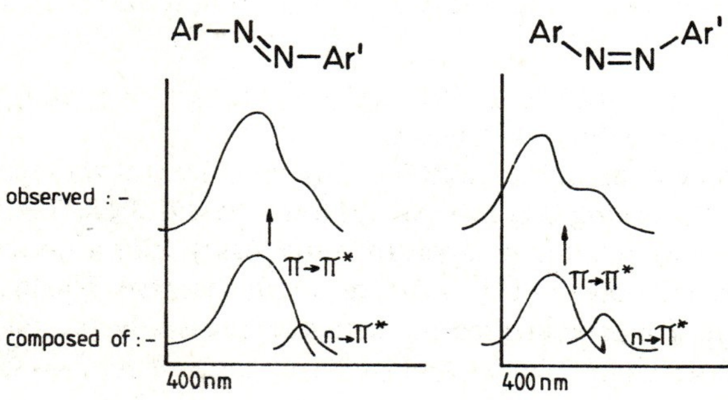

Fig. 6.9. The origin of photochromism in azo dyes

hydroxyazo dye being in equilibrium (though to a small extent) with the hydrazone form. Once in the hydrazone form rapid rotation about the N—N *single* bond occurs and the dye reverts to the *trans* form (Scheme 6.12).

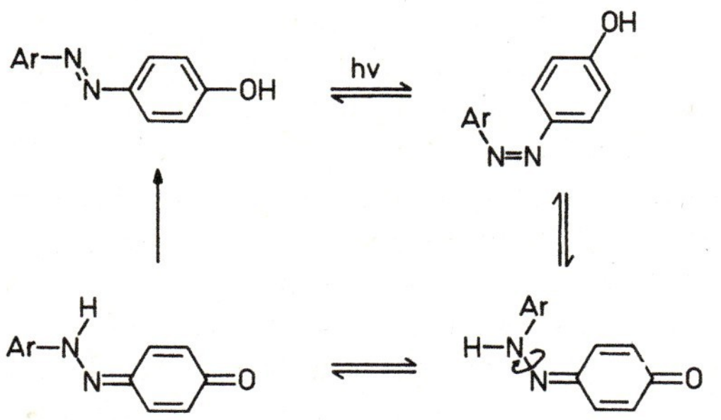

Scheme 6.12

Photochromism is not just limited to azo dyes; other dye classes are also affected. Indigo itself is not photochromic possibly due to the strong hydrogen-bonding, *e.g.* (64), but several thioindigoid dyes and N-substituted indigoids do exhibit photochromism. Once again isomerism about the central carbon-carbon double bond is thought to be responsible for the effect, although this time a hypsochromic shift is usually observed, *e.g.* 60 nm for thioindigo. In contrast, photochromic cyanine dyes undergo a bathochromic shift.

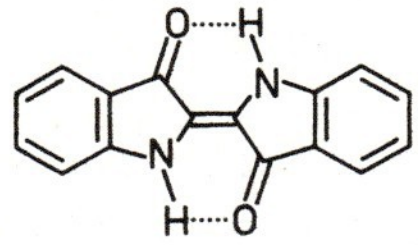

(64)

In some cases the photochromic behaviour is caused by a chemical change. An example is the yellow vat dye flavanthrone which abstracts a hydrogen atom to generate the *leuco* form on cellulose. Oxidation in the dark restores the original colour.

6.6 Heat Fastness of Dyes

There are two major dye types in which heat fastness is a problem — cationic dyes on polyacrylonitrile and disperse dyes on polyester.

On polyacrylonitrile the problem manifests itself as a loss in colour of the dyed fabric on exposure to heat, *viz.* the dyeing becomes paler. However, the effect is exacerbated in shades produced from a mixture of dyes (the usual case) since a noticeable shade change occurs if one component of the mixture is heat sensitive. The heat source is either steam (to simulate a process used for removing creases in the dyed fabric) or dry heat (~180 °C) which simulates carpet backing treatments that the dye may have to withstand. Cationic dyes deficient in heat fastness also tend to hydrolyse in the dyebath.

The problem is particularly pronounced in blue diazahemicyanine dyes such as the market leader, CI Basic Blue 41 (65). It has been established that heat and dyebath instability are a chemical problem. Thus, the treatment of (65) in either boiling water or cold dilute alkali results in the formation of the red merocyanine dye (66). The proposed mechanism is shown in Scheme 6.13. As mentioned in Sect. 5.5.4, resonance form (65A) makes a major contribution to the ground state of the molecule. Hence, the molecule may be more accurately represented as (65B). It can be seen that the aromatic carbon atom (*) is more susceptible to nucleophilic attack than might be expected from the representation in (65).

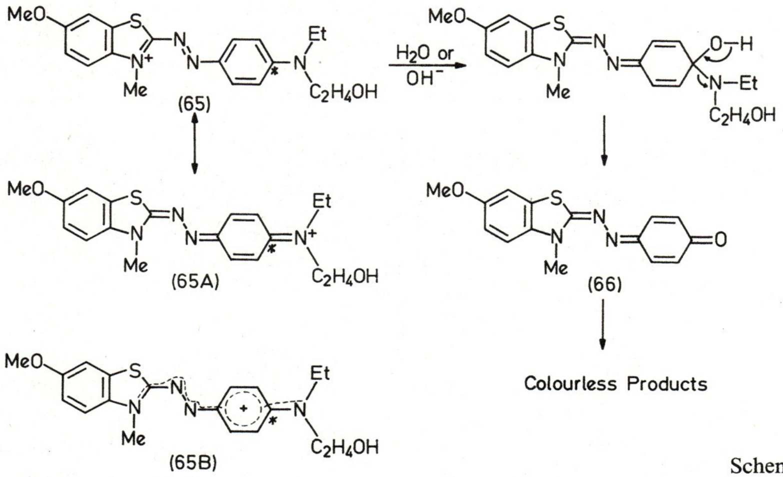

MeO, S, N+, Me, N, N, Et, C₂H₄OH (65) → H_2O or OH^-

(65A) (65B) (66) Colourless Products

Scheme 6.13

More hypsochromic diazahemicyanine dyes such as CI Basic Red 22 (67) are less prone to degradation. Presumably, the contribution of resonance form (67A) to the ground state is lower than for (65A) because of the weaker electron-withdrawing power of the triazolium ring relative to a benzothiazolium ring.

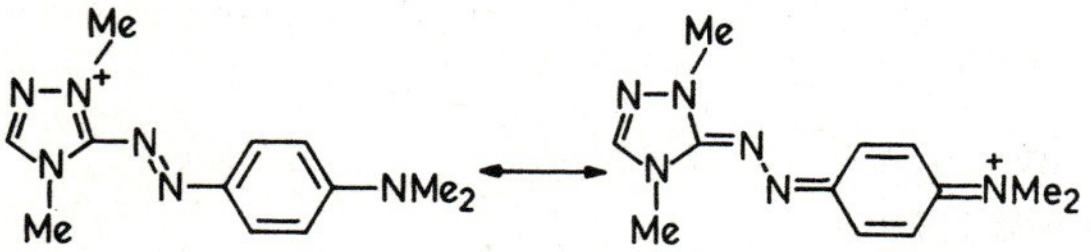

(67) (67A)

Hemicyanine dyes (see Sect. 5.5.3) such as CI Basic Red 14 (68) do not form the merocyanine compound (69). Therefore, they have better heat and dyebath stability than diazahemicyanine dyes.

In contrast to diazahemicyanine dyes, the heat fastness problem with disperse dyes is a physical one. As we have already seen disperse dyes must be small, non-

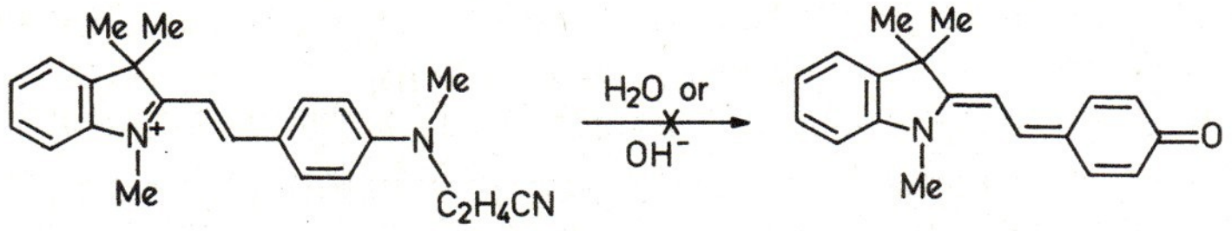

(68) (69)

ionic molecules of low molecular weight. As such, they often exhibit a significant vapour pressure. Therefore, if heat treatments during fixation, or subsequently, *e.g.* ironing, are involved, the dye may sublime and cause contamination of equipment or adjacent undyed material.

Vapour pressure decreases with the mass and also the polarity of the dye molecule. Since molecular size is limited by fibre accessibility, the search for improved sublimation fastness has centred on increasing the polarity of the dye molecules by introducing polar substituents. The common feature of the many substituents introduced for this purpose is to confer increased hydrophilic properties on the dye molecule, but often a compromise has to be achieved with the concomitant decrease in fibre affinity.

Two important groups for improving the sublimation fastness of a dye are the 2-hydroxyethyl and acetylamino groups — Fig. 6.10.

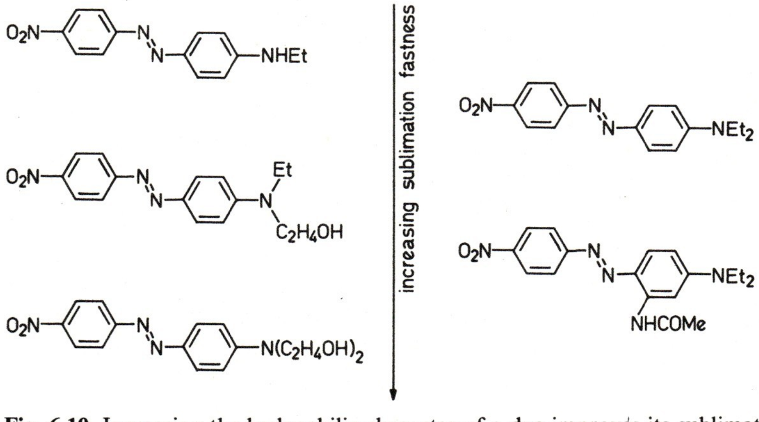

Fig. 6.10. Increasing the hydrophilic character of a dye improves its sublimation fastness

6.7 Bleach Fastness of Dyes

Certain articles need to withstand bleaching treatments; for example, swimwear, towelling, overalls, *etc.* Consequently, the dyes used for these articles must be resistant to bleach.

In general, vat dyes, because of their lack of functional groups, exhibit excellent fastness to bleach. In contrast, most azo dyes are decolourised by bleach. A recent[15]

[15] Gregory, P., Stead, C. V.: J. Soc. Dyers and Colourists *94*, 402 (1978).

study on the effect of bleach (*i.e.* sodium hypochlorite) on azo dyes has shown that they are oxidised, the oxidation products being a diazonium salt and a quinone. Where azo-hydrazone tautomerism is possible (see Sect. 3.3), the hydrazone form is attacked more rapidly than the azo form. The effective oxidising agent is the chloronium ion ($Cl^{\oplus}$) and the probable reaction mechanism is outlined in Scheme 6.14. The products obtained are similar to those observed in the oxidative photo-fading of azo dyes (see Sect. 6.4.4).

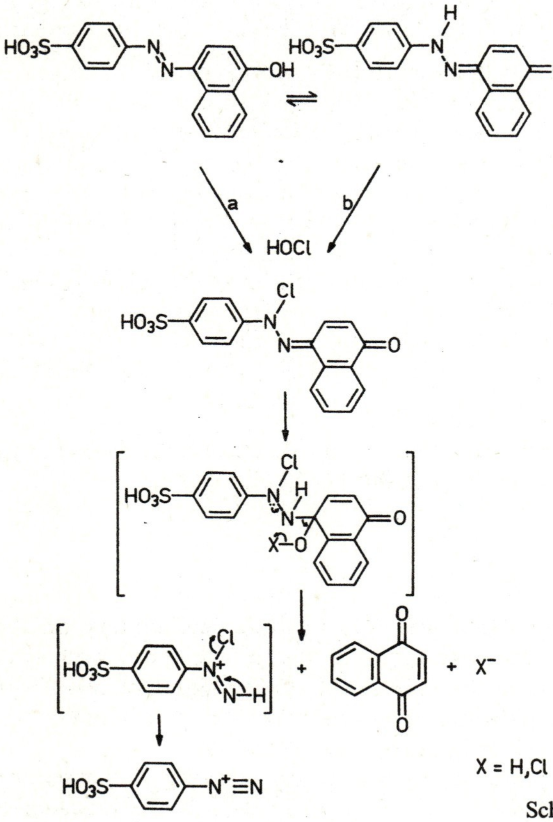

X = H,Cl

Scheme 6.14
Studies on the *N*—Me and *O*—Me compounds show that reaction *b* is much faster than reaction *a*

6.8 Metamerism

As for photochromism, metamerism is considered to be a defect in textile dyes although by careful matching of dyes the problem can often be resolved. The problem is caused by the different light output of various light sources which causes the colour of a dye to appear different depending on the light source.

6.9 Solvatochromism

Solvatochromism is simply the effect of a solvent upon the colour of a dye. Generally, positive solvatochromism is encountered. Here, a dye undergoes a bathochromic shift with increasing solvent polarity. This indicates that the first excited state of the dye is more polar than the ground state. Negative solvatochromism is relatively rare and is observed when the first excited state is less polar than the ground state. It is usually observed with dyes that contain a delocalised charge, *e.g.* CI Basic Yellow 28 (70), or in certain merocyanine dyes such as (71) in which the ground state is polar due to a large contribution from the canonical form (71A).

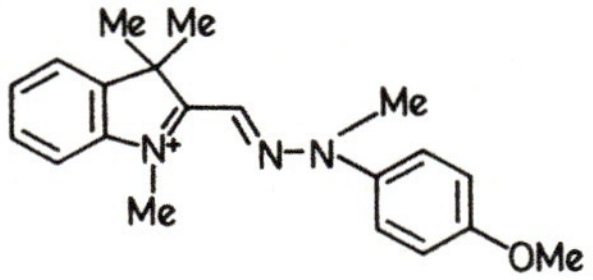

(70)

$\lambda_{max}^{CHCl_3}$ 457 nm λ_{max}^{DMF} 446 nm

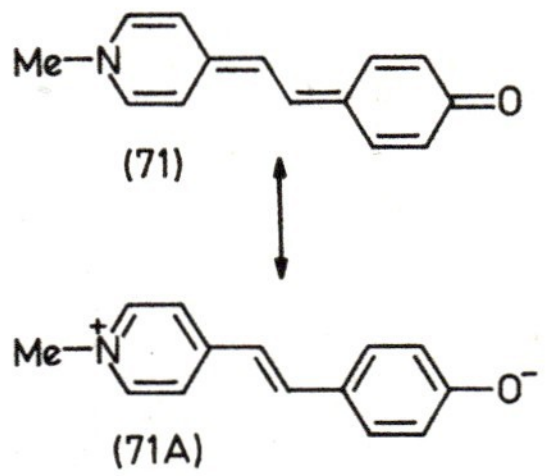

(71)

(71A)

$\lambda_{max}^{CHCl_3}$ 628 nm $\lambda_{max}^{HCONH_2}$ 500 nm

More detailed information on the solvatochromism of individual dye types may be found in the appropriate chapters of this book.

6.10 Summary

Textile fibres are polymers. Until well into the twentieth century all the fibres were of natural origin, *e.g.* wool and cotton. Semi-synthetic fibres such as cellulose acetate were introduced during the 1920's. The three major synthetic fibres in use today, polyamide (nylon), polyester (*e.g.* Terylene) and polyacrylonitrile (acrylic) were developed and marketed during the period 1930–1950.

The diversity of dye types necessitates a variety of dyeing processes to transfer the dye on to the fibre. However, the simplest of these dyeing procedures is still that of boiling the dye and the fibre in water. A dye is retained within a fibre by either physical effects, such as the formation of insoluble aggregates within the fibre structure, or by chemical bonding — this may be either ionic or covalent.

Over the years much attention has been focused on the light fastness of dyes. The azo dyes have been studied in some depth and it appears that the mechanism of fading on protein fibres is reductive whereas on non-protein fibres it is oxidative. Mechanisms have been put forward to account for these observations. The other dye classes have been studied to a lesser extent although here an oxidative mechanism appears to predominate, especially for dealkylative processes.

Many other factors have an effect on the overall light fastness of a dye and it appears that conditions of high humidity and low aggregation of dye are to be avoided if dyes of high light fastness are required. Other atmospheric contaminants are also responsible for accelerating the fading of dyes, such as agents which form peroxides easily, ozone, and oxides of sulphur and nitrogen.

Obviously there are many other treatments which a dyed fibre must withstand: two of the more important are heat and bleach.

The single most important mechanism to account for photochromism is *trans* → *cis* isomerism about a double bond. This mechanism successfully explains the photochromism in a variety of dye classes from azo dyes to indigoids.

The effect of the solvent (solvatochromism) and the light source (metamerism) on the colour of a dye are also important.

6.11 Bibliography

A. *Textile Fibres*

1. Skelly, J. K.: Chem. and Ind. *1965*, 1525
2. Venkataraman, K. (ed.): The chemistry of synthetic dyes, Vol. II, pp. 1253–1264. New York: Academic Press 1952
3. Roberts, W. J.: Fibers, chemical. In: Encyclopedia of chemical technology, Kirk-Othmer (eds.), 3rd ed., Vol. 10, pp. 148–166. New York, Chichester, Brisbane, Toronto: Wiley-Interscience 1980
4. Allen, R. L. M.: Colour chemistry, pp. 68–76. London: Nelson 1971

B. *Application and Wet Fastness of Dyes*

1. Ref. A1
2. Ref. A4.: pp. 77–102, 150–151, 270–277 and 238–240
3. Ramsay, D. W.: J. Soc. Dyers and Colourists *97*, 102 (1981)
4. Dolby, P. J.: Amer. Dyestuff Reporter *1976*, 28
5. Vickerstaff, T.: The physical chemistry of dyeing, 2nd ed. London: ICI and Oliver and Boyd 1954
6. Ref. A2.: pp. 1264–1303
7. Beech, W. F.: Fibre-reactive dyes. London: Logos 1970

C. *Light fastness (and references cited therein)*

1. Griffiths, J.: Rev. Prog. in Colouration: In press
2. Giles, C. H., McKay, R. B.: Textile Research Journal *33*, 528 (1963)
3. Evans, N. A., Stapleton, I. W.: Structural factors affecting the light fastness of dyed fibres. In: The chemistry of synthetic dyes, Venkataraman, K. (ed.). Vol. VIII, pp. 221–276. New York: Academic Press 1978

D. *Photochromism (and references cited therein)*

1. Ref. C 1–3
2. Griffiths, J.: Chem. Soc. Reviews *11*, 481 (1972)
3. Ross, D. L., Blanc, J.: Photochromism by *cis-trans* isomerization. In: Techniques of chemistry, Vol. III, Brown, G. H. (ed.), pp. 471–557. New York: Wiley-Interscience 1971

Appendix I

Some useful references to colour physics

A. *Molecular Orbital Theory and Quantum Mechanics*

1. Dewar, M. J. S.: The molecular orbital theory of organic chemistry. New York: McGraw-Hill 1969
2. Dewar, M. J. S., Dougherty, R. C.: The PMO theory of organic chemistry. New York: Plenum Press 1975
3. Borden, W. T.: Modern molecular orbital theory for organic chemists. Englewood Cliffs, N. J.: Prentice-Hall 1975
4. Zimmerman, H. E.: Quantum mechanics for organic chemists. New York: Academic Press 1975
5. Pople, J. A., Beveridge, D. L.: Approximate molecular orbital theory. New York: McGraw-Hill 1970
6. Jorgensen, W. L., Salem, L.: The organic chemist's book of orbitals. New York: Academic Press 1973
7. Roberts, J. D.: Notes on molecular orbital calculations. New York: W. A. Benjamin 1961
8. Streitwieser, A. Jr.: Molecular orbital theory for organic chemists. New York: Wiley 1961
9. Liberles, A.: Introduction to theoretical organic chemistry. New York: Macmillan 1968

B. *Colour and its Measurement*

1. Jaffé, H. H., Orchin, M.: Theory and applications of ultraviolet spectroscopy. New York: Wiley 1962
2. Wright, W. D.: The measurement of colour, 4th ed. London: Adam Hilger 1969
3. Murray, H. D. (ed.): Colour in theory and practice. London: Chapman and Hall 1952
4. Wyszecki, G., Stiles, W. S.: Colour science. New York: Wiley 1967
5. LeGrand, T.: The measurement of colour. London: Adam Hilger 1957
6. Birren, F. (ed.): Munsell: A grammar of color. New York: Van Nostrand Reinhold 1969

C. *Colour Perception and Vision*

1. Evans, R. M.: The perception of color. New York: Wiley-Interscience 1974
2. Birren, F.: Principles of color. New York: Van Nostrand Reinhold 1969
3. Ref. B3
4. Ref. B5.: pp. 358–381
5. Abbot, A. G.: The color of life. New York: McGraw-Hill 1947
6. Wilson, R. F.: Colour and light at work. London: Seven Oaks Press 1953
7. Henderson, S. T.: Daylight and its spectrum. London: Adam Hilger 1970
8. Graham, C. H.: Vision and visual perception. New York: Wiley 1965
9. Hering, E.: Outlines of a theory of the light sense. Harvard University Press 1964
10. Mueller, C. G., Rudolph, M.: Light and vision. Netherlands: Time-Life International 1967
11. Dratz, E. A.: Science of photobiology. Plenum 1977
12. Waaler, G. H. M.: J. Oslo City Hosp. *27*, 137 (1977)
13. Menger, E. L. (ed.): Special issue on the chemistry of vision. Accounts Chem. Res. *8*, (3) (1975)
14. Sheppard, J. J.: Human color perception: A critical study of the experimental foundation. New York: Elsevier 1968

Appendix II

The Colour Index (3rd Edition)

Throughout this book it will have been noticed that many dye structures have a CI number associated with them. This number refers to the entry in the Colour Index corresponding to that dye. The Colour Index (published by the Society of Dyers and Colourists in six volumes) contains much useful information regarding the properties and the structural type of the dyes contained therein. In many cases the full structure of the dye may be ascertained via the Colour Index.

Author Index*

* Thanks are extended to Mr. G. Craig for his help with this section

Subject Index